MOLECULAR MECHANISMS FOR REPAIR OF DNA

Part A

BASIC LIFE SCIENCES
Alexander Hollaender, General Editor

The University of Tennessee
Knoxville
and Associated Universities, Inc.
Washington, D. C.

1973: Volume 1 ● GENE EXPRESSION AND ITS REGULATION
Edited by F. T. Kenney, B. A. Hamkalo, G. Favelukes,
and J. T. August

Volume 2 ● GENES, ENZYMES, AND POPULATIONS
Edited by A. M. Srb

1974: Volume 3 ● CONTROL OF TRANSCRIPTION
Edited by B. B. Biswas, R. K. Mandal, A. Stevens,
and W. E. Cohn

Volume 4 ● PHYSIOLOGY AND GENETICS OF REPRODUCTION
(Parts A and B)
Edited by E. M. Coutinho and F. Fuchs

1975: Volume 5 ● MOLECULAR MECHANISMS FOR REPAIR OF DNA
(Parts A and B)
Edited by P. C. Hanawalt and R. B. Setlow

Volume 6 ● ENZYME INDUCTION
Edited by D. V. Parke

A Continuation Order Plan is available for this series. A continuation order
will bring delivery of each new volume immediately upon publication. Volumes
are billed only upon actual shipment. For further information please contact
the publisher.

MOLECULAR MECHANISMS FOR REPAIR OF DNA

Part A

Edited by

PHILIP C. HANAWALT
Department of Biology
Stanford University

and

RICHARD B. SETLOW
Biology Department
Brookhaven National Laboratory

PLENUM PRESS • NEW YORK AND LONDON

Library of Congress Cataloging in Publication Data

Main entry under title:

Molecular mechanisms for repair of DNA.

(Basic life sciences; v. 5)
"Based upon a workshop conference on molecular mechanisms for repair of DNA
at Squaw Valley, California, February 25-March 1, 1974."
Includes bibliographies and index.
1. Deoxyribonucleic acid repair—Congresses. I. Hanawalt, Philip C., 1931-
II. Setlow, Richard Burton.

QP624.M64   574.8'732   75-17731

ISBN 0-306-36593-6 (v. 5A)

First half of the Proceedings of a workshop conference on Molecular
Mechanisms for Repair of DNA, held in Squaw Valley, California,
February 25-March 1, 1974

©1975 Plenum Press, New York
A Division of Plenum Publishing Corporation
227 West 17th Street, New York, N.Y. 10011

United Kingdom edition published by Plenum Press, London
A Division of Plenum Publishing Company, Ltd.
Davis House (4th Floor), 8 Scrubs Lane, Harlesden, London, NW10 6SE, England

Printed in the United States of America

This volume is dedicated to the memory
of Dr. Ruth Frances Hill

Ruth Frances Hill 1917–1973

Dr. Ruth Frances Hill died suddenly of a cerebral hemorrhage on November 3, 1973, in Toronto, where she was Professor of Biology at York University. Ruth Hill won an important and secure place in the history of biology by virtue of her discovery in 1958 of the first radiation-sensitive mutant bacterium, strain B_{s-1} of *Escherichia coli.* The isolation of this mutant came as a surprise to radiation geneticists, who up to that time had been more concerned with the isolation of radiation-resistant strains. Furthermore, the extent to which the mutant alleles in B_{s-1} affected sensitivity, especially to ultraviolet light, was vastly greater than had ever been observed before for other physical, chemical, or biological modifiers of radiobiological action. It was not surprising that the very existence of such a mutant should immediately have attracted the interest of radiation biologists, most of whom were accustomed to dealing with three- or fourfold changes in radiosensitivity but certainly not with sensitization factors of 100 or more. The many theories and experiments stimulated by Ruth's discovery of *E. coli* B_{s-1} have, to say the least, changed the complexion of cellular radiobiology. But even more significantly, comparative studies of B_{s-1} and wild-type strains, designed to elucidate the macromolecular basis of wild-type resistance, led to the discovery in 1964 of DNA excision-repair, one of the few fundamental processes in molecular biology that was not foreseen by the clairvoyant pioneers of that field.

Robert H. Haynes

Foreword

An "age" has passed in the 40 years since we first observed recovery from radiation damage in irradiated bacteria. During the early 1930s, we had been discussing the possibility of rapid changes after radiation exposure with Farrington Daniels, Benjamin Duggar, John Curtis, and others at the University of Wisconsin. After working with living cells, we had concluded that organisms receiving massive insults must have a wide variety of repair mechanisms available for restoration of at least some of the essential properties of the cell. The problem was how to find and identify these recovery phenomena.

At that time I was working on a problem considered to be of great importance—the existence of the so-called mitogenetic rays. Several hundred articles and a score of books had already appeared dealing with mitogenetic rays, a type of radiation that was thought to exist in the shorter ultraviolet region. Our search for mitogenetic rays necessitated the design of experiments of greatest sensitivity for the detection of ultraviolet. It was vital that conditions be kept as constant as possible during exposure. All the work was done at icewater temperature (3–5°C) during and after exposure. We knew that light was an important factor for cell recovery, so all our experiments were done in dim light, with the plated-out cells being covered with dark cloth. Our statements on the effect of visible light stimulated Kelner to search for "photoreactivation' (as it was later called). We attempted to exhaust all possible means for the detection of these elusive "rays" before expressing an opinion. Under carefully controlled conditions (this work is described in Monograph No. 100 of the U.S. National Academy of Sciences), we were not able to confirm the existence of this radiation.

However, the strictly controlled experiments we designed made the eventual discovery of "recovery" inevitable. I realized there must be some enzyme that stimulated recovery, and I spent a few weeks in Northrop's laboratory (then called Rockefeller Institute) outside Princeton, New Jersey, working with Kunitz in an effort to isolate the stimulating enzyme. But with my limited knowledge of the recovery process our experiments were not successful.

In the 1940s, we attempted to interest other investigators in the problem of recovery, but to no avail. It was not until the many new developments of the early 1950s and 1960s that interest in this problem was revived. Then began the tremendous progress in understanding the structure and function of nucleic acid and protein chemistry. These observations of 30 years ago opened up the possibilities that led to the development of studies in molecular biology, and formed the basis for many new types of investigations in the field of repair— investigations that are so well discussed in this book.

Alexander Hollaender

Associated Universities, Inc.
and
University of Tennessee

Preface

This book is the outgrowth of a workshop conference on Molecular Mechanisms for Repair of DNA held at Squaw Valley, California, in February 1974. Nearly 200 researchers and students gathered to discuss this timely subject and to present their recent results. Constructive interactions took place among geneticists, biochemists, and radiation biologists as DNA repair models were debated at the conference.

We felt that a book on DNA repair to document that discussion would be useful to researchers in this field and in related fields—especially since a comprehensive treatment of this subject has not recently been published. The ideas and current knowledge in the field of DNA repair are of interest to workers in many areas (e.g. replication of nucleic acids, mutagenesis, environmental biology, carcinogenesis). Although the organization of this book by sections follows closely the organization of the meeting by sessions, the book does not represent direct proceedings. Rather, it is an attempt to compile a group of short articles that serve to define our current state of knowledge. This includes articles by each of the invited participants at the meeting as well as short communications from others. The session chairmen have each prepared introductory articles for the sections that give overviews of the respective subtopics. The manuscripts were assembled in May 1974.

A number of the manuscripts have been severely pared—otherwise, in the interest of prompt publication in this rapidly moving field we have not done much editorial tampering with individual manuscripts. We regret that we were not able to accommodate all of the contributed articles in the book. For historical interest, we reprint an early note (1935) in which Alexander Hollaender hinted that the damage produced by radiation in cells might be subject to repair. This may have been the first suggestion of repair—7 years after Gates had shown that the action spectrum for killing of bacteria by UV paralleled the absorption spectrum for nucleic acid.

The Squaw Valley conference was supported in part by a contract with the U.S. Atomic Energy Commission. We are also indebted to the American Cancer Society for a grant to assist with travel of invited participants. We express our appreciation to Fred Fox and his assistants Charlotte Miller and Fran Stusser for administering the meeting through the ICN-UCLA Squaw Valley Conferences on Molecular Biology. We also acknowledge the assistance of Pat Seawell, who helped with the arrangements during the week of the meeting.

P. C. Hanawalt

Stanford University

R. B. Setlow

Brookhaven National Laboratory

An Early Suggestion of DNA Repair[*]

Effect of Sublethal Doses of Monochromatic Ultraviolet Radiation on Bacteria in Liquid Suspensions

Alexander Hollaender and John T. Curtis
(Introduced by B. M. Duggar)

From the Laboratory of Plant Physiology
University of Wisconsin

In the course of a quantitative investigation[1] of the lethal effect of monochromatic ultraviolet radiation below 3,000 A on stirred saline suspensions of bacteria *(Escherichia coli)* by a method similar to one described in a previous publication,[2] we observed the following phenomena:

If the bacteria are plated out in nutrient agar on Petri dishes subsequent to radiation, the colonies formed by the bacteria which survived the irradiation appear later than the colonies formed by the bacteria from the control suspension treated in the identical manner but protected from the ultraviolet radiation.

A careful study of the growth of the irradiated and control suspensions in a relatively poor medium (3 gm. NaCl, 0.2 gm. KCl, 0.2 gm. $CaCl_2$, 0.04 gm. Difco beef broth powder, 1,000 cc. distilled water, ca 10,000 bacteria/cc.) was made by determining the number of viable, colony-forming, organisms at proper intervals and several interesting observations were made:

1. The retarded growth (lag) phase of the irradiated bacteria was extended considerably over the control. This extension apparently depends on the energy applied to the suspension. When the bacteria had completed their growth, the suspension with the control and that with the irradiated bacteria contained the same number of organisms.

---

[*]Reprinted from the *Proceedings of the Society for Experimental Biology and Medicine* **33**, 61–62 (1935).
[1]Hollaender and Claus, *J. Gen. Physiol.* **19**, 753–765 (1936).
[2]Duggar, B. M., and Hollaender, Alexander, *J. Bact.* **27**, 219, 241 (1934).

2. A careful determination of the lag phase revealed that where the control culture changed little in number of bacteria during this phase, the culture which had survived irradiation increased in number quite rapidly in the earlier part of the lag phase and then slowed down more or less for a certain time before it came into the log phase, thus producing a modified extension of the lag phase. The growth curve of the control rising in the beginning of the lag phase actually crossed the curve of the exposed culture and in the end an extension of the lag phase was induced by the radiation. The total number of viable bacteria of the exposed culture had increased during the lag phase. These increases were so pronounced that in the time during which the control culture showed a change in number of not more than 10 to 15%, the exposed culture had increased up to, or more than, 100%.

Tests have shown that the increased growth in the lag phase of the exposed culture was not produced by the decomposition products of the dead bacteria or by the excretion of any substance by the irradiated organisms. Stimulation is suggested, but the possibility of recovery of the irradiated bacteria is not entirely excluded.

The observed phenomena do not refer to the effects of very small amounts of energy, but only to energy values which kill some of the bacteria and allow the rest of the organisms to survive.

A detailed description of the experiments will be published later. The authors wish to thank Prof. B. M. Duggar for the help and advice he has given them in the course of this investigation.

Contents

I. Repairable Damage in DNA

II. Enzymatic Photoreactivation

III. Dark Repair in Bacteriophage Systems

IV. Enzymology of Excision-Repair in Bacteria

V. Repair by Genetic Recombination in Bacteria

Contents
of Part B

VII. Repair Models and Mechanisms

VIII. Repair Processes in Diverse Systems

I

Repairable Damage in DNA

1

Repairable Damage in DNA: Overview

Peter A. Cerutti

Department of Biochemistry
University of Florida
Gainesville, Florida 32610

Due to the central functional role of DNA and the fact that each cell contains only one or at the most a few copies of each chromosome, damage to DNA has more severe implications for the functional integrity of the cell than does damage to most other cellular components. The chemical makeup and the large target size of chromosomal DNA make it particularly susceptible to attack by exogenous chemical and physical agents. Most, if not all, reactions of exogenous agents with the sugar residues of the DNA backbone result in strand breakage. While the continuity of the sugar-phosphate backbone usually remains intact for reactions involving the heterocyclic bases, such reactions may cause local distortion of the DNA conformation and of the native structure of eukaryotic chromatin. The preservation of the unique three-dimensional structure of the double-stranded DNA helix appears to be a prerequisite for its unimpaired biological activity. DNA base damage and concomitant helix distortion may lead to inhibition of replication and transcription and to a deterioration of the fidelity and a breakdown of the regulation of these processes. Restoration of the structural integrity of the DNA by repair processes is, therefore, a vital function of every cell.

DNA repair phenomena are often classified according to the exogenous agents by which the damage was induced: ultraviolet repair, X-ray repair, repair of alkylation damage, etc. However, most physical and chemical agents reacting with DNA induce a large variety of products. In many cases products produced by different agents are structurally related and are expected to have similar effects on the local conformational structure of the DNA helix—e.g. pyrimidine photohydration (formation of 6-hydroxy-dihydropyrimidines and pyrimidine ring saturation by ionizing radiation under formation of products of the 5,6-dihydroxy-dihydrothymine type); base elimination as a consequence of exposure of DNA to alkylating agents, ionizing radiation, or heat; interstrand cross-linking by nitrogen or sulfur mustards or by psoralen plus light. *It appears more likely that classes of lesions with common structural features are repaired by common repair pathways than that specific repair enzymes exist for each individual lesion or for a specific damaging agent.* Structurally related lesions are also expected to have similar biological effects regardless of the agent responsible for their formation. The criteria for the present classification of DNA base damage are, therefore, the chemical characteristics and the extent of the local helix distortion caused by a given type of base modification. A prerequisite for such a classification is, of course, the structural identification of the lesions formed in the DNA *in situ* in the living cell. This goal has not been reached for most of the biologically important DNA-damaging agents. Considerably more is known about the lesions introduced by ultraviolet light and certain alkylating agents than those produced by ionizing radiation or heat treatment.

While the effect of certain types of base modifications on DNA structure has been studied by physical biochemical techniques, such information is relatively scarce. In many cases the effect of a given type of lesion on DNA conformation can only be guessed from data obtained with the mononucleotide, by analogy with structurally related modifications of synthetic and natural polydeoxynucleotides, and from model building. Further uncertainty arises from the recognition that damage is not always introduced randomly into the DNA. Product distribution may depend on the primary sequence, the conformational fine structure, and the electronic properties of the DNA (see e.g. Patrick and Rahn, 1975). Clustering of damage in native DNA has been observed for cyclobutane-type photodimers (see Brunk, this volume; Shafranovskaya et al., 1973). This phenomenon could be of general importance. Reactions at the helix periphery (e.g. damage of Class I; see below) for which the transition-state and final products have steric requirements that are compatible with the native conformation of the helix may occur randomly. Nonrandom distribution of damage may result from reactions which occur at sites toward the helix center and introduce considerable steric strain. Accessibility of the reaction sites and the reactivity of the individual nucleotide residues in eukaryotic cells are undoubtedly further influenced by chromosomal proteins. Chromosomal proteins may act as radical scavengers in reactions induced by ionizing radiation (see e.g. Roti Roti et al., 1974) and as sensitizers or quenchers in photochemical reactions. Results obtained with free DNA therefore should be evaluated with caution as to their relevance to the *in vivo* situation. Product distribution is expected to influence the extent and the character of the helix distortion and the repairability of the damage (see e.g. Grossman, 1974). High priority should be given in future work to the *chemical* identification of the lesions produced by the various damaging agents in the DNA *in situ* in the living cell, to physical biochemical studies of the effects of characteristic lesions on DNA conformation, to studies of the distribution of the lesions in DNA introduced under *in vivo* conditions, and to the preparation of DNA substrates containing only a single type of lesion.

It is clear then that a classification of DNA base damage on the basis of chemical characteristics and the expected effect of the damage on DNA conformation must be based, at the present time, on many assumptions and much indirect evidence. Only a few examples of typical lesions for each category of damage are discussed here, and no attempt is made to be comprehensive. Undoubtedly the classification proposed below will have to be revised at a later date. It is hoped, however, that it may serve as a framework for organization in this highly complex field. It should be noted that *Class II* base damage may mostly elicit "ionizing type" repair and *Class III* "ultraviolet type repair" in mammalian cells according to the criteria of Regan and Setlow (1973).

STRUCTURAL CLASSIFICATION OF DNA BASE DAMAGE

I. Monofunctional Lesions Causing Negligible Helix Distortion

Products with essentially unaltered base-repairing and base-stacking capacities (e.g. 7-alkylguanine formed by most alkylating agents; 5-hydroxymethyl- or 5-hydroperoxymethylthymine formed by ionizing radiation).

II. Monofunctional Lesions Causing Minor Helix Distortion

A. Products with only slightly altered base-pairing and base-stacking properties; ring-saturation products (e.g. pyrimidine photohydrates; products of the 5,6-dihydroxy-dihydrothymine type induced by ionizing radiation).
B. Ring-contraction, ring-fragmentation, and ring-elimination products (e.g. ring-fragmentation products of pyrimidines and purines produced by ionizing radiation; ring elimination leading to apurinic and apyrimidinic sites induced by alkylating agents, ionizing radiation, or heat).

III. Lesions Causing Major Helix Distortion

A. Monofunctional lesions with bulky substituents or substantially altered base-pairing properties; intercalation damage [e.g. substitution products with *N*-acetoxy-acetylaminofluorine; intercalating polycyclic aromatic ring systems; alkylation products (6-alkoxyguanine, 3-alkyladenine);

arylalkylation products (exocyclic amin-substituted purines and pyrimidines)] .

B. Difunctional lesions
1. Intrastrand cross-linking of neighboring bases (e.g. cyclobutane-type
photodimers; ultraviolet-induced spore products; products formed by
nitrogen and sulfur mustards).
2. Interstrand cross-linking of bases [e.g. products formed by nitrogen
and sulfur mustards (diguaninyl products); psoralen-photosensitized
cross-links; cross-links induced by ionizing radiation; cross-linking as a
consequence of base elimination)] .
3. Cross-linking between DNA and proteins (e.g. radiation-induced cross-
links).

DESCRIPTION OF THE MAJOR CLASSES OF BASE DAMAGE

I. Monofunctional Lesions Causing Negligible Helix Distortion

The least effect on the base-pairing and base-stacking properties is expected
from monofunctional reactions involving an exocyclic substituent or a position
on the heterocyclic ring that does not participate in Watson-Crick base pairing;
aromaticity of the heterocyclic moiety has been retained, no major changes in
the electron density distribution of the ring have occurred, and no bulky
substituents have been introduced by the modification reaction. Such damage
may in many cases be ignored by repair enzymes and tolerated in the DNA
without substantial effects on its functional integrity. We have recently proposed
the term "functionally tolerable damage" for this type of lesion (Cerutti, 1974).
A typical lesion of this class is 7-alkylguanine, which is formed by most
alkylating agents regardless of whether they react by an $S_N 1$ or $S_N 2$ mechanism.
All available evidence indicates that 7-alkylguanine is not removed from the
DNA by excision-repair in bacterial or mammalian cells (see e.g. Strauss et al.,
this volume; Lawley, this volume) and does not miscode in the *in vitro* transcription of copolymers of uridylic acid and 7-methylguanylic acid by *Escherichia
coli* RNA polymerase (Ludlum, 1970). The thymine-methyl group in free DNA
(Swinehart et al., 1974), in isolated chromatin (Roti Roti et al., 1974), and in
DNA *in situ* in bacterial and mammalian cells (Roti Roti and Cerutti, 1974) was
found to be highly reactive to the attack by hydroxyl radicals formed by
homolysis of water by ionizing radiation. The location of the thymine-methyl
groups on the helix periphery may render them particularly susceptible to the
attack by hydroxyl radicals. Monofunctional products formed in this reaction,
e.g. 5-hydroxymethyl- and 5-hydroperoxymethyluracil (Latarjet et al., 1963),
may well represent "functionally tolerable damage" and may be bypassed by the
DNA-repair enzymes. It should be kept in mind, however, that even minor base

modifications of the type described here may have some effect on the helix parameters. A difference in sedimentation properties between DNA containing thymine and DNA from a dinoflagellate in which thymine is replaced by 5-hydroxymethyluracil was observed by Rae (1973). Repair enzymes have recently been described in bacterial and mammalian cells which recognize uracil residues in DNA and accomplish their removal (Smets and Cornelis, 1971; Carrier and Setlow, 1974; Lindahl, 1974).

II. Monofunctional Lesions Causing Minor Helix Distortion

A. Products with Only Slightly Altered Base-Pairing and Base-Stacking Properties

Monofunctional ring-saturation products are representative lesions of this type. Ring saturation, which is best known for pyrimidine nucleotides, results from the addition of hydroxyl radicals, oxygen and hydrogen atoms to the 5,6 double bond upon exposure to ionizing radiation (Ekert and Monier, 1959; Cadet and Téoule, 1971) or from the nucleophilic addition of a molecule of water to the 5,6 double bond upon irradiation with ultraviolet light (pyrimidine photohydration) (see e.g. Setlow and Setlow, 1972). Ring saturation without consecutive fragmentation of the 5-membered ring of adenine by the addition of hydroxyl radicals to carbon-8 under formation of 6-amino-8-hydroxy-7,8-dihydropurine has been described by VanHemmen (1971), but it is not known whether this reaction occurs in DNA. The presence of saturated pyrimidine nucleotides in polyribonucleotides has been shown to decrease their capability to form an ordered single-strand structure (Swinehart et al., 1972) or a regular double-stranded helix with a complementary strand (Cerutti et al., 1966). Local disruption of the native conformation of DNA by 5,6-dihydropyrimidine derivatives is mostly due to an increase in the pK_A by 2 to 3 units upon ring saturation, the loss of ring aromaticity and ring planarity which results in a decrease in the forces involved in base stacking. Uridine photohydrates, 6-hydroxy-5,6-dihydrouridine, were shown to be responsible for the ultraviolet inactivation of single-stranded RNA phage R17 (Remsen et al., 1970) and to miscode as cytidine in an *in vitro* translation system (Remsen and Cerutti, 1972). Deoxycytidine photohydrates, 6-hydroxy-5,6-dihydrodeoxycytidine, are formed by ultraviolet light in DNA (see e.g. in Patrick and Rahn, 1975), but their biological significance is not clear at the present time.

Products of the 5,6-dihydroxy-dihydrothymine type represent major lesions formed by γ-rays *in situ* in living cells (see e.g. Cerutti, 1974) and are removed from the DNA by an excision process in bacterial and mammalian systems (Hariharan and Cerutti, 1972, 1974a-c and this volume; Mattern et al., 1973). While the endonuclease responsible for the incision step has not yet been identified, recent results show that the exonucleotytic removal of these products from the DNA *in vitro* is accomplished by the 5'→3' exonuclease activity accompanying polymerase I of *E. coli* (Hariharan and Cerutti, 1974*a*), and that the last strand-rejoining step is carried out by polynucleotide kinase (Hariharan and Cerutti, 1974*b*). Enzymes in *E. coli* which may be responsible for the incision step in the repair of ring-saturation products are endonuclease II (Goldthwait et al., this volume), γ-endonuclease from *E. coli* (see Strniste and Wallace, this volume) and from *Micrococcus luteus* (Setlow and Carrier, 1973), and apurinic site enzyme (Verly and Paquette, 1972; Verly, this volume). The *uvrA uvrB* endonuclease of *E. coli* is not required for the excision of products of the 5,6-dihydroxy-dihydrothymine type (Hariharan et al., this volume and 1974*c*). Endonuclease activities which may recognize this type of damage have been described in crude HeLa-cell extracts (Bacchetti et al., 1972; Brent, 1973).

B. Ring-Contraction, Ring-Fragmentation, and Ring-Elimination Products

Characteristic examples for this type of damage are the ring contractions occurring upon irradiation of cytosine, thymine, thymidine, and uridylic acid with ionizing radiation in the presence of oxygen. The major products formed are 1-carbamylimidazolidone from cytosine (Hahn et al., 1973) and 5-hydroxy-hydantoin derivatives from thymine (Cadet and Téoule, 1971), thymidine (Cadet and Téoule, 1972), and uridylic acid (Ducolomb et al., 1974). Ring fragmentation under formation of deoxyribosylformamide (Cadet and Téoule, 1972) has been observed for thymidine. The central 4,5-double bond is the major reaction site of the purine bases. Addition of hydroxyl radicals leads to the saturation of both ring systems and is followed by the fragmentation of the five-membered ring (see Alexander and Lett, 1968). It is expected that these radiation-chemical reactions would lead to local distortion of the DNA helix, but it is not known whether they occur in DNA *in vitro* or *in vivo*.

Ring elimination under formation of apyrimidinic and apurinic sites, observed upon exposure of DNA to ionizing radiation (Téoule et al., 1974; Dunlap and Cerutti, 1975) or heat (see Lindahl and Ljungquist, this volume), represents an important secondary reaction following the formation of 7-alkylguanine by various alkylating agents (see Strauss et al., this volume; Lawley, this volume). No data are available on the effect of "aheterocyclic" sites on local DNA conformation. Specific endonucleases have been isolated from bacterial, plant, and mammalian sources which recognize apurinic sites and incise in their proximity (see Goldthwait et al.; Verly; Lindahl and Ljungquist, all in this volume). The relationship of the apurinic site enzyme from *E. coli* to endonuclease II and γ-endonuclease from the same organism has not yet been clarified.

III. Lesions Causing Major Helix Distortion

A. Monofunctional Lesions with Bulky Substituents or Substantially Altered Base-Pairing Properties; Intercalation Damage

Major local helix distortion is expected from ring substitution with large substituents which cannot be accommodated in the native DNA structure because of their steric requirements or because they intercalate. An example is the substitution of carbon-8 of guanine and adenine with the bulky acetylaminofluorine (AAF) group upon reaction of DNA with the carcinogen N-acetoxy-AAF (see Miller and Miller, 1969; Grunberger et al., 1974). Similar effects on DNA conformation may result from the intercalation of other polycyclic aromatic ring systems, such as the acridine dye ICR-170. Arylalkylation of the exocyclic amino groups of the purine bases and of cytosine by the carcinogen 7-bromomethylbenz[α]anthracene undoubtedly leads to local denaturation of the helix. Since bulky substituents are introduced at sites which are normally involved in Watson-Crick hydrogen bonding, the disruptive effects are expected to be particularly severe in this case.

Substantial changes in the electron density distribution and the tautomeric equilibria of the purine ring systems result from the substitution of nitrogen-3 in adenine and guanine by alkylating agents such as N-methyl-N-nitrosourea (MNUA), N-methyl-N'-nitro-N-nitrosoguanidine (MNNG), and methyl methanesulfonate (MMS). Alkylation of oxygen-6 of guanine by MNUA, MNNG, or MMS changes the base-pairing properties by fixing the ring in the unnatural lactim form (see Lawley, 1972).

Monofunctional lesions of the type described here are repairable by excision mechanisms in bacterial and mammalian cells (see Lawley, this volume; Strauss et al., this volume). The uvr and exr gene products have been implicated in their excision in $E.\ coli$ (Lawley and Orr, 1970; Venitt and Tarmy, 1972). Excision-repair of damage produced by N-acetoxy-AAF and ICR-170 is deficient in fibroblasts from patients with xeroderma pigmentosum (see Regan and Setlow, 1973).

B. Difunctional Lesions

1. Intrastrand Cross-Linking of Neighboring Bases. Intrastrand cross-links between neighboring bases are introduced into DNA both by ultraviolet radiation (see e.g. Setlow and Setlow, 1972; Patrick and Rahn, 1975) and by difunctional reagents such as sulfur and nitrogen mustards. Substantial conformational rearrangement is necessary to allow their formation, and the resulting lesions cause considerable local distortion of the DNA helix. Cyclobutane-type photodimers were found to be introduced preferentially into pyrimidine-rich regions of the DNA by a cooperative mechanism (see Brunk, this volume). The presence of cyclobutane-type photodimers was found to facilitate deoxycytidine photohydration in native DNA (Cerutti and Vanderhoek, 1974). Under certain conditions ultraviolet light introduces other types of intrastrand cross-links, such as the adduct 6-4'- [pyrimidine-2'-on]-thymine (Varghese and Patrick, 1969) and spore product 5-thyminyl-5,6-dihydrothymine (Donellan and Setlow, 1965; Varghese, 1970). However, cyclobutane-type photodimers are the most representative lesions of this type, and their formation in model systems, and in DNA *in vitro* and *in vivo,* as well as the pathways for their repair, have been studied more successfully and in more detail than any other type of DNA lesion (see the articles in this volume and recent reviews by Setlow and Setlow, 1972; Cleaver, 1974; Grossman, 1974; Patrick and Rahn, 1975). As are most lesions of Class III, intrastrand cross-links are recognized by the *uvrAuvrB* endonuclease of *E. coli.*

2. *Interstrand Cross-Linking of Bases.* Interstrand cross-links are introduced by difunctional chemicals such as sulfur and nitrogen mustards under the formation of a diguaninyl bridge (Lawley and Brookes, 1967; Lawley et al., 1969), by mitomycin C (Boyce and Howard-Flanders, 1964), by the photosensitized reaction of DNA with psoralen (see Howard-Flanders et al., this volume; Cole and Sinden, this volume), by ionizing radiation (see Smith, 1974), and at apurinic sites formed, e.g. by the elimination of 7-alkyl-guanine (Verly, 1974). Considerable helix distortion is undoubtedly introduced by such lesions. Strand separation cannot occur at the site of a cross-link. Interstrand cross-links introduced by mustards and the photosensitized reaction of DNA with psoralen are recognized by the *uvrAuvrB* endonuclease of *E. coli.* Psoralen cross-links are removed from the DNA in *E. coli* by a combination of prereplication excision-repair and recombinational repair (see Howard-Flanders et al., this volume; Cole and Sinden, this volume).

3. *Cross-Linking Between DNA and Proteins.* Radiation-induced cross-linking of DNA and proteins, both by ultraviolet light and by ionizing radiation, has been demonstrated under both *in vitro* and *in vivo* conditions. Little is known about the structure and the biological implications of such lesions. This topic has recently been reviewed (Smith, 1975).

ACKNOWLEDGMENT

This work was supported by a grant from the National Institutes of Health and a contract from the U.S. Atomic Energy Commission.

REFERENCES

Alexander, P. and Lett, J. (1968). *Comp. Biochem.* 27, 267–356.

Bacchetti, S., Van der Plaas and Veldhuisen, G. (1972). *Biochem. Biophys. Res. Commun.* 48, 662–668.

Boyce, R. and Howard-Flanders, P. (1964). *Z. Vererbungsl.* 95, 345–350.

Brent, T. (1973). *Biophys. J.* 13, 399–401.

Cadet, J. and Téoule, R. (1971). *Biochim. Biophys. Acta* 238, 8–26.

Cadet, J. and Téoule, R. (1972). *Tetrahedron Lett.* 31, 3225–3228.

Carrier, W. and Setlow, R. (1974). *Fed. Proc.* 33, 1599.

Cerutti, P. (1974). *Naturwissenschaften* 61, 51–59.

Cerutti, P. and Vanderhoek, J. (1975). In *Photochemistry and Photobiology of Nucleic Acids* (Wang, S. Y. & Patrick, M., eds.). Gordon Breach, New York (in press).

Cerutti, P., Miles, H. and Frazier, J. (1966). *Biochem. Biophys. Res. Commun.* 22, 466–472.

Cleaver, J. (1974). *Advan. Radiat. Biol.* 4, 1–75.

Donnellan, J. and Setlow, R. (1965). *Science* 149, 308–310.

Ducolomb, R., Cadet, J. and Téoule, R. (1974). *Int. J. Radiat. Biol.* 25, 139–149.

Ekert, B. and Monier, R. (1959). *Nature* 184, BA58–BA59.

Grossman, L. (1974). *Advan. Radiat. Biol.* 4, 77–126.

Grunberger, D., Blobstein, S. and Weinstein, I. (1974). *J. Mol. Biol.* 83, 459–468.

Hahn, B., Wang, S., Flippen, J. and Karle, I. (1973). *J. Amer. Chem. Soc.* 95, 2711–2712.

Hariharan, P. and Cerutti, P. (1972). *J. Mol. Biol.* 66, 65–81.

Hariharan, P. and Cerutti, P. (1974a). *Biochem. Biophys. Res. Commun.* 61, 971–976.

Hariharan, P. and Cerutti, P. (1974b). *Biochem. Biophys. Res. Commun.* 61, 375–379.

Hariharan, P. and Cerutti, P. (1974c). *Proc. Nat. Acad. Sci. U.S.A.* 71, 3532–3536.

Latarjet, R., Ekert, B. and Demerseman, P. (1963). *Radiat. Res.* 3(Suppl.), 247–256.

Lawley, P. (1972). In *Topics in Chemical Carcinogenesis* (Nakahara, W., Takayama, S., Sugimura, T. and Odashima, S., eds.), p. 237. University Park Press, Baltimore; and University of Tokyo Press, Tokyo.

Lawley, P. and Brookes, P. (1967). *J. Mol. Biol.* 25, 143–160.

Lawley, P. and Orr, D. (1970). *Chem.-Biol. Interact.* 2, 154–157.

Lawley, P., Lethbridge, J., Edwards, P. and Shooter, K. (1969). *J. Mol. Biol.* 39, 181–198.

Lindahl, T. (1974). *Proc. Natl. Acad. Sci. U.S.A.* 71, 3649–3653.

Ludlum, D. (1970). *J. Biol. Chem.* 245, 477–482.

Mattern, M., Hariharan, P., Dunlap, B. and Cerutti, P. (1973). Nature New Biol. 245, 230–232.

Miller, E. and Miller, J. (1969). *Ann. N.Y. Acad. Sci.* 163, 731–750.

Patrick, M. H. and Rahn, R. O. (1975). In *Photochemistry and Photobiology of Nucleic Acids (Wang, S. Y. and Patrick, M., eds.)*. Gordon Breach, New York (in press).

Rae, P. M. (1973). *Proc. Nat. Acad. Sci. U.S.A.* 70, 1141–1145.

Regan, J. and Setlow, R. (1973). In *Chemical Mutagens* (Hollaender, A., ed.), vol. 3, pp. 151–170. Plenum Press, New York.

Remsen, J. and Cerutti, P. (1972). *Biochem. Biophys. Res. Commun.* 48, 430–436.

Remsen, J., Miller, N. and Cerutti, P. (1970). *Proc. Nat. Acad. Sci. U.S.A.* 65, 460–466.

Roti Roti, J. and Cerutti, P. (1974). *Int. J. Radiat. Biol.* 35, 413–417.

Roti Roti, J. and Stein, G. and Cerutti, P. (1974). *Biochemistry* 13, 1900–1904.

Setlow, R. and Carrier, W. (1973) *Nature New Biol.* 241, 170–172.

Setlow, R. and Setlow, J. (1972). *Ann. Rev. Biophys. Bioeng.* 1, 293–346.

Shafranovskaya, N., Trifonov, E., Lazurkin, Y. and Frank-Kamenetskii, M. (1973). *Nature New Biol.* 241, 58–60.

Smets, L. and Cornelis, T. (1971). *Int. J. Radiat. Biol.* **19**, 445–457.

Smith, K. (1975). In *Photochemistry and Photobiology of Nucleic Acids* (Wang, S. Y. and Patrick, M., eds.). Gordon Breach, New York (in press).

Swinehart, J., Bobst, A. and Cerutti, P. (1972). *FEBS Lett.* **21**, 56–58.

Swinehart, J., Lin, W. and Cerutti, P. (1974). *Radiat. Res.* **58**, 166–175.

Téoule, R., Bonicel, A., Bert, C., Cadet, J. and Polverelli, M. (1974). *Radiat. Res.* **57**, 46–58.

VanHemmen, J. (1971). *Nature New Biol.* **231**, 79–80.

Varghese, A. (1970). *Biochem. Biophys. Res. Commun.* **38**, 484–490.

Varghese, A. and Patrick, M. (1969). *Nature* **223**, 299–300.

Venitt, S. and Tarmy, E. (1972). *Biochim. Biophys. Acta* **287**, 38–51.

Verly, W. (1974). *Biochem. Pharmacol.* **23**, 3–8.

Verly, W. and Paquette, Y. (1972). *Can. J. Biochem.* **50**, 217–224.

2

The Nature of the Alkylation Lesion in Mammalian Cells

B. Strauss, D. Scudiero, and E. Henderson

Department of Microbiology
The University of Chicago
Chicago, Illinois 60637

ABSTRACT

Methylating agents may produce as many as nine alkylated purine and pyrimidine adducts in DNA, as well as forming phosphotriesters and inducing apurinic sites and strand breaks. Although some of these products are formed in proportionately small amounts, there are sufficient sites affected in the DNA of a mammalian cell to make even the most minor product of potential biological significance. It is not possible to specify the exact reaction sites resulting in biological damage, but it is possible to quantitate the excision-repair of such damage both in the bulk of the DNA and at DNA growing points. Excision-repair can be measured in the bulk of the DNA by determining the specific activity of the NaCl eluate of a benzoylated naphthoylated DEAE-cellulose column of extracts of cells after treatment and incubation in the presence of hydroxyurea and labeled thymidine. The average number of nucleotides inserted per methyl methanesulfonate-induced methyl group is 0.1, per apurinic site is 9. Repair in growing-point regions after methyl methanesulfonate treatment occurs to approximately the same extent as in the bulk of the DNA.

Alkylating agents are electrophilic reagents which can combine with DNA and other cellular macromolecules. Different alkylating agents have specific biologi-

Table 1. Reported Sites of Attack by Electrophilic Reagents on DNA

Compound	Reaction sites[a]					Reaction Type	Reference
	Adenine	Cytosine	Guanine	Thymine	Phosphate		
N-acetoxy-acetyl-aminofluorene (N-acetoxy-AAF)	C-8 (1)	—	C-8 (100)	—	—	S_N1	Miller and Miller (1969)
7-Bromomethylbenz-[α]anthracene (7-BMBA)	N^6 (6-amino)	N^4 (4-amino) (probable)	N^2 (2-amino) (major product)			S_N1	Dipple et al. (1971)
N-Methyl-N-nitroso-urea (MNUA)	N-1 (1.4) N-3 (11.2) N-7 (2.5)	N-3 (0.6)	N-3 (1.1) N-7 (75.7) 0-6 (7.3)	N-3 (0.3) 0-4 (0.1)	(18)	S_N1	Lawley and Shah (1973); Lawley, 1973
N-Methyl-N'-nitro-N-nitrosoguanidine (MNNG)	N-1 (1) N-3 (12)	N-3 (2)	N-3 (2) N-7 (67) O-6 (7)			S_N1	Lawley and Thatcher (1970)
Ethyl Methane-sulfonate (EMS)					(15)	$S_N1 + S_N2$	Bannon and Verly (1972)
Methylmethane-sulfonate (MMS)	N-1 (5) N-3 (9)	N-3 (1)	N-3 (0.68) N-7 (86) O-6 (0.34)		(1)	S_N2	Lawley and Shah (1972)

[a]Values in parentheses give approximate relative amounts of reaction products. In some cases these values are compiled from different publications and therefore do not add up to 100%. An entry means that the product has been reported. Lack of entry means only that the reference given does not record that product.

cal properties; for example, N-methyl-N'-nitro-N-nitrosoguanidine (MNNG) produces a different pattern of mutations and is more carcinogenic than methyl methanesulfonate (MMS) (Malling and de Serres, 1969). This specificity of biological effect is paralleled by a specificity of chemical reaction, since each different alkylating agent gives a different but characteristic "mix" of reaction products, particularly of minor reaction products. The major reaction product of methylating agents with DNA is 7-methylguanine, and the properties which the alkylating agents have in common are likely to be due in some way to formation of this substance. However, alkylation of all purines and pyrimidines can be demonstrated, and most of the organic bases react at several sites (Table 1). Phosphotriester formation also occurs as a result of alkylation (Bannon and Verly, 1972) in both RNA and DNA but with different consequences. The phosphotriesters are stable in DNA but result in chain breakage in RNA (Lawley, 1973; Shooter et al., 1974). There is a specificity of reaction, so that 15% of the total alkylations after treatment with ethyl methanesulfonate are phosphotriesters, compared with only 1% after treatment with MMS (Bannon and Verly, 1974).

Although 7-methylguanine is the major product of methylating agents reacting with DNA, this adduct is ignored by the bacterial and mammalian repair enzymes (Lawley and Orr, 1970; Lawley, this volume) and is replicated without problem (Prakash and Strauss, 1970). However, 7-methylguanine deoxynucleotide is inherently unstable in DNA and depurinates, leaving sites which hydrolyze both spontaneously (Strauss and Hill, 1970) and as a result of endonuclease attack (Ljungquist and Lindahl, 1974; Ljungquist et al., 1974; Hadi and Goldthwait, 1971), leading to single-strand interruptions which are accessible to the excision-repair system. Therefore, although the initial formation of 7-methylguanine is not recognizable "damage," the consequence of the formation of this product is important since apurinic sites will eventually appear at the positions of 7-methylguanine formation.

Not all alkylations produce unstable products. Reaction of N-acetoxy-acetylaminofluorene (N-acetoxy-AAF) at the C-8 of adenine or guanine or of 7-bromomethylbenz[α]anthracene (7-BMBA) at the amino groups of cytosine, guanine, or adenine produces products which should be chemically stable. Combination with amino groups at these positions is likely to affect the hydrogen-bonding characteristics of the bases and/or make the DNA a substrate for the repair endonuclease because of the deforming effect of the adduct on the structure of DNA (Grunberger et al., 1974).

MNNG reacts with DNA to form a larger proportion of the chemically stable 6-methoxyguanine than does MMS, and this special reactivity reflects the different mechanism of alkylation by each compound. MNNG reacts with unimolecular kinetics via a methyldiazonium ion ($S_N 1$ type) intermediate, in contrast to MMS, which reacts via the $S_N 2$ mechanism, in which a transition complex is formed with the nucleophile by a reaction with bimolecular kinetics (Lawley,

1966). In general, $S_N 1$-type compounds give higher yields of 6-methoxyguanine. Methoxyguanine is probably recognized and removed by the UV endonuclease (Lawley and Orr, 1970). The finding that different agents produce different "mixes" of minor reaction products makes it tempting to ascribe differences in the biological activity of different compounds to the particular differences in the reaction products which such compounds form. Thus, the special mutagenic and carcinogenic effects of MNNG might be attributed to the higher proportion of 6-methoxyguanine formed by this substance, although there is no direct evidence indicating how 6-methoxyguanine produces its biological effects.

Notwithstanding the difference in the amount of 6-methoxyguanine formed by different methylating agents, it is important to realize that all of the alkylating agents produce a variety of products, not only in the DNA but also in other macromolecules. Furthermore, the degradation of some adducts permits secondary reactions. For example, low pH results in the formation of DNA cross-links in otherwise untreated DNA (Verly, 1974) as a result of the depurination which occurs under acid conditions. Since alkylation leads to the formation of 7-methyldeoxyguanosine residues at which depurination occurs, methylation should eventually produce cross-links in the DNA, although such a reaction has not been reported. We can estimate the number of cross-links expected after alkylation from the data presented by Verly (1974), who gives 1 cross-link per 140 apurinic sites as an estimate. A rough quantitative description of the different alkylation products to be expected in a mammalian cell as a result of treatment with either MMS or MNNG is given in Table 2.

Roberts et al. (1971) have measured the reactivity of cellular DNA with a number of alkylating agents, and their data are the basis for our calculations (Table 2). However, there is a serious problem in the use of these data for biological comparisons because of the different sensitivities of cell lines. Roberts' HeLa line gives 37% survival at a dose of 0.125 mM MMS for 1 hr. Our experiments have been done with a heteroploid human cell line, HEp.2, which gives 37% survival at an MMS dose of 2.2 mM for 1 hr (Coyle et al., 1971). MNNG is much more toxic for Roberts' HeLa line than is MMS per methyl group fixed (Table 2); on the other hand, MMS and MNNG are almost equitoxic for a particular Chinese hamster cell line used in their laboratory. The chemical data reported by Roberts et al. (1971) agree almost perfectly with our own measurements in HEp.2 cells. At a dose of 0.125 mM MMS for 1 hr, Roberts et al. (1971) report 3.5×10^{11} methyl groups fixed per μg of DNA. We find an average of 4.5×10^{11} methyl groups fixed per μg of DNA after 2 hr of alkylation at 0.1 mM MMS (Table 4), equivalent to 2.8×10^{11} methyl groups under Roberts' conditions. Considering the approximations in our experiments, this is excellent agreement. The finding that different mammalian cell lines have very different sensitivities to the same amount of alkylation and different ratios of response to different sets of compounds implies that there are different repair mechanisms in the different strains and that products lethal for one cell type

Table 2. Expected Products of the Reactions of MMS and MNNG at Equitoxic Doses with a Human Cell Line, HeLa

| Product | Number of alkylated products | | | |
| | Per cell | | Per replicon (30 μm) | |
	MMS	MNNG	MMS	MNNG
1-Methyladenine	2.4×10^5	1.17×10^3	1.6	.0078
3-Methyladenine	4.57×10^5	1.40×10^4	3.0	0.093
3-Methylcytosine	5.08×10^4	2.34×10^3	0.335	0.0156
3-Methylguanine	3.45×10^4	2.34×10^3	0.23	0.0156
7-Methylguanine	4.37×10^6	7.84×10^4	29.1	0.52
O^6-Methylguanine	1.7×10^4	8.54×10^3	0.11	0.056
Phosphotriesters	$5. \times 10^4$	$(1.8 \times 10^4)a$	0.335	(0.119)
Apurinic sites (1st hr)	5.3×10^4	1.01×10^3	0.35	0.067
Cross-links	3.8×10^2	7.2	0.003	4.8×10^{-5}
Breaks (spontaneous)	69.0	1.3	0.0004	8.8×10^{-6}
Total	$5.08 \times 10^{6}b$	$1.17 \times 10^{5}b$	35.06	0.894

[a] Based on an estimate of 15% of reaction products for S_N1 reagents. [b]
[b] Roberts et al. (1971), Figs. 1, 3, 5.
MMS–0.125 mM for 1 hr gives 37% survival.
MNNG–0.057 μM for 1 hr gives 37% survival.
Methyl groups fixed at a dose giving 37% survival: MMS–0.586 μmole/g DNA: MNNG–0.0135 μmole/g DNA (Roberts et al., 1971).
DNA per Hela Cell: 14.4×10^{-12} g.
Av. M.W. per nucleotide = 3.27×10^2; 1 μmole = 2.94×10^3 base pairs (Sober, 1970).

may not be lethal for a second. The precise numbers of Table 2 are useful only insofar as they give an idea of the relative numbers of different products to be expected after alkylation at a biologically meaningful dose.

Values of k_1 and k_2 in the series

$$\text{Alkylated DNA} \xrightarrow{k_1} \text{apurinic DNA} \xrightarrow{k_2} \text{single-strand breaks}$$

are available for reaction at 70°C (Strauss and Hill, 1970; Lindahl and Nyberg, 1972), and Lawley and Brooks (1963) and Lindahl and Nyberg (1972) have published data making possible a calculation of the activation energy of each reaction and therefore the ks at 37°C. The k_1 used in our calculation is an overall measured value for total depurination (Strauss and Hill, 1970). It includes depurination at both 7-methyldeoxyguanosine and 3-methyldeoxyadenosine sites. Although the ratio with which these compounds are formed by methylating agents is about 10:1 (Table 1), the ratio of their rates of depurination is 1:5. Lawley (this volume) gives half-lives of 30 hr for 3-methyladenine and 150 hr for 7-methylguanine, corresponding to ks of 3.8×10^{-4} min^{-1} and $7.7 \times$

10^{-5} min^{-1} at pH 7 and 37°C. An apurinic site is only twice as likely to form in the first hour after treatment as a result of the depurination of methylguanine as methyladenine, and we have therefore used the measured overall rate for our calculations. With a k_1 of 1.84×10^{-4} min^{-1} and a k_2 of 2.17×10^{-5} at 37°C we would expect 1.089×10^{-2} of the 7-methylguanine and 3-methyladenine sites (assuming these to be the only unstable reaction products) to become apurinic within the first hour after treatment, and 1.3×10^{-3} of the apurinic sites to be hydrolyzed. This corresponds to 5.3×10^4 apurinic sites and 6.9×10^1 single-strand breaks per hour. Since there are 2.65×10^{10} nucleotides in a cell with 14.4 pg of DNA, these figures correspond to one apurinic site per 2.5×10^5 base pairs (1.6×10^8 daltons, 85 μM) and about one break per 3.8×10^8 nucleotides plus one cross-link per 7×10^7 nucleotides. The number of breaks and cross-links is small, but the number of apurinic sites produced per hour corresponds to about 0.3 per replicon per hour. Since apurinic sites are rapidly converted *in vivo* to single-strand breaks by the action of endonuclease II (Ljungquist et al., 1974; Hadi and Goldthwait, 1971), it might seem reasonable to consider the apurinic site as the effective toxic alkylation lesion in cells.

The assumption that spontaneously produced apurinic sites are the significant lesions in alkylated cells requires that new damage be continually produced. However, the data of Scudiero and Strauss (1974), Coyle et al. (1971), and Buhl and Regan (1973) show that cells recover from the effects of treatment with MMS by the seventh hour after treatment. Since almost as many apurinic sites would be expected to be produced in the seventh as in the first hour after treatment, it seems necessary to suppose that all of the damage is induced by the time treatment with MMS has ceased. At least two possible known alkylation reactions could give this result. Lawley and Shah (1972) have demonstrated that 0.36% of the alkylation products resulting from treatment of DNA with MMS are O^6-methyldeoxyguanosine residues. We calculate that of 5.1×10^6 methyl groups, 1.7×10^4 would be in O^6 derivatives or about one per 10 replicons of 30 μM (Table 2). An additional, stable reaction product would be the phosphotriester, estimated by Bannon and Verly (1972) for MMS as 1% of the total reaction products or 5×10^4 methyl groups, equivalent to 0.3 phosphotriester per replicon; i.e., about the same number of phosphotriesters and apurinic sites. It is therefore impossible to specify which reaction product is *the* one responsible for a biological effect; indeed, it is difficult to conclude that a given biological effect is solely due to a specific reaction product. The calculations also illustrate that even a very minor reaction product may produce a sufficient number of altered sites in the DNA to result in a major biological effect.

Although it is impossible with most of the known reagents to produce only a single lesion by alkylation (although N-acetoxy-AAF appears relatively specific in its products), we still need to understand the mode of biological action of the different compounds. Furthermore, since it has been reported that one of the methylating agents, MNNG, is particularly mutagenic at DNA growing points

(Cerdà-Olmedo et al., 1968), it would be particularly interesting to compare the reaction of known mutagenic agents with replicating and nonreplicating regions of the DNA. Even without specifying the reaction products excised, it is possible to measure the extent of repair resulting from alkylation damage in the growing-point region and in the bulk of the DNA. We have not as yet measured the changes due to MNNG but we do have data on the repair of DNA alkylated with MMS. Our methodology permits the separation of growing points from the bulk of the DNA because of the single-stranded nature of a portion of the growing-point region (Scudiero and Strauss, 1974). This single-strandedness makes both newly synthesized and parental DNA in the growing-point region adhere to columns of benzoylated naphthoylated DEAE-cellulose (BND-cellulose). Native DNA is eluted from these columns by washing with 1 M NaCl; growing points and regions of single-stranded DNA are eluted by washing with 50% formamide in 1 M NaCl. (In our previous studies we used 2% caffeine in 1 M NaCl to elute single-stranded DNA. The formamide is more easily dialyzed away, and residual traces do not interfere with studies of ultraviolet absorbancy, as suggested to us by Dr. M. Radman.)

Excision-repair can be measured as the increased $[^3H]$ dThd radioactivity incorporated by cultures in the presence of concentrations of hydroxyurea high enough to prevent normal DNA synthesis (Brandt et al., 1972). The incorporated radioactivity can be separated into repair in the growing-point region and in the bulk of the DNA by passage through a BND-cellulose column. We found that treatment of cells with MMS and incubation with $[^3H]$ dThd in the presence of hydroxyurea resulted in an increase in incorporation into both NaCl and caffeine eluates of the BND column (Scudiero and Strauss, 1974). When the treated cells were incubated with $[^3H]$ BrdUrd after MMS treatment, the NaCl eluate contained radioactivity only in the light region of a CsCl gradient (Fig. 1), indicating non-semiconservative (repair) synthesis. Therefore, the NaCl eluate of the BND-cellulose column can be used to measure repair, and we have compared the specific activities of excision-repair in the bulk of the DNA as a function of the number of induced lesions (measured as inhibition of DNA synthesis) after treatment with N-acetoxy-AAF, 7-BMBA, and MMS (Fig. 2). As expected (Regan and Setlow, 1973), (N-acetoxy-AAF and 7-BMBA give larger values for repair per lesion than does MMS. We have also demonstrated repair in the DNA growing-point region after treatment of cells with 7-BMBA and with N-acetoxy-AAF (Table 3).

If the increased incorporation resulting from treatment with MMS (see Scudiero and Strauss, 1974), N-acetoxy-AAF, and 7-BMBA were solely the result of repair, there would seem to be about as much excision-repair in the growing-point region as in all the rest of the DNA. This conclusion appeared to be contrary to the results of Slor and Cleaver (1973), who reported equal excision-repair activity throughout the growth cycle after treatment of cells with ultraviolet light. We therefore decided to determine the specific activity of the

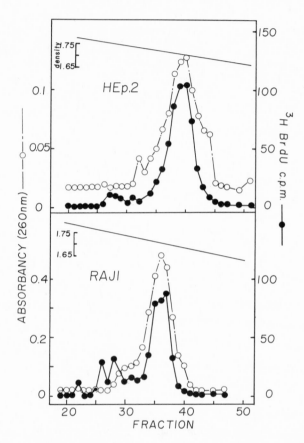

Fig. 1. Non-semiconservative (repair) synthesis in the NaCl eluate from a BND-cellulose column. Cells were preincubated 90 min with 1.5×10^{-5} M BrdUrd. Hydroxyurea was added and 30 min later [^{3}H]BrdUrd was added to give 10 μCi/ml. MMS was added simultaneously. After 60 min the MMS was removed and the cells were incubated an additional hour at 37°C in fresh medium containing [^{3}H]BrdUrd and hydroxyurea. The cells were then harvested, washed, and lysed, and the sheared lysates were passed through BND-cellulose. After elution with NaCl (Scudiero and Strauss, 1974) and dialysis, the eluate was centrifuged to equilibrium in a neutral CsCl gradient. Upper curve: HEp.2 cells, 2.5 mM MMS, 2.0 mM hydroxyurea. Bottom curve: RAJI (human lymphoblast line), 1.7 mM MMS, 1.0 mM hydroxyurea.

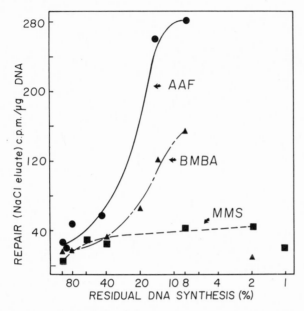

Fig. 2. *DNA repair in RAJI as a function of the residual DNA synthesis after treatment with three inhibitory agents.* Repair is measured as the specific activity of the DNA in the NaCl eluate of a BND-cellulose column. Cells were preincubated with 10 mM hydroxyurea for 30 min, drug was added, and then after 5 min [³H] dThd was added for an additional 60 min before harvest. Inhibition of DNA synthesis by drug was measured in the absence of hydroxyurea. Circles: N-acetoxy-AAF, 2 μg/ml = 45% residual synthesis. Triangles: 7-BMBA, 0.1 μg/ml = 40% residual synthesis. Squares: MMS, 55 μg/ml = 25% residual synthesis.

repaired DNA per induced lesion in DNA isolated from the growing-point region and elsewhere in the DNA. We alkylated HEp.2 cells with radioactive MMS and determined the amount of methylation at growing points and in the bulk of the DNA. We assume that whatever the biologically important lesion(s), the number of such lesion(s) will be proportional to the number of methyl groups fixed. In a separate experiment we then determined the specific repair activity of DNA from alkylated HEp.2 cells treated with MMS and then incubated with [³H]-BrdUrd in the absence of hydroxyurea prior to passage through the column. The BND-cellulose NaCl and formamide eluates were separated and centrifuged in alkaline CsCl. The light, parental strand was then rebanded in neutral CsCl, and the number of nucleotides inserted by repair synthesis per μg of DNA was determined from the radioactivity and from the absorbancy at 260 nm (Table 4). We used the specific activities of the radioactive compounds and their counting efficiencies to calculate the number of CH_3 groups fixed per μg of DNA. The estimated number of apurinic sites was calculated using the data in Table 2.

There is only slightly more repair per unit of DNA in the growing-point region than in the bulk of the DNA (Table 4), as reported by Slor and Cleaver (1973). Although all of the activity in the BND-cellulose NaCl eluate is the result of excision-repair (Fig. 1), the caffeine (or formamide) eluate, insofar as it must contain newly synthesized regions, includes some newly synthesized material formed by semiconservative synthesis. Incorporation of isotope into this fraction after alkylation can never be solely due to repair synthesis. An overestimate of repair in the growing-point region will result unless the column eluate is purified in alkaline CsCl. We think this factor accounts for the difference between the data indicating a preference for repair in the growing-point region (Table 3) and the results of Slor and Cleaver (1973).

The data of Table 4 illustrate the impossibility, at the present time, of identifying the biologically significant lesion after MMS treatment. Regan and Setlow (1973) report that MMS-induced repair is similar to that induced by X-rays, in which patches of about 3 nucleotides are inserted (Painter and Young, 1972). We calculate that the average number of nucleotides inserted by repair per MMS-methyl group fixed is about 0.1, confirming the observations that most of the methyl groups are ignored by the repair system. The average number of bases inserted per (calculated) apurinic site is about 9, which is somewhat too large. Some other methylation-induced lesion or lesions may be excised and repaired, but the data do not permit their identification.

Table 3. Effect of N-Acetoxy-AAF on [^{3}H] dThd Incorporation by RAJI Cells Incubated in the Presence of 10 mM Hydroxyurea[a]

Treatment	Cpm	1.0 M NaCl		50% formamide	
		(cpm	increase)	(cpm	increase)
10 mM HU (control)	10,504	1867		8197	
10 mM HU + 0.5 µg/ml AAAF	5,559	767		4535	
10 mM HU + 2.0 µg/ml AAAF	7,306	1743	976	5410	875
10 mM HU + 10.0 µg/ml AAAF	13,446	5777	4034	7557	2147
10 mM HU + 100.0 µg/ml AAAF	3,442	1047		1904	

[a]10 mM HU permits 1.22% residual thymidine incorporation. Protocol: 0 min, add HU; 30 min, add AAAF; 90 min, add [^{3}H] dThd; 150 min, harvest.

Table 4. Comparison of Repair per MMS-Induced Lesion in the Growing Point
and in the Bulk DNA of HEp.2 Cells[a]

BND-cellulose eulate	Repair (nucleotides inserted per μg of DNA)	Alkylation		Nucleotides inserted	
		CH_3 groups added per μg of DNA	Apurinic sites after 1 hr	Per CH_3 group	Per apurinic site
NaCl	4.09×10^{11}	3.25×10^{12}	4.1×10^{10}	0.125	10.0
		4.25×10^{12}	5.4×10^{10}	0.096	7.6
50% formamide	4.91×10^{11}	4.25×10^{12}	5.4×10^{10}	0.115	9.1
		1.09×10^{13}	1.25×10^{11}	0.045	3.9

[a]Repair .was measured as described in the text after treatment with 2.5 mM MMS for 1 hr. Alkylation was measured by treatment with 0.1 mM [^{14}C]MMS for 2 hrs. In the table, the alkylation values given are those measured multiplied by $2.5/0.1 \times 2 = 12.5$ to make the values comparable. Repair was measured with a mixture of BrdUrd and [^{3}H]dThd at 1.03 mCi/mmole assuming no discrimination between dThd and BrdUrd as has been demonstrated for HEp.2 cells The amount of methylation was determined relative to the amount of DNA determined by uniform labeling with [^{3}H]dThd at a specific activity of 3 Ci/mmole. The [^{14}C]MMS had a specific activity of 46 mCi/mmole. Our calculations are based on a counting efficiency of 0.5 for ^{14}C and 0.2 for ^{3}H.

If the number of nucleotides inserted per lesion by excision-repair is very small, then bromodeoxyuridine may be a bad tracer for the study of alkylation-induced damage. Since it is likely that either adenine or guanine is replaced, the use of bromodeoxyuridine requires that at least two bases be replaced and that thymine be a nearest neighbor of guanine or adenine. Although our data are presented to show that repair of MMS-induced damage occurs about equally well in the growing point and in the rest of the DNA, the results raise questions about the nature of the lesions and about the suitability of the experimental system often used for the study of alkylation-induced excision-repair.

ACKNOWLEDGMENT

The experimental work reported herein was supported by grants from the National Institutes of Health (GM 07816), from the University of Chicago Cancer Research Center (CA14599-01) and from the American Cancer Society (VC 112). Dominic Scudiero is a trainee in a genetics training program (GM 00090) and Earl Henderson is a trainee in a microbiology training program (GM 00603), both supported by the National Institutes of Health.

NOTE ADDED IN PROOF

We have now determined that in contrast to the result obtained with MMS, treatment of cells with MNNG results in a fourfold higher specific activity of repair at the growing point as compared with the bulk of the DNA.

REFERENCES

Bannon, P. and Verly, W. (1972). *Eur. J. Biochem.* **31**, 103–111.

Brandt, W., Flamm, W. and Bernheim, N. (1972). *Chem.-Biol. Interact.* **5**, 327–339.

Buhl, S. and Regan, J. (1973). *Mutat. Res.* **18**, 191–197.

Cerdà-Olmedo, E., Hanawalt, P. and Guerola, N. (1968). *J. Mol. Biol.* **33**, 705–719.

Coyle, M., McMahon, M. and Strauss, B. (1971). *Mutat. Res.* **12**, 427–440.

Dipple, A., Brookes, P., Mackintosh, D. and Rayman, M. (1971). *Biochemistry* **10**, 4323–4330.

Grunberger, D., Blobstein, S. and Weinstein, I. B. (1974). *J. Mol. Biol.* **83**, 459–468.

Hadi, S. and Goldthwait, D. (1971). *Biochemistry* **10**, 4986–4994.

Lawley, P. (1966). *Progr. Nucleic Acid Res. Mol. Biol.* **5**, 89–131.

Lawley, P. (1973). *Chem.-Biol. Interact.* **7**, 127–130.

Lawley, P. and Brookes, P. (1963). *Biochem. J.* **89**, 127–138.

Lawley, P. and Orr, D. (1970). *Chem.-Biol. Interact.* **2**, 154–157.

Lawley, P. and Shah, S. (1972). *Chem.-Biol. Interact.* **5**, 286–288.

Lawley, P. and Shah, S. (1973). *Chem.-Biol. Interact.* **7**, 115–120.

Lawley, P. and Thatcher, C. (1970). *Biochem. J.* **116**, 693–707.

Lindahl, T. and Nyberg, B. (1972). *Biochemistry,* **11**, 3610–3618.

Ljungquist, S., Andersson, A. and Lindahl, T. (1974). *J. Biol. Chem.* **249**, 1536–1540.

Ljungquist, S. and Lindahl, T. (1974). *J. Biol. Chem.* **249**, 1530–1535.

Malling, H. and de Serres, F. (1969). *Ann. N.Y. Acad. Sci.* **163**, 788–800.

Miller, E. and Miller, J. (1969). *Ann. N.Y. Acad. Sci.* **163**, 731–750.

Painter, R. and Young, B. (1972). *Mutat. Res.* **14**, 225–235.

Prakash, L. and Strauss, B. (1970). *J. Bacteriol.* **102**, 760–766.

Regan, J. and Setlow, R. (1973). In *Chemical Mutagens* (Hollaender, A., ed.), vol. 3, pp. 151–170. Plenum Press, New York.

Roberts, J., Pascoe, J., Smith, B. and Crathorn, A. (1971). *Chem.-Biol. Interact.* **3**, 29–47.

Scudiero, D. and Strauss, B. (1974). *J. Mol. Biol.* **83**, 17–34.

Shooter, K., Howse, R., Shah, S. and Lawley, P. (1974). *Biochem. J.* **137**, 303–312.

Slor, H. and Cleaver, J. (1973). *Genetics* 74(Suppl.), S258.

Sober, H. (ed.). (1970). *Handbook of Biochemistry,* 2nd ed. Chemical Rubber Company, Cleveland.

Strauss, B. and Hill, T. (1970). *Biochim. Biophys. Acta* **213**, 14–25.

Verly, W. (1974). *Biochem. Pharmacol.* **23**, 3–8.

3

"Excision" of Bases from DNA Methylated by Carcinogens *in Vivo* and Its Possible Significance in Mutagenesis and Carcinogenesis

P. D. Lawley

Chester Beatty Research Institute
Institute of Cancer Research
Royal Cancer Hospital
London SW3 6JB, England

The methylating carcinogens can be classified into two broad groups according to their chemical reactivities—those typifying Ingold's S_N2 alkylating agents, such as methyl methanesulfonate and dimethyl sulfate, and those such as N-methyl-N-nitrosourea and N-methyl-N'-nitro-N-nitrosoguanidine, envisaged to methylate through the highly reactive methyldiazonium ion, often described as S_N1 agents. These differences in reaction mechanism are reflected in different ratios of methylation products formed in DNA. It should be noted that dimethylnitrosamine requires metabolic activation to yield the same methylating species as generated by hydrolysis of N-methyl-N-nitrosourea. The principal distinction of the S_N1 agents is their ability to alkylate O-atom sites in nucleic acids, including the O-6 atom of guanine (reviewed by Lawley, 1972) and phosphodiester groups (Shooter et al., 1974).

The principal effects of methylation on DNA templates have been indicated by studies with bacteriophage. Inactivation of the template can result from hydrolytic depurination of 3- and 7-methylpurines (for a recent review see Verly, 1974). At neutral pH and 37°C, the half-lives of the principal products are for 3-methyladenine about 30 hr and for 7-methylguanine about 150 hr. Mutations, envisaged to result from miscoding of modified bases in the DNA template, have been attributed mainly to anomalous base-pairing of O^6-methyl-

Table 1. "Excision" of Bases from DNA Methylated by Carcinogens in Vivo[a]

Agent	Cells	Initial extent of alkylation, mmole alkyl/mole DNA-P	Bases indicated to be "excised"	Half-life of base in DNA (hr)	Bases indicated to be stable to "excision"	References
N-Methyl-N'-nitro-N-nitrosoguanidine	E. coli B/r	1.9	3-Methyladenine	$\ll 1$	7-Methylguanine	Lawley and Orr (1970)
			O^6-Methylguanine Other (methylated pyrimidines, phosphotriesters)	<1		
Dimethylnitrosamine	Rat liver	ca. 0.17	3-Methyladenine	<10		O'Connor et al. (1973)
			O^6-Methylguanine	13		
			7-Methylguanine (?)*	72	7-Methylguanine (?)	Capps et al. (1973) Craddock (1973b)
	Rat liver	ca. 1	3-Methyladenine	7		
			O^6-Methylguanine	24		
			7-Methylguanine (?)*	24		
Methyl methanesulfonate	Rat liver	ca. 0.16	3-Methyladenine	3		Margison and C'Connor (1973)
			7-Methylguanine (?)*	72		
N-Methyl-N-nitrosourea	Mouse liver	ca. 0.5	3-Methyladenine	3	O^6-Methylguanine	Maitra and Frei (1975)
	Mouse bone marrow	ca. 0.2	3-Methyladenine		7-Methylguanine O^6-Methylguanine 7-Methylguanine	Frei and Lawley (1975)

[a]"Excision" denotes losses of methylation products from DNA at rates significantly greater than expected for spontaneous chemical hydrolyses at neutral pH and 37° C (i.e. half-lives of about 30 hr for 3-methyladenine and for 7-methylguanine of about 150 hr; O^6-methylguanine is stable under these conditions in vitro). The data for 7-methylguanine marked with * probably do not indicate active "excision."

guanine with thymine (Loveless, 1969; Loveless and Hampton, 1969). It then follows that mutagenic efficiency of methylating agents will depend on their mechanism of reaction, and that the $S_N 1$ agents will be more potent mutagens (reviewed by Lawley, 1974).

Apart from the losses of methylated bases from DNA attributable to spontaneous chemical hydrolysis, evidence has been obtained that "excision" of these bases *in vivo* could occur much more rapidly than expected from the known rates of hydrolytic loss. This was found firstly for bacteria (Lawley and Orr, 1970) and subsequently for mammals (see Table 1). Such "excisions" were attributed to enzymatic action (Lawley and Orr, 1970), and it appeared that the enzyme system involved was not identical with that which removed cross-links from DNA of bacteria treated with difunctional alkylating agents, which latter closely resembled the *uvr* system (Lawley and Brookes, 1968). It now seems probable that the relatively rapid depurinations of methylated bases are mediated by an enzyme, endonuclease II, isolated by Goldthwait and coworkers (Kirtikar and Goldthwait, 1974). Repair of methylated DNA is indicated to involve depurination and conversion of depurinated sites to single-strand breaks, with subsequent stages then possibly analogous to those involved in repair of radiation-induced breaks by DNA ligase.

One notable feature of the "excisions" observed is that O^6-methylguanine, a base stable to hydrolytic depurination at neutral pH, can be removed. As pointed out by Lawley and Orr (1970), excision of this base before replication of methylated DNA would remove a source of mutation by miscoding. Conversely, if methylation occurred in a region of the template undergoing replication, removal of this promutagenic base might be inhibited. This could be envisaged as a possible cause of preferential mutation of the "growing point" of DNA by methylating mutagens, as observed by Cerdá-Olmedo et al. (1968) for N-methyl-N'-nitro-N-nitrosoguanidine.

There is also increasing evidence that tumor induction by methylating carcinogens may depend not only on specific chemical damage but also on the occurrence of cell division before repair of such damage. With regard to the first factor, ability of methylating agents to induce O^6-methylguanine residues in DNA of target issues has been implicated as an essential *in vivo* reaction, for example, for tumorigenesis in rat kidney (Swann and Magee, 1968) and for induction of thymic lymphoma in mice, where the target tissues are bone marrow and thymus (Frei, 1971). With regard to the second factor, from studies on the induction of carcinoma in rat liver by a single dose of dimethylnitrosamine Craddock (1973*a*) has obtained evidence that cell replication, stimulated by partial hepatectomy, is necessary for carcinogenesis.

In mice treated with N-methyl-N-nitrosourea, the rate of "excision" of O^6-methylguanine from DNA of various organs including liver and bone marrow has been found to be much slower than for rat liver (see Table 1), no loss being detected up to 18 hr after treatment with the carcinogen.

In summary, therefore, the relatively rapid losses of methylated bases from DNA methylated by mutagens and carcinogens *in vivo* are thought to result from enzymatic excision, as the first stage toward removal of either potentially template-inactivating or promutagenic groups. Dependences of such excision on various factors, such as species, organ, and possibly the DNA replication cycle, are indicated by the available data, and it is reasonable to speculate that these factors may be important as determinants of biological effects of methylation. Thus mutagenesis and carcinogenesis may require persistence of promutagenic bases, notably O^6-methylguanine, in the DNA template until replicated to cause anomalous base-pairing.

REFERENCES

Capps, M. J., O'Connor, P. J. and Craig, A. W. (1973) *Biochim. Biophys. Acta,* 331, 33.

Cerdà-Olmedo, E., Hanawalt, P. C. and Guerola, N. (1968). *J. Mol. Biol.* 33, 705.

Craddock, V. M. (1973*a*). *Nature* 245, 386.

Craddock, V. M. (1973*b*). *Biochim. Biophys. Acta* 312, 202.

Frei, J. V. (1971). *Int. J. Cancer* 7, 436.

Frei, J. V. and Lawley, P. D. (1975). *Chem.-Biol. Interact.* (in press).

Kirtikar, D. M. and Goldthwait, D. A. (1974). *Proc. Nat. Acad. Sci. U.S.A.* 71, 2022.

Lawley, P. D. (1972). In *Topics in Chemical Carcinogenesis* (Nakahara, W., Takayama, S., Sugimura, T. and Odashima, S., eds.), p. 237. University Park Press, Baltimore; and University of Tokyo Press, Tokyo.

Lawley, P. D. (1974). *Mutat. Res.* 23, 283.

Lawley, P. D. and Brookes, P. (1968). *Biochem. J.* 109, 433.

Lawley, P. D. and Orr, D. J. (1970). *Chem.-Biol. Interact.* 2, 154.

Loveless, A. (1969). *Nature* 223, 206.

Loveless, A. and Hampton, C. L. (1969). *Mutat. Res.* 7, 1.

Maitra, S. C. and Frei, J. V. (1975). *Chem.-Biol. Interact.* 10, 285.

Margison, G. P. and O'Connor, P. J. (1973). *Biochim. Biophys. Acta* 331, 349.

O'Connor, P. J., Capps, M. J. and Craig, A. W. (1973). *Brit. J. Cancer* 27, 153.

Shooter, K. V., Howse, R., Shah, S. A. and Lawley, P. D. (1974). *Biochem. J.* 137, 303.

Swann, P. F. and Magee, P. N. (1968). *Biochem. J.* 110, 39.

Verly, W. G. (1974). *Biochem. Pharmacol.* 23, 3.

4

Alkali-Labile Lesions in DNA from Cells Treated with Methylating Agents, 4-Nitroquinoline-*N*-oxide, or Ultraviolet Light

I. G. Walker and R. Sridhar

Department of Biochemistry
University of Western Ontario
London, Ontario, Canada N6A 3K7

ABSTRACT

Methylating agents, 4-nitroquinoline-*N*-oxide, and ultraviolet radiation produce alkali-labile structures in the DNA of mammalian cells. The number of these structures does not diminish during a 5-hr posttreatment incubation.

We demonstrated previously (Walker and Ewart, 1973*a*) that after L-cells were treated with low doses of methylating agents the DNA sedimented more slowly in an alkaline sucrose gradient than did DNA from untreated cells. Evidence was presented to show that an alkali-labile structure had been formed in the DNA, most probably a phosphotriester, and that the single-strand breaks in the DNA were produced when the cells were lysed in the highly alkaline lysing reagent and were not the result of a cellular event. This study has now been extended to other agents which interact with DNA—namely, 4-nitroquinoline-*N*-oxide and ultraviolet light. These agents were also found to induce single-strand breaks in the cell's DNA, as indicated by sedimentation analysis in alkaline sucrose gradients.

Using HeLa cells as well as L-cells, the former having a much greater capacity for repair synthesis (Walker and Ewart, 1973*b*), we examined the effect of posttreatment incubation on the occurrence of these single-strand breaks resulting from treatment with N-methyl-N-nitrosourea, 4-nitroquinoline-N-oxide, or ultraviolet light. Posttreatment incubation of the cells in growth medium for periods of up to 5 hr did not lead to restitution of the damaged DNA. It is likely that all three agents produce alkali-labile structures which lead to single-strand breaks only after exposure to high pH. These structures do not appear to be recognized and removed by the cell's enzymatic machinery.

ACKNOWLEDGMENTS

This work was supported by the Medical Research Council of Canada.

NOTED ADDED IN PROOF

In subsequent work with another strain of HeLa cells we have found that single-strand breaks produced by 4-nitroquinoline-N-oxide, but not the other agents, could be repaired.

REFERENCES

Walker, I. G. and Ewart, D. F. (1973a). *Mutat. Res.* **19**, 331–341.
Walker, I. G. and Ewart, D. F. (1973b). *Can. J. Biochem.* **51**, 148–157.

5

Apurinic and Apyrimidinic Sites in DNA

Tomas Lindahl and Siv Ljungquist

Department of Chemistry
Karolinska Institute
Stockholm 60, Sweden

ABSTRACT

An outline review is given of the chemical properties of apurinic and apyrimidinic sites in DNA, the rate of introduction of such sites under different conditions, secondary lesions such as chain breaks, and the properties of endonucleases that specifically attack DNA at apurinic sites.

The selective release of free purines from DNA by hydrolysis in acid is well known and has been employed for structural chemical work on DNA for many years. In contrast, it is not widely recognized that depurination is also a result of alkali-catalyzed DNA degradation, and that depurination and depyrimidination of both single-stranded and double-stranded DNA proceed at appreciable rates at neutral pH. Further, apurinic sites and/or apyrimidinic sites are found in DNA after treatment with alkylating agents or ionizing radiation. The recent discovery of endonucleases that specifically attack DNA at such sites raises the possibility that apurinic sites are removed by an excision-repair mechanism *in vivo*.

MECHANISM OF CLEAVAGE OF THE BASE-SUGAR BOND

The most detailed work on this subject has been done with model compounds. The acid-catalyzed hydrolysis of purine deoxyribonucleosides occurs by reversible protonation of the purine ring to give a cation, followed by rate-limiting cleavage of the glycosidic bond (Zoltewicz et al., 1970). This mechanism is now definitely established, and alternative models involving e.g. transient protonation and ring-opening of the sugar residue have been ruled out. The reaction thus proceeds by specific acid-catalyzed hydrolysis, so the rate is the same in different buffers of the same pH. Pyrimidine deoxynucleosides are also slowly hydrolyzed by the same reaction mechanism in acid and neutral solution, and in addition the unprotonated form of thymidine is cleaved by a pH-independent reaction with water (Shapiro and Kang, 1969).

Deoxynucleotides are hydrolyzed in the same fashion as deoxynucleosides but at a slightly lower rate, probably because of small conformational changes in the sugar ring (Shapiro and Danzig, 1973). Cleavage of glycosidic linkages in DNA also occurs by the same reaction mechanism (Lindahl and Nyberg, 1972), although for unknown reasons these bonds are hydrolyzed 10–100 times more slowly in single-stranded DNA than in the corresponding deoxynucleosides (Shapiro and Danzig, 1973; Lindahl and Karlström, 1973). Above pH 8.5, hydroxyl ion attack causes an increase in the rate of cleavage of the glycosidic bond of deoxyadenosine with increasing pH, but the detailed reaction mechanism has not yet been established (Garrett and Mehta, 1972b).

DEPURINATION AS A FUNCTION OF pH AND TEMPERATURE

The rate of depurination of DNA at neutral pH and high temperature has been determined by measuring the amount of free purines released on incubation of calf thymus DNA (Greer and Zamenhof, 1962) or *Bacillus subtilis* DNA radioactively labeled in the purine residues (Lindahl and Nyberg, 1972). In 1 M NaOH, depurination of single-stranded DNA is approximately 100 times faster than for denatured DNA at pH 7.4 (Ullman and McCarthy, 1973); at 70°C, the rate of depurination of denatured DNA is estimated to be $k = 1 \times 10^{-6}$ sec^{-1} in alkali, and $k = 1 \times 10^{-8}$ sec^{-1} at neutral pH. At pH 7.4, depurination of native DNA occurs 4 times slower than for denatured DNA (Lindahl and Nyberg, 1972). Thus, the double-helical structure offers only partial protection against hydrolysis of the glycosidic bonds in DNA. While deoxynucleosides are hydrolyzed at similar rates at both low and high ionic strength, depurination of double-stranded DNA proceeds more rapidly at low ionic strength. Guanine is released 1.5 times faster than adenine from DNA at both acid and neutral pH (Tamm et al., 1952; Greer and Zamenhof, 1962), but in alkaline solution dAMP is cleaved more rapidly than dGMP (Jones et al., 1966).

Hydrolytic depyrimidination of DNA also occurs at neutral pH, but 20 times slower than depurination (Lindahl and Karlström, 1973). Both cytosine and thymine are released at very similar rates. As previously observed for deoxynucleosides (Shapiro and Danzig, 1972), depyrimidination takes place in a completely analogous fashion to depurination.

The activation energies for the hydrolysis of different deoxynucleosides have consistently been found to fall within the range 95–150 kJ/mole under widely varying solvent conditions (Zoltewicz et al., 1970; Shapiro and Danzig, 1972; Garrett and Mehta, 1972a). The values obtained for DNA agree well with those for deoxynucleosides, as depurination of both native DNA and denatured DNA in neutral and weakly acidic buffers is associated with an activation energy of 130 ± 10 kJ/mole (Lindahl and Nyberg, 1972). Thus, if the temperature of a solution of native DNA is decreased by 10°C, depurination will proceed 3–4 times slower.

ALKYLATED DNA

Treatment of DNA with a monofunctional alkylating agent primarily causes purine alkylation. Such modified bases are released by hydrolysis much more rapidly than normal purines (Lawley and Brookes, 1963; Strauss and Hill, 1970). The large difference between the rates of release in 7-methylguanine and guanine from DNA does not signal any change in reaction mechanism, but simply reflects the increased basicity of the alkylated base. Guanine residues are almost entirely present in uncharged form at neutral pH and should be protonated prior to cleavage of the glycosidic bond, while 7-methylguanine residues are already protonated under such solvent conditions (Zoltewicz et al., 1970).

It is interesting to note that the rate of release of 7-methyl guanine residues from DNA is practically the same *in vivo* as in neutral buffers of physiological ionic strength, as found both for phage T7 (Lawley et al., 1969) and for normal as well as regenerating rat liver (Margison and O'Connor, 1973; Capps et al., 1973). At 37°C, 7-methylguanine is released from DNA with a half-life of 3–6 days both *in vivo* and *in vitro*. This indicates that there are no special mechanisms operative in living cells to protect the DNA from hydrolytic depurination. On the other hand, there is also no evidence for active excision of 7-methylguanine from DNA *in vivo* (Prakash and Strauss, 1970). This is in agreement with the observation that endonucleases reported to act on alkylated DNA seem to act only on partly depurinated DNA (Verly et al., 1973; Ljungquist et al., 1974). However, some minor alkylation products, e.g. 3-methyladenine and O^6-methylguanine, are released from the DNA *in vivo* at a faster rate than *in vitro*, and they might be enzymatically excised (Lawley and Orr, 1970; Margison and O'Connor, 1973).

Treatment of DNA with nitrous acid also causes depurination. The mechanism in this case involves deamination of guanine residues to xanthine, followed by hydrolysis of the labile xanthine-deoxyribose bond (Schuster, 1960).

EFFECTS OF IRRADIATION

On γ-irradiation of mononucleotides, glycosidic bond cleavage is a major lesion (Ducolomb et al., 1974). Bond rupture occurs partly by attack of a hydroxyl radical at the C-1 of the sugar, leading to immediate cleavage of the glycosidic bond and release of an unaltered base, and partly by initial damage to the base residue, leading to weakening of the glycosidic bond. On γ-irradiation of double-stranded DNA in neutral solution, lesions of both these types are apparently introduced (Ljungquist et al., 1974). Apurinic and/or apyrimidinic sites are present in DNA immediately after γ-irradiation, and on incubation of the irradiated DNA in neutral solution a further increase in the number of such sites is observed. The apurinic/apyrimidinic sites[1] were in this case identified as alkali-labile sites in DNA, also sensitive in neutral solution to attack by an endonuclease acting at apurinic sites. Under the experimental conditions employed (irradiation under air in tris buffer), apurinic/apyrimidinic sites were as common lesions as directly introduced single-strand breaks. Apyrimidinic sites have recently also been identified by other techniques as important lesions after γ-irradiation of DNA (Téoule et al., 1974; P. Cerutti, personal communication). On the other hand, irradiation of double-stranded DNA with high doses of ultraviolet light does not seem to cause direct sensitization of the DNA to the endonuclease acting at apurinic sites, and only small numbers of enzyme-sensitive sites appear on subsequent incubation of the irradiated DNA (Ljungquist et al., 1974).

CHAIN BREAKAGE AT APURINIC SITES IN DNA

The deoxyribose residue at an apurinic site is no longer fixed in the furanose form, and an equilibrium is established between this form and an open form with a free aldehyde group. In alkaline solution, chain cleavage due to β-elimination therefore occurs at apurinic sites in DNA (Bayley et al., 1961). The DNA chain is split at the 3′ side of the apurinic site, generating a 3′-hydroxy and a 5′-phosphate group. In neutral solution, chain breakage at apurinic sites takes place by the same mechanism (Lindahl and Andersson, 1972).

The chain cleavage at apurinic sites in DNA in alkali does not occur

---

[1] It seems likely that at apurinic sites in DNA directly introduced by γ-irradiation the sugar residue is present in an oxidized form, as an alkali-labile 3′,5′-disubstituted deoxyribonic acid residue (Rhaese and Freese, 1968). Another type of structural modification of the sugar at an apurinic site is that the free aldehyde form of the deoxyribose can be reduced with sodium borohydride to the corresponding alcohol. Such reduced apurinic sites in DNA are alkali stable but remain susceptible to endonucleases acting at apurinic sites (Hadi and Goldthwait, 1971).

instantaneously. For quantitative analysis of apurinic sites in DNA by alkaline sucrose gradient centrifugation, it is therefore necessary to preincubate the DNA in alkali. We use the following procedure: an equal volume of 2 M glycine-NaOH, pH 13.1, is added to a DNA solution to be analyzed, followed by incubation at 25°C for 3–4 hr prior to centrifugation. The average lifetime of the sensitive phosphodiester bond at an apurinic site in DNA is 32 min under these conditions (Lindahl and Andersson, 1972).

In buffered 0.15 M KCl, pH 7.4, chain breakage at an apurinic site in double-stranded DNA occurs after approximately 500 hr at 37°C (Lindahl and Andersson, 1972). A variety of compounds, including polyamines, basic proteins, monoamines, and divalent cations, promote such chain breakage. Nevertheless, under solvent conditions similar to those present *in vivo*, cleavage at apurinic sites in DNA is still a relatively slow process requiring 20–100 hr at 37°C in the absence of additional factors such as endonucleases specifically attacking such sites.

Slow chain breakage occurs on incubation of intact DNA chains, initially free from apurinic sites, in alkaline solution. From the data reviewed here, it may be estimated that the rate of alkali-catalyzed depurination of DNA in 0.3 M NaOH should be $k = 10^{-10}$ to 10^{-9} sec^{-1} at 25°C. The lower estimate, which corresponds to 1 chain break per 10^9 daltons of DNA per hour at 25°C, is in good agreement with the rate of chain scission of *Escherichia coli* DNA and phage λ DNA in 0.3 M NaOH (Hill and Fangman, 1973). It would therefore appear that depurination and subsequent chain breakage are a major cause of cleavage of DNA chains in alkaline solution. As noted by Hill and Fangman (1973), "analysis of the single-strand molecular weight of large DNA molecules such as those found in eukaryotic chromosomes will need to be carried out under conditions which avoid prolonged exposure to alkali." Unfortunately, due to artifacts of speed dependence it is not possible to circumvent this problem in sedimentation velocity studies by simply centrifuging for short times at high rotor speeds (McBurney et al., 1971).

Heat-induced chain breakage of intact DNA in neutral solution is also primarily a consequence of depurination. For double-stranded DNA there is good agreement between the rate of depurination, as directly measured by the release of purines from DNA, and the combined rate of introduction of apurinic sites and strand breaks in phage PM2 DNA on incubation in neutral buffers at 70 or 80°C. Similarly, for single-stranded DNA, depurination and subsequent β-elimination probably cause most chain breaks, but in addition a second mechanism of hydrolytic chain scission seems to be operative, accounting for about one-third of the total cleavage and presumably depending on the destruction of sugar residues or the direct cleavage of phosphodiester bonds. Thus, while adenine is released 10- to 20-fold more rapidly than cytosine from DNA by hydrolysis of glycosidic bonds at pH 7.4, chain breakage is only 3 times more rapid in poly(dA) than in poly(dC) at 70 and 80°C (T. Lindahl, unpublished).

CROSS-LINKS AT APURINIC SITES

The aldehyde group at an apurinic site in double-stranded DNA can cause cross-linking to the complementary DNA chain, and the amount of such cross-linking increases with decreasing pH (Freese and Cashel, 1964). DNA treated with monofunctional alkylating agents at neutral pH also contains small numbers of cross-links, introduced as a consequence of depurination. After treatment of DNA with methyl methanesulfonate at pH 6.8, followed by incubation at $50°C$, one cross-link per 140 apurinic sites is found (Burnotte and Verly, 1972). Cross-linking of DNA treated with nitrous acid also depends on depurination (Burnotte and Verly, 1971).

REPAIR OF APURINIC SITES

The spontaneous hydrolysis of glycosidic bonds in DNA and the ability of cells to survive large amounts of DNA depurination caused by treatment with monofunctional alkylating agents indicate that DNA repair occurs *in vivo* at apurinic sites. A mammalian cell probably loses 5000–10,000 purines and 200–500 pyrimidines by spontaneous hydrolysis within a 20-hr generation time (Lindahl and Nyberg, 1972; Lindahl and Karlström, 1973). After treatment of mammalian cell lines with large but nonlethal doses of methyl methanesulfonate, $10^6–10^7$ alkylated purine residues are released from the DNA in the same time period in each cell (Strauss et al., 1975). Similar estimates have been made for the amount of depurination of DNA in liver cells after administration of methyl methanesulfonate to rats (Margison and O'Connor, 1973).

Enzymes that would directly add missing purine residues to partly depurinated DNA have not been found. The observation that single-strand breaks are rapidly introduced at apurinic sites in DNA by crude cell extracts suggests instead that an excision-repair mechanism is operative *in vivo* (Lindahl and Andersson, 1972; Verly and Paquette, 1972). Endonucleases that specifically catalyze the formation of single-strand breaks at apurinic sites in double-stranded DNA have so far been found in all cells investigated, and the enzymes from calf thymus (Ljungquist and Lindahl, 1974) and from *E. coli* (Verly et al., 1973) have been studied in detail.

The *E. coli* endonuclease II discovered by Goldthwait and coworkers attacks partly depurinated DNA (Hadi and Goldthwait, 1971) and appears to be identical to the *E. coli* endonuclease acting at apurinic sites investigated by Verly et al. (1973). There is presently some disagreement on the substrate specificity of this enzyme, as in addition to its action at apurinic sites it has been reported to introduce a few single-strand breaks in native DNA from phage T4 and T7, to degrade the polydeoxynucleotide poly[d(A-T)], and to attack alkylated DNA (Hadi et al., 1973). On the other hand, the mammalian endonuclease appears to be absolutely specific for apurinic sites in double-stranded DNA. Thus, the calf thymus enzyme of this kind has no detectable effect on native DNA from phage PM2 or T7, denatured T7 DNA, single-stranded DNA containing apurinic sites ("apurinic acid"), RNA, or DNA containing 7-methylguanine residues or pyrimidine dimers at enzyme concentrations 100- to 1000-fold higher than that sufficient for quantitative chain cleavage at apurinic sites in double-stranded DNA (Ljungquist and Lindahl, 1974; Ljungquist et al., 1974). Further, poly-[d(A-T)] is completely resistant to high concentrations of the enzyme, as judged from alkaline sucrose gradient experiments (S. Ljungquist, unpublished). This endonuclease therefore can be used as a reagent for the detection of small numbers of apurinic sites in double-stranded DNA. The enzyme has a broad pH optimum at pH 8.5, is strongly stimulated by Mg^{2+} but is not absolutely dependent on addition of divalent metal ions, has an approximate molecular weight of 32,000, and apparently cleaves the DNA chain at the 3' side of the apurinic site.

While the existence of endonucleases acting at apurinic sites indicates that such sites are removed by excision-repair *in vivo,* the amount of spontaneous repair replication in normal, unirradiated mammalian cells is too low to be detectable. Thus, fewer than 6000 bromodeoxyuridine molecules per cell are incorporated by repair replication in 2 hr at 37°C (Gautschi et al., 1972). These findings can be reconciled with the estimated spontaneous depurination rate, which would predict the loss of 500–1000 purines in the same time period (Lindahl and Nyberg, 1972), and the hypothesis of excision-repair at apurinic sites if the excised region at each apurinic site in DNA were smaller than 20 nucleotides. Such short-patch repair has previously been found to occur in mammalian cells after X-irradiation (Painter, 1972).

Depurination involves loss of genetic information, but the available evidence indicates that it is not mutagenic. There is unfortunately no method to specifically introduce large numbers of apurinic sites in DNA *in vivo,* and consequently the structure of DNA synthesized from a template containing apurinic sites is not known. It seems likely that apurinic and apyrimidinic sites in DNA are repaired so effectively that they are rarely expressed as lethal or mutagenic events.

ACKNOWLEDGMENT

The authors' work was supported by the Swedish Natural Science Research Council and the Swedish Cancer Society.

REFERENCES

Bayley, C. R., Brammer, K. W. and Jones, A. S. (1961). *J. Chem. Soc.* 1903–1917.
Burnotte, J. and Verly, W. G. (1971). *J. Biol. Chem.* **246,** 5914–5918.

Burnotte, J. and Verly, W. G. (1972). *Biochim. Biophys. Acta* 262, 449–452.

Capps, M. J., O'Connor, P. J. and Craig, A. W. (1973). *Biochim. Biophys. Acta* 331, 33–40.

Ducolomb, R., Cadet, J. and Téoule, R. (1974). *Int. J. Radiat. Biol.* 25, 139–149.

Freese, E. and Cashel, M. (1964). *Biochim. Biophys. Acta* 91, 67–77.

Garrett, E. R. and Mehta, P. J. (1972a). *J. Amer. Chem. Soc.* 94, 8532–8541.

Garrett, E. R. and Mehta, P. J. (1972b). *J. Amer. Chem. Soc.* 94, 8542–8547.

Gautschi, J. R., Young, B. R. and Painter, R. B. (1972). *Biochim. Biophys. Acta* 281, 324–328.

Greer, S and Zamenhof, S. (1962). *J. Mol. Biol.* 4, 123–242.

Hadi, S. M. and Goldthwait, D. A. (1971). *Biochemistry* 10, 4986–4994.

Hadi, S. M., Kirtikar, D. and Goldthwait, D. A. (1973). *Biochemistry* 12, 2747–2754.

Hill, W. E. and Fangman, W. L. (1973). *Biochemistry* 12, 1772–1774.

Jones, A. S., Mian, A. M. and Walker, R. T. (1966). *J. Chem. Soc.* 692–695.

Lawley, P. D. and Brookes, P. (1963). *Biochem. J.* 89, 127–138.

Lawley, P. D. and Orr, D. (1970). *Chem.-Biol. Interact.* 2, 154–157.

Lawley, P. D., Lethbridge, J. H., Edwards, P. A. and Shooter, K. V. (1969). *J. Mol. Biol.* 39, 181–198.

Lindahl, T. and Andersson, A. (1972). *Biochemistry* 11, 3618–3623.

Lindahl, T. and Karlström, O. (1973). *Biochemistry* 12, 5151–5154.

Lindahl, T. and Nyberg, B. (1972). *Biochemistry* 11, 3610–3618.

Ljungquist, S. and Lindahl, T. (1974). *J. Biol. Chem.* 249, 1530–1535.

Ljungquist, S., Andersson, A. and Lindahl, T. (1974). *J. Biol. Chem.* 249, 1536–1540.

Margison, G. P. and O'Connor, P. J. (1973). *Biochim. Biophys. Acta* 331, 349–356.

McBurney, M. W., Graham, F. L. and Whitmore, G. F. (1971). *Biochem. Biophys. Res. Commun.* 44, 171–177.

Painter, R. B. (1972). In *Molecular and Cellular Repair Processes* (Beers, R. F., Jr., Herriott, R. M. and Tilghman, R. C., eds.), pp. 140–146. The Johns Hopkins University Press, Baltimore.

Prakash, L. and Strauss, B. (1970). *J. Bacteriol.* 102, 760–766.

Rhaese, H. J. and Freese, E. (1968). *Biochim. Biophys. Acta* 155, 476–490.

Schuster, H. (1960). *Z. Naturforsch.* 15b, 298–304.

Shapiro, R. and Danzig, M. (1972). *Biochemistry* 11, 23–29.

Shapiro, R. and Danzig, M. (1973). *Biochim. Biophys. Acta* 319, 5–10.

Shapiro, R. and Kang, S. (1969). *Biochemistry* 8, 1806–1810.

Strauss, B. and Hill, T. (1970). *Biochim. Biophys. Acta* 213, 14–25.

Strauss, B., Scudiero, D. and Henderson, E. (1975) this volume, p. 13.

Tamm, C., Hodes, E. and Chargaff, E. (1952). *J. Biol. Chem.* 195, 49–63.

Téoule, R., Bonicel, A., Bert, C., Cadet, J. and Polverelli, M. (1974). *Radiat. Res.* 57, 46–58.

Ullman, J. S. and McCarthy, B. J. (1973). *Biochim. Biophys. Acta* 294, 396–404.

Verly, W. G. and Paquette, Y. (1972). *Can. J. Biochem.* 50, 217–224.

Verly, W. G., Paquette, Y. and Thibodeau, L. (1973). *Nature New Biol.* 244, 67–69.

Zoltewicz, J. A., Clark, D. F., Sharpless, T. W. and Grahe, G. (1970). *J. Amer. Chem. Soc.* 92, 1741–1750.

6

Maintenance of DNA and Repair of Apurinic Sites

Walter G. Verly

Department of Biochemistry
University of Montreal
Montreal, Quebec, Canada

ABSTRACT

Escherichia coli cells contain an enzyme which hydrolyzes a phosphodiester bond near each apurinic site in double-stranded DNA. This endonuclease is specific for apurinic sites; it has no effect on normal DNA, and its action on alkylated DNA is restricted to apurinic sites. *In vitro* incubation with the endonuclease for apurinic sites, DNA polymerase I, and ligase permits repair of DNA containing apurinic sites. The endonuclease for apurinic sites might thus play a role in cell survival after a treatment with alkylating agents; as DNA spontaneously loses purines, the enzyme might also play a role in the maintenance of a normal DNA in every cell. Indeed, an endonuclease for apurinic sites has been found not only in bacteria but also in animal and plant cells; it is very active in thermophilic bacteria.

Delayed inactivation of T7 coliphage treated with monofunctional alkylating agents is due to depurination (Lawley et al., 1969; Brakier and Verly, 1970). This clearly indicates that an apurinic site is much more toxic than an alkylated purine in DNA. Repair mechanisms increasing the organism's survival after a treatment with monofunctional alkylating agents might be directed mostly against the toxic apurinic sites. If the repair of apurinic sites is of the excision type, the cell must contain an endonuclease that recognizes the apurinic sites in DNA.

Starved *Escherichia coli* B41 cells were treated with ethyl methanesulfonate (EMS) or methyl methanesulfonate (MMS) to alkylate their DNA; as controls, isolated bacterial DNA samples were treated in parallel to have the same levels of alkylation (Verly and Paquette, 1970, 1972a). After removal of the alkylating agent, the bacteria and the DNA treated *in vitro* were incubated at 37°C in order to eliminate part of the alkylated purines. The DNA was then isolated from the bacteria, and the DNA treated *in vitro* was submitted to the same manipulations. The DNA alkylated and incubated *in vitro* had a slower sedimentation velocity in neutral sucrose gradients after denaturation by NaOH than after denaturation by formamide at neutral pH. Denaturation by formamide at neutral pH has no effect on apurinic sites (Strauss and Robbins, 1968), while denaturation by NaOH produces the hydrolysis of a phosphodiester bond near each apurinic site (Tamm et al., 1953). The conclusion is that, as expected, the DNA alkylated and incubated *in vitro* contained apurinic sites. By contrast, the DNA alkylated and incubated within the bacteria had the same sedimentation velocity in neutral sucrose gradients whether denatured by NaOH or formamide at neutral pH: no apurinic site can be demonstrated by this technique in the isolated DNA when the DNA is depurinated within the cell (Fig. 1) (Verly and Paquette, 1970, 1972a).

Although other explanations are possible, the *in vivo* disappearance[1] of apurinic sites may be due to the action of a specific endonuclease. To prove this hypothesis, DNA alkylated and depurinated *in vitro* was incubated with the proteins of *E. coli* B41; this led to the disappearance of the apurinic sites. *E. coli* B41 thus contains an endonuclease for apurinic sites (Verly and Paquette, 1970, 1972a); since *E. coli* B41 is a mutant lacking endonuclease I (Duerwald and Hoffman-Berling, 1968), the latter enzyme cannot be involved in the process.

A quantitative method was developed to assay the endonuclease for apurinic sites. DNA labeled with tritium was alkylated with 0.3 M MMS for 1 hr at 37°C, then partially depurinated by heating 6 hr at 50°C. This depurinated DNA, which is still in the native form, is completely precipitated in 5% perchloric acid, but hydrolysis of phosphodiester bonds yields acid-soluble fragments whose radioactivity can be measured. After treatment with NaOH, the substrate DNA gives about 40% acid-soluble radioactivity. The acid-soluble radioactivity is nearly proportional to the amount of enzyme as long as it does not exceed 50% of what is given by NaOH (Paquette et al., 1972).

To purify the *E. coli* B41 endonuclease for apurinic sites, the method used by Goldthwait and his colleagues (Friedberg et al., 1969) for endonuclease II was followed. The crude extract, obtained by sonication of the bacterial cells followed by centrifugation, was treated with streptomycin sulfate, submitted to fractional precipitation with ammonium sulfate, and chromatographed on

---

[1] Disappearance means only that the apurinic sites are no longer demonstrable by the centrifugation technique. They might still be near a strand break.

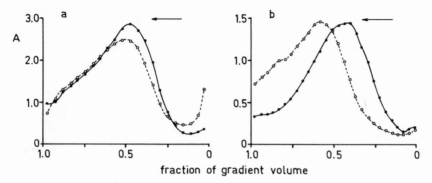

Fig. 1. In vivo instability of apurinic sites. (*a*) *E. coli* cells were alkylated with 0.1 M EMS for 1 hr at 37°C, then incubated for 24 hr at 37°C (*in vivo* treatment); the DNA was extracted from the bacteria. (*b*) *E. coli* DNA was alkylated with 0.1 M EMS for 1 hr at 37°C, then incubated for 24 hr at 37°C (*in vitro* treatment); the DNA was subsequently treated as in *a.* Both DNA samples were denatured by NaOH (●) or by formamide (○) and centrifuged (about 200 μg) in sucrose gradients at 45000 rpm for 3 hr; the bottom of the tube was then punctured and 10-drop fractions were collected. The curves show the absorbance at 260 nm of these fractions versus the volume collected; the direction of sedimentation is indicated by the arrow. From Verly and Paquette (1972a).

DEAE-cellulose and phosphocellulose. On phosphocellulose, the peak of enzyme activity was well separated from the bulk of proteins (Paquette et al., 1972).

Using labeled DNA and the determination of the acid-soluble radioactivity for the assay, it was shown that the purified enzyme has no action on normal DNA, native or denatured (Fig. 2). This lack of action was confirmed by Crine with a very sensitive technique: labeled T7 phage DNA was incubated for 1 hr at 37°C with and without enzyme before NaOH denaturation and centrifugation on alkaline sucrose gradients; the sedimentation profiles were the same in both cases (Paquette et al., 1972).

The purified *E. coli* B41 enzyme has some action on alkylated DNA, but when the alkylated DNA has been partially depurinated it becomes a much better substrate (Fig. 2). Alkylated DNA always contains apurinic sites resulting from the spontaneous loss of alkylated purines, and the action of the purified *E. coli* B41 enzyme on the alkylated DNA is restricted to those apurinic sites. Labeled alkylated DNA was incubated with and without enzyme. At intervals, aliquots were denatured by NaOH before determination of the acid-soluble radioactivity. At any time, the results were the same whether the enzyme was present or not. This shows that enzyme and NaOH cut the DNA strands at the same spots, i.e. near the apurinic sites (Fig. 3) (Paquette et al., 1972). The *E. coli* B41 enzyme is thus an endonuclease that seems entirely specific for apurinic sites. It is also a repair enzyme analogous to the UV endonuclease.

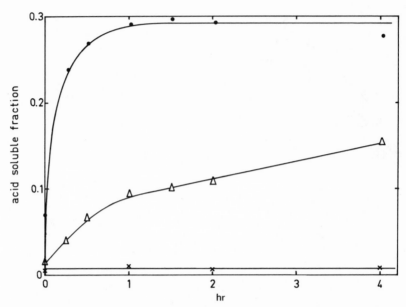

Fig. 2. Activities of the purified enzyme on labeled normal, alkylated, or depurinated DNA. The enzyme was mixed with an equal volume of labeled native untreated DNA (×), alkylated DNA (Δ), or depurinated DNA (•) in SSC. The mixtures containing 0.01 M MgCl$_2$ were incubated at 37°C. Aliquots were taken at increasing time intervals (hr), and the acid-soluble radioactivity was measured and plotted as fraction of the total radioactivity in the aliquot. From Paquette et al. (1972).

Gossard introduced a few apurinic sites in labeled T7 phage DNA by a short incubation at pH 4, which kept the DNA in native form. This slightly depurinated DNA was incubated with the purified *E. coli* B41 endonuclease for apurinic sites, DNA polymerase I and the four deoxyribonucleoside triphosphates, T4 phage ligase, and its coenzyme ATP. After 3 hr at 37°C, the DNA was analyzed by denaturation with NaOH or formamide, followed by centrifugation on neutral sucrose gradients. In each case, the number of breaks per DNA strand was deduced from the sedimentation profile, using the computer technique of Gillespie et al. (1972); from these data, the average number of apurinic sites and the average number of single-strand breaks per strand in the original DNA were calculated. Table 1 shows that the substrate DNA contained, after NaOH denaturation, around 3 breaks per strand, about 2 of which were derived from apurinic sites. When this depurinated DNA was incubated with the three enzymes, no apurinic site was left, and the number of breaks per strand was reduced to 0.8 in one experiment and to 0.5 in the other. The apurinic sites were thus repaired. Omission of any one of the three enzymes, in particular, omission of the endonuclease for apurinic sites, prevents the repair of apurinic sites. This is illustrated in Fig. 4, which presents the sedimentation profiles after NaOH denaturation of a control without enzyme (*A*), the sample incubated with the three enzymes (*C*), and a sample where the nuclease for apurinic sites had been omitted (*B*) (Verly et al., 1974). All the enzymes used in this *in vitro* work are present in *E. coli*. It is thus possible that this system plays an important role in DNA repair and bacterial survival after treatment with a monofunctional alkylating agent.

In a very careful work, Lindahl and Nyberg (1972) showed that DNA spontaneously loses purines at a rate which is not negligible at physiological pH and temperature. There is also a spontaneous depyrimidination, although at a slower rate (Lindahl and Karlström, 1973); the result is indistinguishable—a base has been lost. It is possible that, in the absence of any physical or chemical aggression, the repair system for apurinic sites has a role in the maintenance of normal DNA within *E. coli* cells. Indeed, such a system for keeping intact the genetic information ought to exist in every cell (Verly et al., 1973). Rat liver contains a highly active endonuclease for apurinic sites; when purified, it is entirely specific for apurinic sites (Verly and Paquette, 1972*b*, 1973). Lindahl and Andersson (1972) found a nuclease for apurinic sites in calf thymus; it is also completely specific (Ljungquist and Lindahl, 1974*a,b*). L. Thibodeau

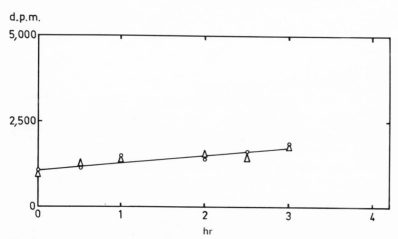

Fig. 3. Localization of the sites of action of the purified enzyme on the alkylated DNA. A 100-μl aliquot of labeled alkylated DNA in SSC was incubated with 100 μl of enzyme preparation (Δ) or SSC ($\circ$). At various time intervals (hr), 10-μl aliquots (6640 dpm) were taken from both assay and control, mixed with 100 μl of calf thymus DNA in SSC (2 mg/ml) and 100 μl of 0.4 N NaOH, incubated for 15 min at 37°C, and precipitated with perchloric acid; the acid-soluble radioactivity (dpm) was measured. From Paquette et al. (1972).

Table 1. Average Number of Breaks and Apurinic Sites per Strand in
Depurinated T7 DNA Incubated with Various Combinations
of Endonuclease for Apurinic Sites, DNA Polymerase I, and Ligase
(Verly et al., 1974)[a]

3 hr incubation at 37°C	Enzymes	Average number of breaks per T7 DNA strand		Average number of apurinic sites per T7 DNA strand
		NaOH	Formamide	
−		3.2	1.1	2.1
+	None	3.1	1.5	1.6
+	Endonuclease, polymerase, ligase	0.8	0.8	0.0
+	polymerase, ligase	2.7	1.5	1.2
−		2.4	0.5	1.9
+	None	2.4	0.8	1.6
+	Endonuclease, polymerase, ligase	0.5	0.5[b]	0.0[b]
+	Endonuclease, polymerase	2.5	2.8	0.0
+	Endonuclease, ligase	2.0	2.2	0.0
+	Endonuclease	2.8	2.8	0.0

[a]Freshly prepared T7 phage DNA containing an average of 0.5 apurinic site or break per strand was submitted to a limited depurination; two preparations were made and each was used in a different set of experiments. The partially depurinated DNA was incubated with various combinations of the three enzymes: endonuclease specific for apurinic sites, DNA polymerase I, and ligase. The DNA was then denatured by NaOH or formamide before being centrifuged on a neutral linear 5–20% sucrose gradient. In each case, the average number of breaks per DNA strand has been calculated from the sedimentation profile by the computer method of Gillespie et al. (1972). Alkaline treatment hydrolyzes a phosphodiester bond near each apurinic site (Tamm et al., 1953), while formamide treatment leaves them intact (Strauss and Robbins, 1968). Consequently, centrifugation after formamide denaturation yields the average number of breaks per strand; centrifugation after NaOH denaturation yields the sum of breaks and apurinic sites; the difference between the values from the two methods gives the average number of apurinic sites per DNA strand.

[b]Estimated values because the sample denatured by formamide was lost.

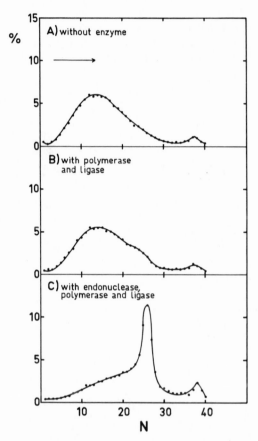

Fig. 4. In vitro repair of apurinic sites in DNA. T7 phage DNA (3 μg) containing apurinic sites and labeled with tritium (150,000 dpm per μg) was incubated without enzyme (*A*), with DNA polymerase I and ligase (*B*), or with the endonuclease specific for apurinic sites, DNA polymerase I, and ligase (*C*). After a 3-hr incubation at 37°C, the 100-μl solution was brought to 1 ml with NaCl-EDTA and dialyzed against the same solvent. To 0.2 ml of the dialyzed solution, NaOH was added to a final concentration of 0.2 M, and after an incubation of 15 min at 37°C the pH was adjusted to 7 with 1.4 M H_3PO_4. An aliquot (0.2 ml) of the neutralized denatured DNA solution was then layered on top of a neutral linear 5–20% sucrose gradient in 1 M NaCl, 0.01 M tris-HCl, 0.001 M EDTA, and centrifuged for 90 min at 35,000 rpm in the SW-50 rotor of a Beckman L2-65B centrifuge at 20°C. Fractions of 0.125 ml were then collected and their radioactivity measured in a Triton-toluene scintillation mixture. The sedimentation profile gives the percentage of the total radioactivity (%) found in each fraction (*N* = fraction number). The arrow indicates the direction of sedimentation. From Verly et al. (1974).

showed that the enzyme exists in plant tissues; it was found in leaves, roots, and seed embryos of *Phaseolus aureus* (Verly et al., 1972, 1973). The endonuclease for apurinic sites is very active in the thermophilic bacterium *Bacillus stearothermophilus* (V. Bibor, unpublished). It thus seems that the nuclease for apurinic sites might be present in every cell and that it might be needed to maintain intact DNA and normal cellular lines.

REFERENCES

Brakier, L. and Verly, W. G. (1970). *Biochim. Biophys. Acta* 213, 296.

Duerwald, H. and Hoffmann-Berling, H. (1968). *J. Mol. Biol.* 34, 331.

Friedberg, E. C., Hadi, S. M. and Goldthwait, D. A. (1969). *J. Biol. Chem.* 244, 5879.

Gillespie, C. J., Gislason, G. S., Dugle, D. L. and Chapman, J. D. (1972). *Radiat. Res.* 21, 272.

Lawley, P. D., Lethbridge, J. H., Edwards, P. A. and Shooter, K. V. (1969). *J. Mol. Biol.* 39, 181.

Lindahl, T. and Andersson, A. (1972). *Biochemistry,* 11, 3618.

Lindahl, T. and Karlström, O. (1973). *Biochemistry,* 12, 5151.

Lindahl, T. and Nyberg, B. (1972). *Biochemistry,* 11, 3610.

Ljungquist, S. and Lindahl, T. (1974*a*). *J. Biol. Chem.* 249, 1530.

Ljungquist, S. and Lindahl, T. (1974*b*). *J. Biol. Chem.* 249, 1536.

Paquette, Y., Crine, P. and Verly, W. G. (1972). *Can. J. Biochem.* 50, 1199.

Strauss, B. S. and Robbins, M. (1968). *Biochim. Biophys. Acta* 161, 68.

Tamm, C., Shapiro, H. S., Lipshitz, R. and Chargaff, E. (1953). *J. Biol. Chem.* 203, 673.

Verly, W. G., Gossard, F. and Crine, P. (1974). *Proc. Nat. Acad. Sci. U.S.A.* 71, 2273.

Verly, W. G. and Paquette, Y. (1970). *Can. Fed. Biol. Soc.*, 685.

Verly, W. G. and Paquette, Y. (1972*a*). *Can. J. Biochem.* 50, 217.

Verly, W. G. and Paquette, Y. (1972*b*). *Fed. Proc.* 31(2), 918 (abstr. 4001).

Verly, W. G. and Paquette, Y. (1973). *Can. J. Biochem.* 51, 1003.

Verly, W. G., Paquette, Y. and Thibodeau, L. (1973). *Nature New Biol.* 244, 67.

Verly, W. G., Paquette, Y., Tremblay, C. and Thibodeau, L. (1972). *Can. Fed. Biol. Soc.*, 673.

7

DNA Turnover and Strand Breaks
in *Escherichia coli*

Philip Hanawalt, Anthony Grivell,[1] and Hiroaki Nakayama[2]

Department of Biological Sciences
Stanford University
Stanford, California 94305

ABSTRACT

The extent of DNA turnover has been measured in a *dnaB* mutant of *Escherichia coli*, temperature sensitive for semiconservative DNA replication. At the nonpermissive temperature about 0.02% of the deoxynucleotides in DNA are exchanged per generation period. This turnover rate is markedly depressed in the presence of rifampicin. During thymine starvation strand breaks accumulate in the DNA of *E. coli* strains that are susceptible to thymineless death. Rifampicin suppresses the appearance of these breaks, consistent with our hypothesis that transcription may be accompanied by repairable single-strand breaks in DNA.

DNA turnover is enhanced severalfold in strands containing 5-bromodeoxyuridine in place of thymidine, possibly because the analog (or the deoxyuridine, following debromination) is sometimes recognized and excised.

A low level of "turnover" in *Escherichia coli* DNA was previously reported by Couch and Hanawalt (1967), and it has been suggested that this represents repair

---

[1] Present address: School of Biological Sciences, Flinders University, Bedford Park, South Australia, 5042.
[2] Present address: Department of Biochemistry, Kyushu Dental College, Kokura-Kita, Kitakyushu 803, Japan.

of parental strand scissions introduced in the course of normal DNA replication and/or transcription (Hanawalt et al., 1968). The periodic introduction of single-strand breaks in the duplex DNA would permit free rotation of portions of the bacterial chromosome. Such "swivels" may be obligatory for the unwinding of DNA strands for replication and transcription.

We have examined the quantitative extent of DNA turnover and the frequency of strand breaks in the DNA of *E. coli* under a variety of physiological conditions. Density labeling with ^{15}N, ^{13}C and high-specific-activity radioactive labeling were used to distinguish semiconservative replication from nonconservative DNA turnover as described by Hanawalt and Cooper (1971). Using a *dnaB* mutant temperature sensitive for DNA synthesis, Nakayama et al. (1972) showed that at the nonpermissive temperature DNA turnover occurred to a similar extent in the regions ahead of and behind the replication point(s) in the bacterial chromosome. The quantitative extent of this turnover or background repair synthesis has been measured in *E. coli* strain TG149, isolated from strain CR34/43 (Bonhoeffer, 1966), with the genotype F^- *dnaB*$^-$ *thy*$^-$ *leu*$^-$ *his*$^-$ *thi*$^-$ *lac*$^-$. At the nonpermissive temperature, 0.02% of the nucleotides are exchanged per generation (Grivell and Hanawalt, 1972; Grivell et al., 1974).

When the cells are incubated at the permissive temperature, the turnover rate is enhanced fivefold or more. We suggest that the additional turnover associated with DNA replication may be the result of the repair of parental strand scissions introduced as "swivels." Whatever the origin of these strand breaks, it is clear that some of them require more than the action of polynucleotide ligase for closure. The *uvrB* mutation was shown to have no effect on the extent of turnover, thus ruling out involvement of the endonuclease specific for *uvrA uvrB* damage (see Braun et al., this volume).

The extent of DNA turnover in the *dnaB* mutant at the restrictive temperature was markedly reduced by treatments which inhibited transcription. Rifampicin at 30 μg/ml reduced the extent of turnover by a factor of 2, while naladixic acid enhanced the extent of turnover somewhat at 43°C. Puromycin greatly reduced turnover, and CN$^-$ eliminated it altogether.

It was previously reported by Pauling and Hanawalt (1965) that thymine starvation stimulates the nonconservative repair mode of DNA synthesis in thymine-requiring bacteria. It was suggested that unrepaired single-strand gaps may accumulate in the absence of the thymine needed for their repair. We have been able to follow the appearance of single-strand breaks in the DNA during thymine starvation by examining the distribution of the single-strand molecular weight of the DNA in alkaline sucrose gradient sedimentation (Nakayama and Hanawalt, 1974). We have demonstrated the accumulation of strand breaks in the DNA of many different thymine-requiring strains, including those for which negative results had previously been reported (cf. Walker, 1970; Baker and Hewitt, 1971; Reichenbach et al., 1971; Sedgwick and Bridges, 1971). In particular, we find that the accumulation of strand breaks is markedly enhanced by a mutant allele at the *polA* locus. *PolA* strains are also unusually sensitive to thymineless death (Bendigkeit and Hanawalt, 1968). The accumulation of strand breaks is totally dependent upon an energy supply and also requires conditions that permit transcription. Thus, we find that rifampicin (30 μg/ml) blocks the production of these strand breaks—just as rifampicin was shown to suppress DNA turnover in the presence of thymine. Protein synthesis inhibition also appeared to suppress the strand break accumulation in some strains. When thymine is added back to a starved culture the strand breaks can be repaired, as shown by the return to normal of the DNA molecular weight distribution in alkaline gradients.

On the basis of our observations we propose the following model:

1. Transcription may be accompanied by single-strand breaks in DNA.
2. DNA polymerase I is involved in the efficient repair of these breaks.
3. Thymine deprivation results in the accumulation of unrepaired breaks.
4. Repair mediated by polymerase I is less affected by thymine starvation than are the alternative pathways operating in *polA*⁻ cells, because it closes the breaks with short patches, thus requiring less thymine (cf. Cooper and Hanawalt, 1972).

DNA turnover was examined in strain TG149 after a period of growth with the thymine analog 5-bromouracil in place of thymine in the medium. The cells were then incubated at the restrictive temperature to eliminate further normal replication while permitting repair synthesis. The semiconservatively replicated "hybrid" DNA was separated from the unreplicated parental DNA in a neutral CsCl equilibrium density gradient. The thymine-containing DNA strands were then separated from those containing 5-bromouracil by rebanding of the "hybrid" DNA in an alkaline CsCl gradient. The extent of turnover (incorporation of [^{3}H] thymidine at 43°C) in 5-bromouracil strands was compared to that in thymine strands, and a severalfold enhancement was seen in the former. The explanation for this result could involve a greater extent of repair synthesis per break in 5-bromouracil DNA, or the additional turnover could be due to additional strand breaks in 5-bromouracil DNA. Evidence for the latter possibility is our finding that DNA strands containing 5-bromouracil exhibited about 1.5-fold more strand breaks than did thymine-containing DNA at the time of cell harvest (Grivell et al., 1974). Although photochemical effects on the 5-bromouracil DNA were eliminated in this study by keeping cell cultures in the dark (Hutchinson, 1973), it is conceivable that a low level of debromination may have occurred. If so, then the resulting deoxyuridine in the DNA may have been recognized and excised by the action of DNA polymerase I (Worcha and Warner, 1973) or some specific endonuclease that acts on uracil in DNA. Carrier and Setlow (1974) have recently reported an endonuclease activity in *Micrococcus luteus* that produces strand scissions in uracil-containing DNA. Finally, it is possible that the 5-bromouracil itself is occasionally recognized and excised.

Philip Hanawalt et al.

ACKNOWLEDGMENTS

This work was supported by a contract AT(04-3)326-7 with the U.S. Atomic Energy Commission and a grant GM 09901 from the U.S. Public Health Service. We appreciate the technical assistance of Marlene Grivell and Yasuko Nakayama.

REFERENCES

Baker, M. L. and Hewitt, R. R. (1971). *J. Bacteriol.* **105**, 733–738.
Bendigkeit, H. and Hanawalt, P. C., (1968). *Bacteriol. Proc. Abstr.* G103.
Bonhoeffer, F. (1966). *Z. Vererbungsl.* **98**, 141.
Carrier, W. L. and Setlow, R. B. (1974). *Fed. Proc.* **33**, 5 (abstr.).
Cooper, P. K. and Hanawalt, P. C. (1972). *Proc. Nat. Acad. Sci. U.S.A.* **69**, 1156–1160.
Couch, J. and Hanawalt, P. C. (1967). *Biochem. Biophys. Res. Commun.* **29**, 779–784.
Hanawalt, P. C. and Cooper, P. K. (1971). *Methods Enzymol.* **21D**, 221–230.
Hanawalt, P. C., Pettijohn, D., Pauling, E. C., Brunk, C. F., Smith, D. W., Kanner, L. C. and Couch, J. L. (1968). *Cold Spring Harbor Symp. Quant. Biol.* **33**, 187–194.
Hutchinson, F. (1973). *Quart. Rev. Biophys.* **6**, 201–246.
Grivell, A. R. and Hanawalt, P. C. (1972). *Biophys. Soc. Abstr.* 37a, *Biophys. J.* **12**.
Grivell, A. R., Grivell, M. and Hanawalt, P. C. (1975). *J. Mol. Biol.* (in press).
Nakayama, H. and Hanawalt, P. (1975). *J. Bacteriol.* **121**, 537–547.
Nakayama, H., Pratt, A. and Hanawalt, P. (1972). *J. Mol. Biol.* **70**, 281–289.
Pauling, C. and Hanawalt, P. (1965). *Proc. Nat. Acad. Sci. U.S.A.* **54**, 1728–1735.
Reichenbach, D. L., Schaiberger, G. E. and Sallman, B. (1971). *Biochem. Biophys. Res. Commun.* **42**, 23–30.
Sedgwick, S. G. and Bridges, B. A. (1971). *J. Bacteriol.* **108**, 1422–1423.
Walker, J. R. (1970). *J. Bacteriol.* **104**, 1391–1392.
Worcha, M. G. and Warner, H. R. (1973). *J. Biol. Chem.* **248**, 1746–1750.

8

Excision-Repair of γ-Ray-Damaged Thymine in Bacterial and Mammalian Systems

P. V. Hariharan, J. F. Remsen, and P. A. Cerutti

Department of Biochemistry
University of Florida
Gainesville, Florida 32610

ABSTRACT

The selective excision of products of the 5,6-dihydroxy-dihydrothymine type (t') from γ-irradiated or OsO_4-oxidized DNA or synthetic poly[d(A-T)] was observed with crude extracts of *Escherichia coli* and isolated nuclei from human carcinoma HeLa S-3 cells and Chinese hamster ovary cells.

The results with *E. coli* extracts allow the following conclusion: (1) The *uvrA*-gene product is not required for t' excision. (2) Radiation-induced strand breakage is not required for product excision. (3) Experiments with extracts of *E. coli* polAexl showed that the $5' \rightarrow 3'$ exonuclease associated with polymerase I is responsible for the removal of t'. (4) Experiments with extracts of *E. coli* endo I lig 4 and the ligase inhibitor nicotinamide mononucleotide showed that polynucleotide ligase accomplishes the last strand resealing step in the excision-repair of t'.

Isolated nuclei from HeLa and Chinese hamster ovary cells possess the necessary enzymes for the selective excision of t' from γ-irradiated or osmium tetroxide oxidized DNA. Approximately 25 to 35% of the products were removed from DNA within 60 min. Unspecific DNA degradation was very low. Radiation-induced strand breakage is not required for product removal.

The best evidence for the contribution of DNA base damage to the lethal effect of ionizing radiation has come from work with bacteriophages. The formation of single- and double-strand breaks as a function of dose and irradiation conditions could be accurately determined in these systems by physical biochemical techniques and related to the loss of infectivity. Participation of base damage in the inactivation process was implicated when less than one double-strand break was introduced per genome per biological hit (Freifelder, 1965, 1968; Van der Schans et al., 1973; reviewed by Cerutti, 1975). In our work we have attempted to assess the efficiency of the formation of DNA base damage, in particular of thymine damage, relative to the formation of single-strand breaks in free DNA (Swinehart et al., 1974), in isolated HeLa chromatin (Roti Roti et al., 1974), *in situ* in *Escherichia coli* (Swinehart and Cerutti, 1975), and *in situ* in mammalian cells (Roti Roti and Cerutti, 1974). Our results have recently been summarized (see Cerutti, 1974, 1975). The reductive assay (Hariharan and Cerutti, 1971) or the alkali-acid degradation assay (see Hariharan and Cerutti, 1974a) was used for the determination of products of the 5,6-dihydroxy-dihydrothymine type, while the radiation chemical reactivity of the thymine-methyl[^{3}H] group was assessed by the reaction with OH radicals (see Cerutti, 1974). It should be stressed that only part of the radiation-induced base damage was measured in our experiment and that additional damage undoubtedly occurs at adenine, guanine, and cytosine. It was concluded that base damage is a major type of damage induced by γ-rays *in situ* in bacterial and mammalian cells.

The removal from the DNA of γ-ray-damaged thymine residues during postirradiation incubation has been shown to occur in bacterial and mammalian cells (Hariharan and Cerutti, 1971, 1972; Mattern et al., 1973), but the results obtained so far have not yielded much information concerning the molecular mechanism leading to the release of γ-ray-damaged thymine. In particular, the question remains unanswered whether the removal of damaged thymine residues from DNA is accomplished by a *selective* excision process similar or identical to that effecting the removal of cyclobutane-type photodimers or whether it is a reflection of postirradiation DNA degradation. Recently indirect evidence was presented which favors the first alternative. It has been shown that bacterial (Paterson and Setlow, 1972; Setlow and Carrier, 1973) and mammalian cells (Brent, 1973; Ljungquist et al., 1974; Ljungquist and Lindahl, 1974; Lindahl and Ljungquist, this volume) possess endonucleolytic activities which recognize X-ray- or γ-ray-induced lesions which may involve the heterocyclic bases. In the case of *E. coli* it was found that such "endonuclease-sensitive sites" were rapidly removed from the DNA during postirradiation incubation of the culture (Paterson and Setlow, 1972; Wilkins, 1973). It should be mentioned here that there is considerable indirect evidence for the existence of excision-repair processes in mammalian cells following exposure to ionizing radiation. The radiation-induced insertion of patches of nucleotides into the DNA in γ-irradiated mammalian cells has been observed in the form of repair replication and unscheduled DNA synthesis. These processes are best understood in molecular terms if it is assumed that damaged and undamaged residues are first removed from the DNA in early

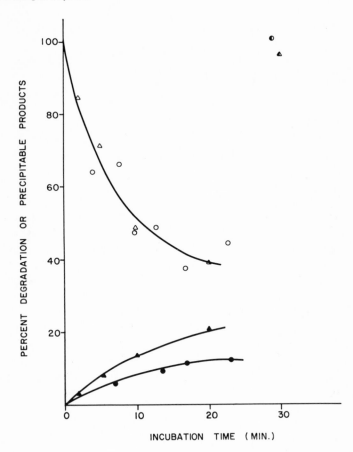

Fig. 1. *Removal of products of the 5,6-dihydroxy-dihydrothymine type (t') and of undam-aged thymine from γ-irradiated poly[d(A-T)] by* E. coli *extracts.* To a solution of γ-irradi-ated poly[d(A-T)]-thymine-methyl[³H] in 350 μl of 0.05 M phosphate buffer (pH 7) containing approximately 1 μg of the polymer and 5–10 × 10⁴ cpm was added 200 μl of crude *E. coli* extract freshly prepared according to Wickner et al. (1972) (irradiation under aerobic, nonprotective conditions; the amount of t' in the irradiated poly[d(A-T)] varied from 0.7 to 1.4%). The samples were incubated at 37°C and the reaction was terminated at various time intervals up to 30 min by the addition of 2.5 ml of cold 7% TCA. The disappearance of products of the 5,6-dihydroxy-dihydrothymine type from the acid-precipitable material was determined by the alkali-acid degradation assay. The amount of acid-soluble material released from the polymer during incubation was determined by standard techniques (see Hariharan and Cerutti, 1974a). (a) ○———○, Removal of t' from acid-precipitable γ-irradiated poly[d(A-T)] by crude extracts of *E. coli* endo I⁻; (b) △____△, removal of t' from acid-precipitable γ-irradiated poly[d(A-T)] by crude extracts of *E. coli* endo I⁻ *uvrA6;* ○, ▲, same as (a) and (b) except that heat-inactivated extracts were used in the incubation mixture; ●———●, acid solubilization of thymine radioactivity by crude extracts of *E. coli* endo I⁻; ▲———▲, acid solubilization of thymine radioactivity by crude extracts of *E. coli* endo I⁻ *uvr A6*.

steps of postirradiation repair (see e.g. Painter, 1970, 1972; Regan and Setlow, 1973).

As mentioned above, ionizing radiation induces damage to all the nucleic acid bases and additionally produces strand breakage, cross-links, apyrimidinic sites (Dunlap and Cerutti, 1975), and possibly apurinic sites. It is, therefore, virtually impossible to relate changes induced in the biological activity of DNA by ionizing radiation to a specific type of lesion. This situation has been improved for the case of thymine base damage by the following approach. The excision of products of the 5,6-dihydroxy-dihydrothymine type from γ-irradiated poly[d(A-T)] or bacteriophage PM2 DNA by crude concentrated *E. coli* extracts or isolated mammalian nuclei has been investigated and compared to results obtained with OsO_4-oxidized poly[d(A-T)]. OsO_4 selectively oxidizes thymine in poly[d(A-T)] to 5,6-dihydroxy-dihydrothymine, which represents a major type of radiation product induced by ionizing radiation. Unlike ionizing radiation, OsO_4 does not induce strand breakage or adenine damage but does produce a small number of apyrimidinic sites. It should be noted that the repair of an exogenous DNA substrate rather than endogenous chromosomal DNA has been investigated.

The following results concerning the molecular steps of excision-repair of products of the 5,6-dihydroxy-dihydrothymine type in *E. coli* have been obtained. As shown in Fig. 1, crude extracts of *E. coli* endo I⁻ and *E. coli* endo I⁻ *uvrA6* prepared according to Wickner et al. (1972) possess the ability to selectively remove ring-damaged thymine residues from γ-irradiated poly[d(A-T)] and PM2 DNA (not shown). Within approximately 20 min of incubation at 37°C, 60% of the products were removed from acid-precipitable polymeric material by the extracts of both *E. coli* strains. The results in Fig. 2 show that extracts of both strains, *E. coli* endo I⁻ and *E. coli* endo I⁻ *uvrA6*, also accomplished the selective removal of 5,6-dihydroxy-dihydrothymine from OsO_4-oxidized poly[d(A-T)]. Maximally 20% of the modified poly[d(A-T)] was rendered acid soluble within the same time period. It is concluded that the *uvrA* gene product is not required for the excision of products of the 5,6-dihydroxy-dihydrothymine type from γ-irradiated poly[d(A-T)] or PM2 DNA or from OsO_4-oxidized poly[d(A-T)].

On average, 8 to 16 undamaged nucleotides are removed from the polymer per ring-damaged thymine residue excised by extracts from both strains and for the γ-irradiated and OsO_4-oxidized substrate (Hariharan and Cerutti, 1974a). It has been speculated that DNA strand breaks induced by ionizing radiation may play a role in the removal of base-damaged residues by serving as starting points for exonucleolytic degradation (Cleaver, 1969; Hariharan and Cerutti, 1972). Our results with OsO_4-oxidized poly[d(A-T)] show that strand breakage is not required for the excision of ring-damaged thymine by *E. coli* extracts. Both γ-irradiated and OsO_4-oxidized DNA substrates contain some apyrimidinic sites.

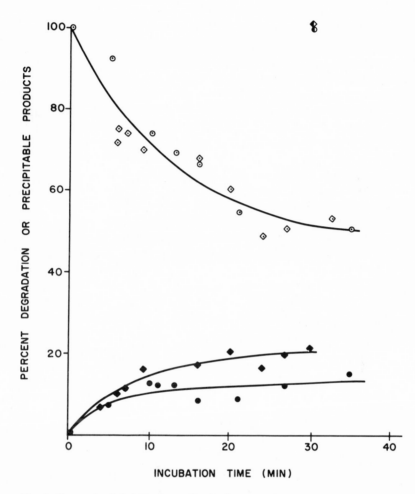

Fig. 2. Removal of 5,6-dihydroxy-dihydrothymine (t') and undamaged thymine from OsO₄-oxidized poly[d(A-T)]. Experimental conditions are given in the caption of Fig. 1. Osmium tetroxide oxidation of poly[d(A-T)] was carried out according to a modification of the procedure of Beer et al., (1966) (see Hariharan and Cerutti, 1974a). (a) o——o, Removal of t' from acid-precipitable OsO₄-oxidized poly[d(A-T)] by crude extracts of *E. coli* endo I⁻; (b) ◊——◊, removal of t' from acid-precipitable OsO₄-oxidized poly[d(A-T)] by crude extracts of *E. coli* endo I⁻ *uvrA6*; ●, ◆, same as (a) and (b) except that heat-inactivated extracts were used in the incubation mixture; ●——●, acid solubilization of thymine radioactivity by crude extracts of *E. coli* endo I⁻; ◆——◆, acid solubilization of thymine radioactivity by crude extracts of *E. coli* endo I⁻ *uvrA6*.

The effect of these lesions on the excision of ring-damaged thymine cannot be assessed at the present time.

Our results raise the question of the identity of the endonuclease recognizing ring-saturated thymine products of the 5,6-dihydroxy-dihydrothymine type in *E. coli*. Possible candidates are endonuclease II (see Goldthwait et al., this volume) and the endonuclease recognizing depurinated DNA (Verly and Paquette, 1972; Verly this volume) of *E. coli*. An endonuclease specific for γ-ray-induced DNA damage other than strand breakage has been purified from *Micrococcus luteus* (Setlow and Carrier, 1973) and *E. coli* (Strniste and Wallace, this volume). While this question remains unanswered, additional information has recently been obtained in our laboratory concerning the subsequent steps in the excision of γ-ray products of the 5,6-dihydroxy-dihydrothymine type. No excision of ring-damaged thymine from γ-irradiated PM2 DNA was observed at the nonpermissive temperature with extracts of *E. coli* endo I$^-$ polA ex1 (Uyemura et al., this volume; Konrad and Lehman, 1974), which is temperature sensitive in the $5' \rightarrow 3'$ exonuclease (exonuclease VI) associated with polymerase I (Hariharan and Cerutti, 1974c). Using extracts of the temperature-sensitive polynucleotide ligase mutant *E. coli* endo I$^-$ lig4 (Gellert and Bullock, 1970) or wild-type extracts with the ligase inhibitor nicotineamide mononucleotide (Campbell et al., 1971), it was demonstrated that polynucleotide ligase is responsible for the strand-resealing step in the *in vitro* excision-repair of 5,6-dihydroxy-dihydrothymine from OsO$_4$-oxidized poly[d(A-T)] (Hariharan and Cerutti, 1974b). The number of strand breaks which remained open in the presence of the ligase inhibitor or at the nonpermissive temperature of the ligase mutant corresponded to the number of ring-saturated thymine residues in the OsO$_4$-oxidized poly[d(A-T)].

Isolated nuclei were used in our experiments of γ-ray excision-repair in mammalian systems, since they have many of the advantages of an "open system": (1) Unirradiated nuclei can be used and the repair of an exogenous damaged DNA substrate can be investigated. (2) Synthetic, modified polydeoxynucleotides (e.g. OsO$_4$-oxidized poly[d(A-T)]) can be used as substrates to study certain questions. (3) Since the repair by isolated nuclei of a highly purified low-molecular-weight DNA specimen can be investigated (e.g. bacteriophage PM2 DNA in our experiments), changes in the molecular weight can readily be followed by standard sedimentation techniques. The problems encountered in the sedimentation analysis of DNA of very high molecular weight are, therefore, avoided. (4) The effect of the addition to the nuclei preparations of cytoplasmic factors or purified "repair enzymes" can be investigated. A major drawback of most isolated nuclei systems is the loss of an unknown fraction of the nuclear content during their preparation. A disadvantage of using an exogenous DNA substrate is that some of the repair enzymes may be tightly bound to the nuclear DNA and, therefore, not available to interact freely with the exogenous DNA. It should be noted, however, that repair of "exogenous DNA"

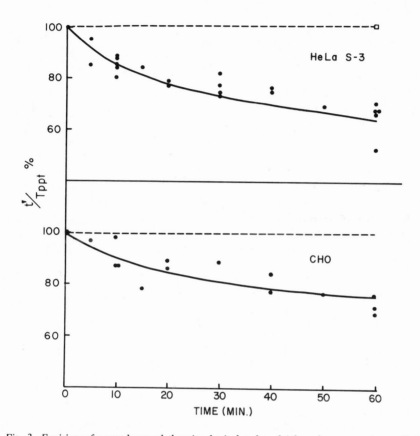

Fig. 3. Excision of γ-ray-damaged thymine by isolated nuclei from human carcinoma HeLa S3 and Chinese hamster ovary cells. The removal of ring-damaged thymine of the 5,6-dihydroxy-dihydrothymine type (t′) from γ-irradiated pseudomonas phage PM2 DNA by isolated nuclei from HeLa S3 and CHO cells was determined (irradiation under aerobic, nonprotective conditions; the amount of t′ in the irradiated DNA was usually about 0.3%). Isolated nuclei were prepared by a modification of the procedure of Berkowitz et al., (1969). A sample of 0.1ml contained approximately 5×10^6 nuclei suspended in a solution of 0.01 M phosphate buffer (pH 7.2), 0.1 M NaCl, 0.001 M dithiothreitol, 15 units pyruvate kinase, 15 mg/ml BSA, 8.24 mg/ml phosphoenolpyruvate, 4.42 mg/ml ATP, and approximately 2.0 mg/ml of each of the four deoxynucleosidetriphosphates. To the nuclei suspension was added, in 25 μl, $0.5–1.5 \times 10^5$ cpm of γ-irradiated PM2 DNA-thymine-methyl-[³H]. The samples were then incubated up to 60 min at 37°C. Some clumping of the nuclei was usually observed after approximately 25 min of incubation. The reaction was terminated by the addition of 7% TCA, and the acid-precipitable and acid-soluble materials were analyzed by the alkali-acid degradation assay (see Hariharan and Cerutti, 1974a). The data are presented as the percent of t′ remaining acid percipitable during incubation with the nuclei at 37°C.

by mammalian cells has been observed in the form of host-cell reactivation (Rabson et al., 1969; Aaronson and Lytle, 1970).

As shown in Fig. 3, isolated nuclei from human carcinoma HeLa S3 and Chinese hamster ovary (CHO) cells prepared by a modification of the procedure of Berkowitz et al. (1969) selectively removed products of the 5,6-dihydroxy-dihydrothymine type from acid-precipitable PM2 DNA. Within 60 min of incubation at 37°C, HeLa nuclei had removed approximately 35% and CHO nuclei 25% of the ring-damaged thymine residues from the DNA. During the same time period only 2–3% of the PM2 DNA had been rendered acid soluble. Similar results have recently been obtained with young and medium-aged diploid human lung WI38 fibroblasts and mouse embryo 3T3 cells (Remsen, Mattern, and Cerutti, unpublished results). It is estimated from these data that only a very few undamaged residues are removed from the DNA per ring-damaged thymine residue, in agreement with the small average patch size inserted by γ-ray-induced repair replication as determined by density gradient analysis (Painter and Young, 1972) or the bromouracil photolysis method (Setlow and Regan, 1973). Under our conditions product removal started to level off after approximately 30 min of incubation. While this could be a peculiarity of the isolated nuclei system (some clumping of the nuclei is observed after 30 min of incubation), similar kinetics have been obtained for the appearance of products of the 5,6-di-hydroxy-dihydrothymine type in acid-soluble form in the cytoplasm and the culture medium during postirradiation incubation of CHO cells (Mattern et al., 1973). It appears then that excision of γ-ray-damaged thymine is considerably faster than that of cyclobutane-type photodimers, which has been shown to continue for several hours in human cells (reviewed by Cleaver, 1974). The characteristics of the excision of γ-ray-damaged thymine by isolated nuclei fit well the criteria for "ionizing-type repair" defined by Regan and Setlow (1973). It should be noted that, in contrast to the excision of γ-ray-damaged thymine, the removal of cyclobutane-type photodimers is rather inefficient in most rodent cells.

ACKNOWLEDGMENTS

This work was supported by a grant from the National Institutes of Health and a contract from the U.S. Atomic Energy Commission.

REFERENCES

Aaronson, S. and Lytle, C. (1970). *Nature,* **228,** 359–361.
Beer, M., Stern, S., Carmalt D. and Mohlenrich, K. (1966). *Biochemistry,* **5,** 2283–2288.
Berkowitz, D., Kakefuda, T. and Sporn, M. (1969). *J. Cell Biol.* **42,** 851–858.
Brent, T. (1973). *Biophys. J.* **13,** 399–401.

Campbell, J., Fleischman, A. and Richardson, C. (1971). *Fed. Proc.* **30**, 1313 (abstr.).

Cerutti, P. (1975*a*). In *Photochemistry and Photobiology of Nucleic Acids* (Wang, S. Y. & Patrick, M., eds.) Gordon Breach, New York (in press).

Cerutti, P. (1974). *Naturwissenschaften*, **61**, 51–59.

Cleaver, J. (1969). *Proc. Nat. Acad. Sci. U.S.A.* **63**, 428–435.

Cleaver, J. (1974). *Advan. Radiat. Biol.* **4**, 1–75.

Dunlap, B. and Cerutti, P. (1975). *FEBS Letters.* **51**, 188–190.

Freifelder, D. (1965). *Proc. Nat. Acad. Sci. U.S.A.* **54**, 128–134.

Freifelder, D. (1968). *Virology*, **36**, 613–619.

Gellert, M. and Bullock, M. (1970). *Proc. Nat. Acad. Sci. U.S.A.* **67**, 1580–1587.

Hariharan, P. and Cerutti, P. (1971). *Nature* New Biol. **229**, 247–249.

Hariharan, P. and Cerutti, P. (1972). *J. Mol. Biol.* **66**, 65–81.

Hariharan, P. and Cerutti, P. (1974*a*). *Proc. Nat. Acad. Sci. U.S.A.* **71**, 3532–3536.

Hariharan, P. and Cerutti, P. (1974*b*). *Biochem. Biophys. Res. Commun.* **61**, 375–379.

Hariharan, P. and Cerutti, P. (1974*c*). *Biochem. Biophys. Res. Commun.* **61**, 971–976.

Konrad, E. and Lehman, I. (1974). *Proc. Nat. Acad. Sci. U.S.A.* **71**, 2048–2051.

Ljungquist S. and Lindahl, T. (1974). *J. Biol. Chem.* **249**, 1530–1535.

Ljungquist, S., Andersson, A. and Lindahl, T. (1974). *J. Biol. Chem.* **249**, 1536–1540.

Mattern, M., Hariharan, P., Dunlop, B. and Cerutti, P. (1973). *Nature New Biol.* **245**, 230–232.

Painter, R. (1970). *Curr. Top. Radiat. Res.* **7**, 45–70.

Painter, R. (1972). *Johns Hopkins Med. J.* **1** (Suppl.), 140–146.

Painter, R. and Young, B. (1972). *Mutat. Res.* **14**, 225–235.

Paterson, M. and Setlow, R. (1972). *Proc. Nat. Acad. Sci. U.S.A.* **69**, 2927–2931.

Rabson, A., Tyrrell, S. and Legallais, F. (1969). *Proc. Soc. Exp. Biol. Med.* **132**, 802–806.

Regan, J. and Setlow, R. (1973). In *Chemical Mutagens* (Hollaender, A., ed.), vol. 3, pp. 151–170. Plenum Press, New York.

Roti Roti, J. and Cerutti, P. (1974). *Int. J. Radiat. Biol.* **35**, 413–417.

Setlow, R. and Carrier, W. (1973). *Nature New Biol.* **241**, 170–172.

Setlow, R. and Regan, J. (1973) Biophys. Soc. Abst. 307a *Biophys. J. 13.*

Swinehart, T. and Cerutti, P. (1975). *Int. J. Radiat. Biol.* **27**, 83–94.

Swinehart, J., Lin, W. and Cerutti, P. (1974). *Radiat. Res.* **58**, 166–175.

Van der Schans, G., Bleichrodt, J. and Blok, J. (1973). *Int. J. Radiat. Biol.* **23**, 133–150.

Verly, W. and Paquette, Y. (1972). *Can. J. Biochem.* **50**, 217–224.

Wickner, R., Wright, M., Wickner, S. and Hurwitz, J. (1972). *Proc. Nat. Acad. Sci. U.S.A.* **69**, 3233–3237.

Wilkins R. J. (1973). *Nature New Biol.* **244**, 269–271.

9

Formation of Dimers in Ultraviolet-Irradiated DNA

Clifford Brunk

Biology Department and Molecular Biology Institute
UCLA
Los Angeles, California 90024

ABSTRACT

Evidence has been obtained that in UV-irradiated native DNA, pyrimidine dimers are preferentially formed in long pyrimidine tracts. This effect is not the result of the existence of more dimerizable thymine in the long pyrimidine tracts. The preferential formation of dimers in long pyrimidine tracts is enhanced as the dose of irradiation is decreased. These results suggest that the formation of dimers in native DNA occurs by a cooperative mechanism. Quite likely, dimers are formed only in regions of native DNA that are locally denatured (breathing), in which the bases can be aligned. The formation of a dimer in such a breathing region would tend to lock it open and could lead to cooperative formation of dimers in such regions. If this mechanism for the cooperative formation of dimers is operative then it should be restricted to double-stranded DNA. The pyrimidine tracts of irradiated single-stranded DNA all show virtually the same dimer content, irrespective of the length of the tract. These results are consistent with the hypothesis that dimers form cooperatively in breathing regions.

The formation of dimers between adjacent pyrimidine residues caused by ultraviolet irradiation has been the subject of intense investigation for almost a decade. The biological significance of such dimers has also been demonstrated directly and is evidenced by the existence of several mechanisms for removing dimers (R. Setlow, 1968; J. Setlow, 1966; Rupp and Howard-Flanders, 1968). Until recently the implicit or explicit assumption has been that dimers are formed randomly along the DNA chain.

Recently we have shown that dimers are formed in a cooperative fashion (Brunk, 1973). Pyrimidine tracts from UV-irradiated double-stranded DNA were prepared and separated according to chain length on a DEAE-cellulose column. The percentage of thymine dimerized for different chain lengths was determined. The dimer content of pyrimidine tracts increases with chain length. This clustering of dimers in longer pyrimidine tracts indicates that the formation of dimers is cooperative. It was also noted that this clustering of dimers in the longer pyrimidine tracts was enhanced at lower UV doses. Appropriate controls demonstrate that this clustering of dimers is a function of the mechanism by which dimers are formed and not a result of the base sequences of the longer pyrimidine tracts.

In this analysis the use of pyrimidine tracts provides a convenient parameter by which the clustering of dimers can be readily observed. More general approaches to determine the distribution of dimers have indicated a relatively random distribution of dimers (Rahn, 1973). Although the selective formation of dimers in long pyrimidine tracts may have profound biological implications. in this study we view it primarily as a sensitive measure of the cooperative formation of dimers.

Having demonstrated that dimers were clustered in the longer pyrimidine

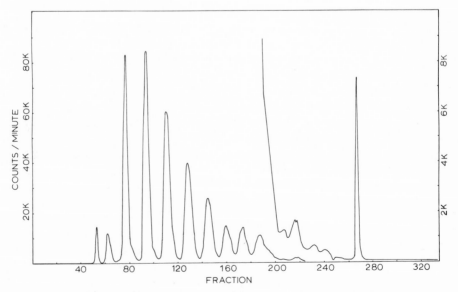

Fig. 1. Pyrimidine tract profile from UV-irradiated M13 DNA prelabeled with [³H] dThd.

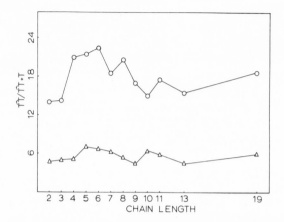

Fig. 2. Dimer content of UV-irradiated single-stranded DNA as a function of chain length. DNA irradiated with (△) 5000 ergs mm⁻² or (○) 30,000 ergs mm⁻².

tracts of UV-irradiated double-stranded DNA, we were interested in the mechanism leading to this distribution. Ingenious experiments have demonstrated that energy absorbed along the DNA chains may be transmitted to other regions of the DNA (Sutherland and Sutherland, 1970). Initially we considered the possibility that the pyrimidine tracts might act as a sink for energy collected along the DNA chains and transmitted to them. Although no direct evidence at present makes this mechanism untenable, it is possible that an even simpler mechanism may be responsible for the clustering of dimers. In order for adjacent pyrimidines to dimerize in double-stranded DNA, the pyrimidines undergo a significant rearrangement from the classical B-structure of DNA. The adjacent pyrimidines must align directly above one another (requiring a $36°$ rotation), and the planes of the bases must approach to a distance only one-half that normally separating the bases in the classical B-structure. Such rearranged bases can then form the two sets of 5,6 double bonds characteristic of cyclobutyl pyrimidine dimers. The rearrangement required for dimerization necessarily creates a locally denatured region in the DNA. Conversely, dimerization occurs in locally denatured regions of double-stranded DNA. This immediately suggests a mechanism for the cooperative formation of dimers. The initially formed dimers occur in regions that are "breathing" (locally denatured). Such "breathing" is common in double-stranded DNA (von Hippel and Wong, 1971). Later dimers are formed either in other "breathing" regions or in the locally denatured regions caused by the initial dimers, which of course leads to the clustering of dimers due to their cooperative formation.

Our initial approach was to investigate the effect of temperature on dimer

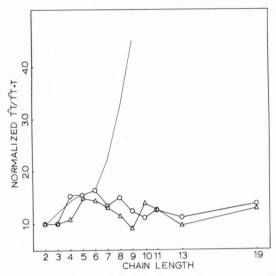

Fig. 3. *The data in Fig. 2 normalized to dimer content of tracts two pyrimidines long.* The steep upper curve shows, for comparison, the dimer content of double-stranded DNA after 4000 ergs mm^{-2}.

formation. We reasoned that at higher temperatures the amount of DNA in locally denatured regions should increase and the yield of dimers for a given dose of UV might also increase. We irradiated DNA at 0, 25, and 60°C with identical doses of UV but were unable to detect significant differences in the amount of dimerization. Others have reached similar conclusions in more detailed studies (Hosszyland Rahn, 1967; Patrick, personal communication). It remains uncertain what degree of increase in dimerization should be directly measurable over this temperature range. The use of a parametric measure of dimer formation such as clustering in pyrimidine tracts should afford a much more sensitive measure of this effect.

Another immediate conclusion of the hypothesis that the clustering of dimers is due to their cooperative formation in locally denatured regions is that such clustering should not occur in irradiated single-stranded DNA. Because the entire chain of DNA is denatured in single-stranded DNA, the formation of a dimer would not be expected to promote the cooperative formation of a dimer adjacent to a dimer formed earlier.

For these experiments we chose to use DNA from the bacteriophage M13. This DNA occurs in the single-stranded form in the phage and has unusually long pyrimidine tracts (up to 19 pyrimidines in length). The DNA was highly labeled with [3H] dThd and irradiated with 254-nm UV, and pyrimidine tracts were prepared as described earlier (Brunk, 1973). The results of this type of separation are shown in Fig. 1. The prominent peak at the right is the pyrimidine tract, 19 long. Each peak was pooled and the dimer content was determined by acid hydrolysis and separation on ion-exchange paper (Brown and Holt, 1967). The results of such determinations for DNA irradiated with two different doses of UV are shown in Fig. 2. The lower and upper plots are the dimer contents of single-stranded DNA irradiated with about 5000 and 30,000 ergs/mm^2 of 254-nm UV, respectively, as a function of chain length of the pyrimidine tract. Fig. 3 shows these data normalized to the dimer content of the tracts two pyrimidines long. Also shown in Fig. 3 is the dimer content of double-stranded DNA irradiated with 4000 ergs/mm^2 (the longer pyrimidine tracts are not present in this double-stranded DNA).

The data clearly show that the dimers found in single-stranded DNA are not clustered. This is completely consistant with the hypothesis that the clustering of dimers observed in double-stranded DNA is a result of the cooperative formation of dimers in locally denatured regions that result from the presence of initial dimers. Although other explanations may be possible, the simplicity of this hypothesis is most attractive.

ACKNOWLEDGMENT

This work was supported by a grant from the National Science Foundation.

REFERENCES

Brown, R. D. and Holt, C. E. (1967). *Anal. Biochem.* **20**, 358.
Brunk, C. F. (1973). *Nature New Biol.* **241**, 74.
Hosszy, J. L and Rahn, R. O. (1967). *Biochem. Biophys. Res. Commun.* **29**, 327.
Rahn, R. O. (1973). *Photophysiology*, 8, 231.
Rupp, W. D. and Howard-Flanders, P. (1968). *J. Mol. Biol.* **31**, 291.
Setlow, J. K. (1966). *Curr. Top. Radiat. Res.* **2**, 195.
Setlow, R. B. (1968). *Progr. Nucl. Acid Res. Mol. Biol.* **8**, 257.
Sutherland, J. C. and Sutherland, B. M. (1970). *Biopolymers* **9**, 639.
Von Hippel, P. H. and Wong, K. Y. (1971). *J. Mol. Biol.* **61**, 587.

10

An Enzymatic Assay for
Pyrimidine Dimers in DNA

Ann Ganesan

Department of Biological Sciences
Stanford University
Standford, California 94305

ABSTRACT

An assay for pyrimidine dimers in DNA employing the T4 endonuclease V is described. The assay can detect approximately 50 dimers per *Escherichia coli* genome, or the effect of 10 ergs/mm^2 ultraviolet irradiation.

The endonuclease V of phage T4 produces single-strand breaks in UV-irradiated but not in unirradiated DNA (Yasuda and Sekiguchi, 1970; Friedberg, 1972; Friedberg and Clayton, 1972). It acts on both single- and double-stranded DNA, apparently causing breaks in the phosphodiester bonds on the 5' side of pyrimidine dimers (Yasuda and Sekiguchi, 1970; Friedberg et al., this volume). Although this endonuclease does not remove dimers from DNA, it renders them susceptible to release by exonuclease (Friedberg et al., 1974). We have used the T4 endonuclease V to estimate the number of dimers in the DNA of irradiated cells of *Escherichia coli* (Ganesan, 1973, 1974). Cells were first treated for 30 min at 0°C with EDTA, lysozyme, and Brij-58. This rendered the cells permeable to enzymes but left the DNA inside the cell sac protected from mechanical shear. The permeable cells were then mixed with T4 endonuclease V and incubated for 15 min at room temperature [see Ganesan (1974) for details of concentration, etc]. The reaction was terminated by layering samples on 5–20% (w/v) alkaline sucrose gradients, the top layer of which consisted of 0.1 ml of 5%

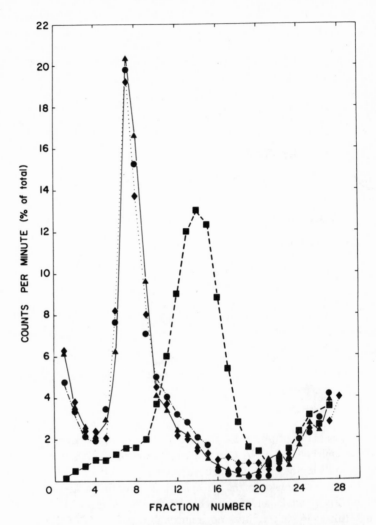

Fig. 1. Effect of T4 endonuclease on DNA of irradiated and unirradiated cells of E. coli. Cells of AB2499 (*uvrB5*), labeled with [³H]thymine, were irradiated with 20 ergs/mm² at 254 nm and treated with lysozyme, Brij-58, and endonuclease (7.5 μg/ml). They were then lysed, and their DNA was analyzed on alkaline sucrose gradients. Unirradiated control cells not treated with lysozyme, Brij, or endonuclease (♦); unirradiated cells treated with lysozyme, Brij, and endonuclease (▲); irradiated cells treated with lysozyme and Brij but not with endonuclease (●); irradiated cells treated with lysozyme, Brij, and endonuclease (■).

Sarkosyl in 0.5 N NaOH. The number of breaks produced by the nuclease, and from this the number of dimers originally present in the DNA, were estimated from the relative rate of sedimentation of the treated DNA in alkaline sucrose compared to that of untreated, control DNA. Using this procedure the effect of 10 ergs/mm^2, or approximately 50 dimers per *E. coli* genome, could easily be detected. A typical result is shown in Fig. 1.

Similar procedures have been developed by other investigators using endo-nuclease preparations from *Micrococcus luteus* to measure dimers in DNA isolated from bacterial (Wilkins, 1973*a*) and mammalian cells (Wilkins, 1973*b*; Paterson et al., 1973).

REFERENCES

Friedberg, E. C. (1972). *Mutat. Res.* **15**, 113–123.
Friedberg, E. C. and Clayton, D. A. (1972). *Nature,* **237**, 99–100.
Friedberg, E. C., Minton, K., Pawl, G. and Verzola, P. (1974). *J. Virol.* **13**, 953–959.
Ganesan, A. K. (1973). *Proc. Nat. Acad. Sci. U.S.A.* **70**, 2753–2756.
Ganesan, A. K. (1974). *J. Mol. Biol.* **87**, 103–119.
Paterson, M. C., Lohman, P. H. M. and Sluyter, M. L. (1973). *Mutat. Res.* **19**, 245–256.
Wilkins, R. J. (1973*a*). *Biochim. Biophys. Acta*, **372**, 33–37.
Wilkins, R. J. (1973*b*). *Int. J. Radiat. Biol.* **24**, 609–613.
Yasuda, S. and Sekiguchi, M. (1970). *Proc. Nat. Acad. Sci. U.S.A.* **67**, 1839–1845.
Proc. Nat. Acad. Sci. U.S.A. **67**, 1839–1845.

II

Enzymatic Photoreactivation

11

Enzymatic Photoreactivation: Overview

Claud S. Rupert

Institute for Molecular Biology
The University of Texas at Dallas
Richardson, Texas 75080

Photoreactivation—a reduction in the effect of ultraviolet irradiation by subsequent exposure to longer wavelengths—stems from at least two different kinds of processes. The first is direct, photoenzyme-mediated repair of ultraviolet radiation damage to DNA, while the second ("indirect photoreactivation") is an enhancement of light-independent repairs due to physiological changes induced in cells by light.[1] These two kinds of processes can be distinguished by their different wavelength and temperature dependences (Jagger and Stafford, 1965). Since indirect photoreactivation is merely one aspect of recovery through mechanisms able to act in the dark, it is best discussed in that context. We are concerned in this section only with the direct photoenzymatic process.

Although there had been earlier indications of recovery from radiation damage (e.g. Hollaender and Claus, 1937), the photoenzyme-mediated repair was the first type to be noted as an explicit, controllable process (Kelner, 1949; Dulbecco, 1949). Largely because of its simplicity and absolute requirement for light, it was also the first to be operated *in vitro* (Rupert et al., 1958), and the first to be characterized with regard to mechanism (Rupert, 1962a,b), establishing beyond question the existence of specific DNA repair. Before its mechanism was understood the phenomenon was reviewed by Dulbecco (1955), and subsequently by Jagger (1958), who gave thorough coverage to the early literature. Later accounts were provided by Rupert (1964) and J. K. Setlow (1966, 1967), and more recent work has been covered by Cook (1970).

---

[1] Indirect recovery can also be observed when light is applied *before* UV irradiation, in which case it is called "photoprotection."

Three general questions might be asked about a biological mechanism in any organism: (1) What does it do? (2) How does it do it? (3) What is its significance to the organism in which it occurs? The first question has been answered for photoenzymatic repair (although there are complications about some uses one would like to make of the information). The second has been answered down to the enzymatic level, but not at the level of chemical mechanism. The third remains uncertain, since the most plausible answer does not seem consistent with the peculiar patterns of the photoenzyme's occurrence and nonoccurrence in the biological world. We can take up each of these points in turn.

With regard to what the process does, a considerable variety of photoproducts can be produced in nucleic acids by ultraviolet radiation, depending on the kind of polynucleotide and the environment in which it is placed (see Varghese, 1972). Among these products are the cyclobutadipyrimidines, or "pyrimidine dimers," formed by addition of one pyrimidine ring to another at both ends of their 5,6 double bonds through a cyclobutane linkage. The rings themselves can be added to each other in either a "doghouse" or a "chair" configuration, as illustrated by sketches (*a*) and (*b*), respectively. Furthermore,

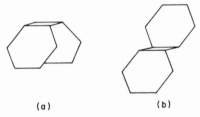

(a)   (b)

in each configuration the two rings can be arranged either "symmetrically" (say, with the No. 1 nitrogen atoms on both rings pointed toward the reader's right) or "antisymmetrically" (with these nitrogens pointed oppositely). Formal nomenclature and abbreviations, both for these dipyrimidine isomers and for other DNA photoproducts, have been summarized by Cohn et al. (1974).

The relative positions of the pyrimidine rings in the "symmetrical doghouse" structure (the *cis-syn* isomer) differ only moderately from the positions of two adjacent bases in the same strand of native double-stranded DNA, so that one might expect preferential formation of this isomer. In fact, it is the only dithymine isomer isolated from such a DNA after irradiation (Blackburn and Davies, 1966; Varghese and Wang, 1967), and presumably also the only isomer of the other cyclobutadipyrimidines formed under the same circumstances. This makes it the major photoproduct type detected after moderate UV doses under typical biological conditions (see R. B. Setlow, 1966). The photoenzymatic repair process removes this isomer from DNA (Wulff and Rupert, 1962; Setlow and Carrier, 1966; R. B. Setlow, 1966), converting it to its constituent pyrimidines (Cook, 1967; Williams et al., 1971). The "symmetrical chair" structure (which can be formed to some extent in denatured DNA on account of the greater freedom of base orientation in a random coil) is not removed (Ben-Hur and Ben-Ishai, 1968), nor are other recognized photoproducts (Rahn et al., 1969; Patrick, 1970).[2] Thus, we are fairly sure that any biological restoration effected by the photoenzyme process stems only from the splitting of a particular cyclobutadipyrimidine isomer in DNA.

Knowing this fact allows rather less to be concluded about the nature of the photoproducts responsible for various biological effects in a wild-type cell than one might suppose. The reason is that the "pyrimidine dimers," as major photoproducts in DNA, compete with all the other photoproducts for the general-purpose, light-independent repair mechanisms, and such a competition is relieved if these dimers are removed by photoenzymatic repair. The dark repair of the other products can then proceed more efficiently, even though they themselves have not been repaired by direct photoreactivation. This means that the alleviation of an effect by photoenzymatic repair does not demonstrate unambiguously that it stems from *cis-syn* cyclobutadipyrimidines in any case where other repair processes are also active. The point has been discussed at some length by Witkin (1969, pp. 501–505) in connection with ultraviolet mutagenesis.

The agent performing the photorepair is a single enzyme (EC 4.1.99.3), with the *cis-syn* cyclobutadipyrimidine isomer in DNA as substrate, catalyzing a photochemical reaction in its enzyme-substrate complex. Since enzymes are usually fastidious about their substrates, such a mechanism increases our confidence about the uniqueness of the photoproduct being repaired. Much of the present knowledge about the enzyme's operation is summarized in the following article by Walter Harm (this volume). As he relates, it has been possible by utilizing the peculiarities of this system to determine, both *in vivo* and *in vitro*, the numbers of enzyme molecules present, the rate constants for formation and dissociation of the enzyme-substrate complex, the approximate magnitude of the molar absorption coefficient of the complex, and some information about the quantum yield of the photochemical step. Similar numerical values have also been obtained for these quantities using an entirely different experimental method (Madden and Werbin, 1974).

The binding of this enzyme to its substrate resembles other specific binding of protein to nucleic acids in being endothermic, driven only by a large entropy increase during complex formation (see Rupert et al., 1972, 1974). It is critically dependent on ionic strength (Harm and Rupert, 1970*a*), and this dependence is identical for enzyme from several different sources and for several different monovalent cations (Cook, 1972).[3] The binding, as well as subsequent normal photorepair, will occur with substrate in a single-stranded DNA (Rupert and

---

[2] The "antisymmetrical" dipyrimidine isomers can form photochemically from the isolated bases but are not detected in DNA (see Varghese, 1972).

[3] The enzyme from human leukocytes, however, has a lower ionic strength optimum than others examined (Sutherland, this volume).

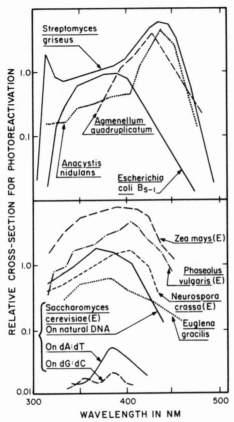

RELATIVE CROSS-SECTION FOR PHOTOREACTIVATION

WAVELENGTH IN NM

Fig. 1. Action spectra for photoenzymatic repair. Adapted and updated from Jagger et al. (1970). Spectra labeled (E) were obtained with extracted enzyme acting *in vitro* on irradiated transforming DNA [the system generally described below by Harm (this volume)]. Absolute magnitudes of different curves are without significance, except that the three spectra for *S. cerevisiae* enzyme (acting on irradiated natural DNA, on irradiated dA:dT, and on irradiated dG:dC) are shown on the same scale. Sources of spectra: *S. griseus* (Jagger et al., 1970); *A. quadruplicatum* (blue-green alga) (Van Baalen, 1968); *A nidulans* (Saito and Werbin, 1970); *E. coli* B_{s-1} (Jagger et al., 1969); *Z. mays* (Ikenaga et al., 1974); *P. vulgaris* (Saito and Werbin, 1969); *N. crassa* (Terry and Setlow, 1967)—very similar to the action spectrum for photorecovery of irradiated conidia (Terry et al., 1967); *E. gracilis* (Schiff et al., 1961); *S. cerevisiae* enzyme on natural DNA (Harm and Rupert, 1970b)—very similar to the action spectrum for photorecovery of yeast cells (H. Harm, personal communication); *S. cerevisiae* enzyme on dA:dT and dG:dC (Rupert et al., 1974). Fine structure in spectra for *E. coli* B/r (Jagger and Latarjet, 1956) and for the *S. cerevisiae* and *E. coli* enzymes *in vitro* (Setlow and Boling, 1963; J. K. Setlow, 1966) were not found in the later spectra shown here—as discussed by Jagger et al. (1969) and by Harm and Rupert (1970b). Although *Agmenellum* photoreactivation is not definitely shown to be direct photoenzymatic repair, the high efficiency at these long wavelengths would be inconsistent with known cases of the indirect mechanism in bacteria. A rough action spectrum (not shown) has been published for *Tetrahymena pyriformis* (Francis and Whitson, 1969).

Harm, 1966), provided that the polynucleotide chain is sufficiently long: in the case of polythymidylic acid a minimum of 9 residues is necessary, and improvement in binding affinity occurs as this number increases up to around 20 (Setlow and Bollum, 1968). Nothing, however, is known about the structure of the complex that is finally formed.

A variety of action spectra have been observed for the enzyme from different species (Fig. 1), all of them extending to wavelengths longer than the major absorption bands of either nucleotides or nucleotide photoproducts.

Because cyclobutadipyrimidines can be split directly by short UV wavelengths which they themselves can absorb (Setlow, 1961; Johns et al., 1962; for summary see Varghese, 1972), and can be induced to split by longer wavelengths in the presence of photochemical sensitizers (Lamola, 1972, plus further references therein), it has seemed plausible that such a mechanism might operate in the complex. One has accordingly pictured an enzyme-borne sensitizer as being responsible for the splitting of the cyclobutadipyrimidines under illumination. In the past, direct evidence on this point has been hard to obtain due to the extremely low concentration of the enzyme in cells (constituting about 10^{-5} of the total extract protein), and the consequent difficulty of purification and analysis of the enzyme itself. A major advance, described by B. M. Sutherland (this volume), has now permitted the enzyme production in *Escherichia coli* cells to be increased as much as 2500-fold above its normal level. The amount of highly purified enzyme made possible by this clever tactic has permitted direct observation of the protein spectrum in the near ultraviolet. Surprisingly, it shows no appreciable absorption band around 360 nm (Sutherland and Sutherland, 1972)—a wavelength where the molar absorption coefficient of the enzyme-substrate complex is greater than 10^4 liter·mole^{-1}·cm^{-1} (see Harm, this volume).

Studies of the action spectrum of the yeast enzyme *in vitro* complement this observation. The spectrum for splitting cyclobutadithymine in irradiated dA:dT (where these are the only type of dipyrimidine formed) is distinctly different from the spectrum for splitting cyclobutadicytosine in dG:dC, and both of these are different from the spectrum for splitting cyclobutadipyrimidines in natural DNA (Rupert et al., 1975). It is noteworthy that even the action spectrum for splitting cyclobutadithymines is different, depending on whether these particular products are located in natural DNA or in dA:dT. Thus, it would appear that the absorption spectrum of the complex is generated during its formation, and that it depends on the structure (either helical dimensions or local base sequence) of the nucleic acid in which the "pyrimidine dimer" resides. These facts argue against any simple analogy with the known photochemical sensitizers.

The most obvious biological utility of photoenzymatic repair lies in its removal of cyclobutadipyrimidines induced in cellular DNA by solar ultraviolet. It is easily shown that it performs this function. Bacterial transforming DNA inactivated by exposure to sunlight in a quartz flask can be partly reactivated by adding a small amount of photoreactivating enzyme and continuing the solar exposure in the same flask (Rupert, 1960a). Presence of the enzyme from the start of the exposure markedly reduces the inactivation (Rupert, 1964). Mutant cells deficient in this mechanism not only survive solar exposure less well than wild type (Harm, 1966), but also accumulate more photoreactivable damage than the latter in their DNA (Rupert and Harm, 1966). The simplicity of the enzymatic mechanism and its widespread occurrence (Table 1) would be consis-

Table 1. Occurrence of DNA Photoreactivating Enzyme Activity[a]

Phylum	Species	Tissues possessing enzyme activity[b]	Tissues failing to show activity[b]
Cyanophyta	*Plectonema boreanum* (blue-green alga)	Cells (1)[a]	—
	Anacystis nidulans (blue-green alga)	Cells (1)[b]	—
Schizomycophyta	*Escherichia coli* (colon bacterium)	Cells (1)[c]	—
	Streptomyces griseus (soil actinomycete)	Cells (1)[d]	—
	Serratia marcescens (bacterium)	Cells (4)[e]	—
	Haemophilus influenzae (respiratory tract bacterium)	—	Cells (1)[f]
	Diplococcus pneumoniae (respiratory tract bacterium)	—	Cells (1)[f]
	Bacillus subtilis (soil bacterium)	—	Vegetative cells (4)[g]
	Micrococcus luteus (bacterium)	—	Cells (1)[h]
	Micrococcus radiodurans (bacterium)	—	Cells (1, 4)[i]

Eumycophyta	Saccharomyces cerevisiae (bakers' yeast)	Cells (1)[j]	—
	Schizosaccharomyces pombe (fission yeast)	—	Cells (1)[η]
	Neurospora crassa (bread mold)	Conidia (1)[k]	—
Euglenophyta	Euglena gracilis	Cells (1)[l] (2)[λ]	—
Angiospermae	Phaseolus vulgaris (pinto bean)	Plumule, hypocotyl, cotyledon of dark-grown sprout, young leaves (1)[m]	Radicles, older leaves (1)[m]
	Phaseolus lunatus (lima bean)	Whole sprout (1)[m]	—
	Zea mays (maize)	Harvested pollen (1)[n]	—
	Nicotiana tabacum (tobacco)	Cultured cells (3)[p]	—
	Phaseolus aureus (mung bean)	—	Whole sprout (1)[m]
	Haplopappus gracilis	—	Cultured cells (3)[p]
Gymnostermae	Gingko biloba (gingko tree)	Cultured cells (3)[o]	—

(cont.)

Table 1. Cont.

Phylum	Species	Tissues possessing enzyme activity [b]	Tissues failing to show activity [b]
Protozoa	*Paramecium aurelia* (paramecium)	Cells (3)[q]	—
	Tetrahymena pyriformis	Cells (3)[r]	—
Mollusca	*Physa sp* (pond snail)	Muscle (2)[s]	—
Echinodermata	*Arbacia punctulata* (sea urchin)	Testis, eggs, ovary (1, 2)[t,u]	Sperm[v]
	Echinarachnius parma (sand dollar)	Egg, early embryo (3)[w]	—
Arthropoda	*Anagasta kühniella* (flower moth)	Adult female abdomen (1)[u]	—
	Gecarcinus lateralis (land crab)	Testis, ovary epithelium, somatic and heart muscle (1)[u]	Midgut gland (1)[u]
	Artemia salina (brine shrimp)	Whole nauplii (2)[s]	—
	Homarus americanus (lobster)	Ovary (2)[s]	—

Chordata			
Bony fish			
(Teleost)	*Haemulon sciurus* (blue striped grunt)	Established dorsal fin line (1,2,3)[x]	—
(Teleost)	*Poecilia formosa*	Whole fish (2)[y]	—
(Teleost)	*Pimephales promelas* (fathead minnow)	Established epithelial cell line (1)[s]	—
Amphibian			
	Bufo marinus (cowflop toad)	White blood cells (1)[u]	Blood serum, red blood cells (1)[u]
	Xenopus laevis (African clawed toad)	Established liver-cell line (1)[z]	—
	Rana pipiens (frog)	Skeletal and cardiac muscle, sciatic nerve, brain, liver (1)[u]	—
Reptile			
	Terrapene carolina (box turtle)	Established heart cell line (1)[α]	—
	Iguana iguana (lizard)	Established heart cell line (1)[α]	—
	Gekko gekko (lizard)	Established lung cell line (1)[α]	—

(cont.)

Table 1. Cont.

Phylum	Species	Tissues possessing enzyme activity[b]	Tissues failing to show activity[b]
Chordata (continued)			
Bird	*Gallus gallus* (domestic chicken)	Primary fibroblasts, whole 4-day embryos, adult brain (1)[u]; primary embryo cultures (3,4)[θ]	Kidney, liver, skeletal muscle, egg white, egg yolk (1)[u]
Mammal (Marsupial)	*Didelphis marsupialis* (American opossum)	Brain, liver, kidney, testis, heart, lung (1,2,3)[β]	—
(Marsupial)	*Caluromys derbianus* (South American wooly opossum)	Established kidney cell line (1,2,3)[β]	—
(Marsupial)	*Potorous tridactylis* (Tasmanian rat kangaroo, or potoroo)	Two established cell lines (male and female)(1, 2, 3)[β]	—
(Rodent)	*Mus musculus* (mouse)	—	Skin (1)[u]
(Rodent)	*Rattus novegicus* (rat)	—	Liver, skeletal and heart muscle, ovary, testis, brain, kidney (1)[u]; uterus plus fetus (1)[s]

(Rodent)	*Cricetulus griseus* (Chinese hamster)			Established embryonic cell line (3)(γ)
(Lagomorph)	*Oryctolagus cuniculus* (rabbit)	White blood cells (1)(δ)	—	Skin (1)(u); Primary kidney culture (3, 4)(θ)
(Ungulate)	*Bos taurus* (domestic beef)	White blood cells (1)(δ); marrow (2)(ϵ)	3-cm fetus (1)(u)	
(Primate)	*Homo sapiens* (human)	White blood cells (3)(ϵ)		Cultured skin fibroblasts (3)(ζ)

[a]This table expands and updates an earlier list compiled by J. K. Setlow (1972). Because different methods have been used to detect the presence of photoenzymatic repair, indirect mechanisms are excluded to various degrees in different published works, and negative reports are of varying certainty. The present list includes only cases for which the evidence is relatively simple and direct. The negative list for microorganisms is arbitrarily restricted to a few illustrative cases (see Jagger, 1958, for others).

[b]Literature references are given by superscript letters keyed to the reference list below, and test methods are given by parenthetical numbers according to the following code: (1) light-induced repair of bacterial transforming DNA *in vitro* by cell extracts; (2) light-induced removal of cyclobutadipyrimidines from DNA by cell extracts *in vitro* (by hydrolysis and chromatography of the nucleic acid, except in reference ϵ, where it was detected by restored susceptibility to nuclease attack); (3) light-induced removal of cyclobutadipyrimidines from DNA of irradiated intact cells; (4) analysis of biological photorecovery [photoreactivation of cell survival, distinguishing indirect mechanisms by the Jagger and Stafford (1965) criteria, or photoreactivation of DNA viruses in host cells, or—for negative indications—absence of any observed photorecovery]. References: (a) Werbin and Rupert, 1968; (b) Saito and Werbin, 1970; (c) Rupert et al., 1958; (d) Eker, 1972; (e) Kaplan et al., 1960; (f) Goodgal et al., 1957; (g) Keiner, 1964; (h) Elder and Beers, 1965; (i) J. K. Setlow and M. E. Boling, personal communication; (j) Rupert, 1960b; (k) Terry and Setlow, 1967; (l) Diamond et al., 1969; (m) Saito and Werbin, 1969; (n) Ikenaga et al., 1974; (o) Trosko and Mansour (1969); (p) Trosko and Mansour (1968); (q) Sutherland et al., 1967; (r) Francis and Whitson, 1969; (s) J. S. Cook, personal communication; (t) Cook and Setlow, 1966; (u) Cook and McGrath, 1967; (v) Blum et al., 1967; (y) J. S. Cook and R. W. Hart, personal communication; (z) Regan et al., 1968; (α) Regan et al., 1969; (β) Cook and Regan, 1969; (γ) Trosko et al., 1965; (δ) H. Harm, private communication; (ϵ) Sutherland, this volume; (ζ) Cook, 1970; (η) C. S. Rupert and K. Haefner, unpublished experiment; (θ) Pfefferkorn and Coady, 1968; (λ) Diamond et al. (1975).

tent with its early evolution as a protection against sunlight—although the unaided process cannot preserve a modern *E. coli* cell from serious injury by the sun (Harm, 1969).

The difficulty with this hypothesis about its purpose lies in the number of cell types possessing photoreactivating enzyme for which no significant ultraviolet exposure can be pictured. One may perhaps explain the presence of the mechanism in an intestinal bacterium like *E. coli* on the basis of some advantage during transmittal from one animal to another at a water hole (although its absence from a soil bacterium like *Bacillus subtilis* is then less clear, particularly since it is present in the soil actinomycete *Streptomyces griseus*). But its presence in internal tissues—liver, muscle, brain, lung—of metazoa whose size precludes penetration by ultraviolet light (see Cook, 1970) suggests that some additional dimension of its utility may as yet be unappreciated. Harm et al. (1972) have suggested that the small number of photoenzyme molecules synthesized per cell makes close regulation of their production uneconomical, the necessary regulatory machinery costing more than the unneeded synthesis. From this viewpoint if any cell of the organism benefits from it, certain others might also continue to produce it after differentiation. It cannot be said, however, that the enzyme's presence or absence from various cell types is understood.

Until recently it was believed (largely from the extensive surveys by Cook and his coworkers) that the photoenzymatic repair did not occur above marsupials in the phylogenetic tree. However, the second article by B. M. Sutherland in this volume describes its occurrence in human white blood cells. It has also been found in similar cells of beef and rabbit, and so quite possibly exists in other higher mammals as well (H. Harm, personal communication). The puzzle about its utility therefore extends to our own bodies.

Repair of DNA is evidently important to the cellular economy because of the limited number of copies of genetic information it provides, and the inability of a cell to retrieve this information from any other source. Neither of these statements is true for RNA, whose repair would therefore seem less important. On the other hand, the photoreactivation of irradiated plant viral RNAs after infection of their host cells has long indicated that some sort of RNA repair must exist.

Chemically the difference between DNA and RNA is small, and both natural and synthetic RNAs can form cyclobutadipyrimidines under irradiation (see Varghese, 1972), yet the photoenzyme from yeast fails completely to act on irradiated RNA (Rupert and Harm, 1966). The enzyme will, however, act on irradiated, denatured DNA from the *B. subtilis* bacteriophage SP2, which contains uracil in place of thymine, and which therefore differs from a single-stranded RNA essentially only in the absence of a $2'$-OH group from its sugar moieties. Failure of yeast enzyme to act on dipyrimidines in RNA must consequently stem from this single difference.

Recent work with plant cell extracts and infectious viral RNAs, described in the concluding paper of this section by Gordon, now provides strong indications of an enzymatic photorepair for RNA, somewhat analogous to the one known for DNA. Not only can RNA be repaired by the appropriate photoenzyme, but there is also evidence that it may undergo repair by light-independent processes, suggesting that (at least in some cases) cells actively preserve the integrity of their longer-lived RNA species.

More detail is known about the photoenzymatic repair of nucleic acids than about any other repair process, but, as is often the case, the information itself raises new questions. At the conclusion of the conference session represented in this section, the fundamental chemical mechanism and biological purpose of this enzymatic process still required clarification.

ACKNOWLEDGMENT

Support under Research Grant GM 16547, from the National Institute of General Medical Sciences, U.S. Public Health Service, and from The University of Texas at Dallas Research Fund is gratefully acknowledged.

REFERENCES

Ben-Hur, E. and Ben-Ishai, R. (1968). *Biochim. Biophys. Acta* **166**, 9–15.

Blackburn, G. M. and Davies, R. J. H. (1966). *Biochem. Biophys. Res. Commun.* **22**, 704–706.

Blum, H. F., Robinson, J. C. and Loos, G. M. (1951). *J. Gen. Physiol.* **35**, 323–342.

Cohn, W. E., Leonard, N. J. and Wang, S. Y. (1974). *Photochem. Photobiol.* **19**, 89–94.

Cook, J. S. (1967). *Photochem. Photobiol.* **6**, 97–101.

Cook, J. S. (1970). *Photophysiology* **5**, 191–233.

Cook, J. S. (1972). In *Molecular and Cellular Repair Processes* (Beers, R. F., Herriott, R. M. and Tilghman, R. C., eds.), pp. 79–94. Johns Hopkins Press, Baltimore.

Cook, J. S. and McGrath, J. R. (1967). *Proc. Nat. Acad. Sci. U.S.A.* **58**, 1359–1365.

Cook, J. S. and Regan, J. D. (1969). *Nature* **223**, 1065–1066.

Cook, J. S. and Setlow, J. K. (1966). *Biochem. Biophys. Res. Commun.* **24**, 285–289.

Diamond, J., Schiff, J. A. and Kelner, A. (1969). *Plant Physiol.* **44** (Suppl.), 9.

Diamond, J., Schiff, J. A. and Kelner, A. (1975). *Archives Biochem. Biophys.* **167**, 603–614.

Dulbecco, R. (1949). *Nature* **163**, 949–950.

Dulbecco, R. (1955). In *Radiation Biology* (Hollaender, A., ed.), Vol. II, pp. 455–486. McGraw-Hill, New York.

Eker, A. P. M. (1972). In *Book of Abstracts. VI Int. Congr. Photobiol. (Bochum)*, No. 118.

Elder, R. L. and Beers R. F., Jr. (1965). *J. Bacteriol.* **90**, 681–686.

Francis A. A. and Whitson, G. L. (1969). *Biochim. Biophys. Acta* **179**, 253–257.

Goodgal, S. H., Rupert, C. S. and Herriott, R. M. (1957). In *The Chemical Basis of Heredity* (McElroy, W. D. and Glass, B., eds.), pp. 341–343. Johns Hopkins Press, Baltimore.

Harm, W. (1966). In *The Physiology of Gene Mutation and Mutation Expression* (Kohoutová, M. and Hubáček, J., eds.), pp. 51–59. Academia, Prague.

Harm, W. (1969). *Radiat. Res.* **40**, 63–69.

Harm, H. and Rupert, C. S. (1970a). *Mutat. Res.* **10**, 291–306.

Harm, H. and Rupert, C. S. (1970b). *Mutat. Res.* 10, 306–318.
Harm, W., Rupert, C. S. and Harm, H. (1972). In *Molecular and Cellular Repair Processes* (Beers, R. F., Herriott, R. M. and Tilghman, R. C., eds.), pp. 53–63. Johns Hopkins Press, Baltimore.
Hollaender, A. and Claus, W. D. (1937). *Bull. Nat. Res. Council* 100, 75–88.
Ikenaga, M., Kondo, S. and Fujii, T. (1974). *Photochem. Photobiol.* 19, 109–113.
Jagger, J. (1958). *Bacteriol. Rev.* 22, 99–142.
Jagger, J. and Latarjet, R. (1956). *Ann Inst. Pasteur* 91, 858–873.
Jagger, J. and Stafford, R. S. (1965). *Biophys. J.* 5, 75–88.
Jagger, J., Stafford, R. S. and Snow, J. M. (1969). *Photochem. Photobiol.* 10, 383–395.
Jagger, J., Takebe, H. and Snow, J. M. (1970). *Photochem. Photobiol.* 12, 185–196.
Johns, H. E., Rapaport, S. A. and Delbrück, M. (1962). *J. Mol. Biol.* 4, 104–114.
Kaplan, R. W., Winkler, U. and Wolf-Ellmauer, H. (1960). *Nature* 186, 330–331.
Kelner, A. (1949). *Proc. Nat. Acad. Sci. U.S.A.* 35, 73–79.
Kelner, A. (1964). *Radiat. Res.* 25, 205.
Lamola, A. A. (1972). *J. Amer. Chem. Soc.* 94, 1013–1014.
Madden, J. J. and Warbin, H. (1974). *Biochemistry* 13, 2149–2154.
Patrick, M. H. (1970). *Photochem. Photobiol.* 11, 477–485.
Pfefferkorn, E. R. and Coady, H. M. (1968). *J. Virol.* 2, 474–479.
Rahn, R. O., Setlow, J. K. and Hosszu, J. L. (1969). *Biophys. J.* 9, 510–517.
Regan, J. D. and Cook, J. S. (1967). *Proc. Nat. Acad. Sci. U.S.A.* 58, 2274–2279.
Regan, J. D., Cook, J. S. and Lee, W. H. (1968). *J. Comp. Cell. Physiol.* 71, 173–176.
Regan, J. D., Cook, J. S. and Takeda, S. (1969). In *Hemic Cells in Vitro* (Farnee, P., ed.), p. 162. Williams and Wilkins, Baltimore.
Rupert, C. S. (1960a). In *Comparative Effects of Radiation* (Burton, M. and Kirby-Smith, J. S., eds.), pp. 49–61. Wiley, New York.
Rupert, C. S. (1960b). *J. Gen. Physiol.* 43, 573–595.
Rupert, C. S. (1962a). *J. Gen. Physiol.* 45, 703–724.
Rupert, C. S. (1962b). *J. Gen. Physiol.* 45, 724–741.
Rupert, C. S. (1964). *Photophysiology* 2, 283–327.
Rupert, C. S. and Harm, W. (1966). *Adv. Radiat. Biol.* 2, 1–81.
Rupert, C. S., Goodgal, S. H. and Herriott, R. M. (1958). *J. Gen. Physiol.* 41, 451–471.
Rupert, C. S., Harm, W. and Harm, H. (1972). In *Molecular and Cellular Repair Processes* (Beers, R. F., Herriott, R. M. and Tilghman, R. C., eds.), pp. 64–78. Johns Hopkins Press, Baltimore.
Rupert, C. S., Harm, H. and To, K. (1975). In *Proc. Symp. New Trends in Photobiology, Rio de Janeiro* (in press).
Saito, N. and Werbin, H. (1969). *Photochem. Photobiol.* 9, 389–393.
Saito, N. and Werbin, H. (1970). *Biochemistry* 9, 2610–2620.
Schiff, J. A., Lyman, H. and Epstein, H. T. (1961). *Biochim. Biophys. Acta* 50, 310–318.
Setlow, J. K. (1966). *Curr. Top. Radiat. Res.* 2, 195–248.
Setlow, J. K. (1967). *Comp. Biochem.* 27, 157–209.
Setlow, J. K. (1972). In *Research Progress in Organic, Biological, and Medicinal Chemistry* (Gallo, V. and Santamaria, L., eds.), vol. 3, part I, pp. 335–355. North Holland, Amsterdam-London.
Setlow, J. K. and Boling, M. E. (1963). *Photochem. Photobiol.* 2, 471–477.
Setlow, J. K. and Bollum, F. J. (1968). *Biochim. Biophys. Acta* 157, 233–237.
Setlow, R. B. (1961). *Biochim. Biophys. Acta* 49, 237–238.
Setlow, R. B. (1966). *Science* 153, 379–386.
Setlow, R. B. and Carrier, W. L. (1966). *J. Mol. Biol.* 17, 237–254.
Sutherland, B. M., Carrier, W. L. and Setlow, R. B. (1967). *Science* 158, 1699–1700.

Sutherland, J. C. and Sutherland, B. M. (1972). In *Book of Abstracts VI Int. Congr. Photobiol. (Bochum)*, No. 115.

Terry, C. E. and Setlow, J. K. (1967). *Photochem. Photobiol.* **6**, 799–803.

Terry, C. E., Kilbey, B. J. and Howe, H. B., Jr. (1967). *Radiat. Res.* **30**, 739–747.

Trosko, J. E. and Mansour, V. H. (1968). *Radiat. Res.* **36**, 333–343.

Trosko, J. E. and Mansour, V. H. (1969). *Mutat. Res.* **7**, 120–121.

Trosko, J. E., Chu, E. H. Y. and Carrier, W. L. (1965). *Radiat. Res.* **24**, 667–672.

Van Baalen, C. (1968). *Plant Physiol.* **43**, 1689–1695.

Varghese, A. J. (1972). *Photophysiology* **7**, 207–274.

Varghese, A. J. and Wang, S. Y. (1967). *Nature* **213**, 909–910.

Werbin, H. and Rupert, C. S. (1968). *Photochem. Photobiol.* **7**, 225–230.

Williams, D. L., Hayes, F. N., Varghese, A. J. and Rupert, C. S. (1971). *Biophys. Soc. Abstr.* **11**, 191a.

Witkin, E. M. (1969). *Ann Rev. Microbiol.* **23**, 487–514.

Wulff, D. L. and Rupert, C. S. (1962). *Biochem. Biophys. Res. Commun.* **7**, 237–240.

12

Kinetics of Photoreactivation

Walter Harm

The University of Texas at Dallas
Institute for Molecular Biology
Richardson, Texas 75080

ABSTRACT

This paper summarizes experimental work (most of which is published) in which light flashes were used for an analysis of photoenzymatic repair *in vivo* and *in vitro*. The method permits determination of the reaction rate constants for the formation, dark dissociation, and repair photolysis of enzyme-substrate complexes under various conditions, and estimation of the number of photoreactivating enzyme molecules present. Investigation of these characteristics is basic for understanding of the overall photoreactivation kinetics observed in biological systems, its dependence on experimental parameters, and possibly its biological significance.

Photoenzymatic repair of UV lesions in DNA is the simplest repair process so far known. It employs only one kind of enzyme ("photoreactivating enzyme" or PRE), which in conjunction with light energy specifically repairs cyclobutadipyrimidines ("pyrimidine dimers") by monomerizing ("splitting") them *in situ*. The light dependence of the reaction is evident from the general kinetics of photoreactivation (PR): The survival of a UV-irradiated biological system increases with the time of post-illumination with monochromatic or polychromatic light of the appropriate wavelength(s) (ranging from about 310 to 480 nm) until it attains a maximal level, which is determined by the photorepairable fraction of total lethal lesions. Details of the PR kinetics depend on a variety of parameters: in particular, the spectral composition and intensity of photoreactivating light, the PRE content of the biological system, temperature, pH, ionic strength, presence of inhibitors, and others.

The reaction scheme for PR (Rupert, 1962a) is expressed by

$$E + S \underset{k_2}{\overset{k_1}{\rightleftharpoons}} ES \xrightarrow[k_3]{\text{light}} E + P$$

where E is the PRE, S is its substrate (pyrimidine dimers), ES is the enzyme-substrate complex, and P is the repair product (monomers). It is evident from this scheme that understanding of the overall PR kinetics requires knowledge of the characteristics of the component steps, their rate constants, and their dependence on various experimental parameters, as well as information regarding the quantities of PRE molecules and of repairable UV lesions involved. The following is a brief review of the relevant results and considerations in this respect.

USE OF LIGHT FLASHES FOR THE INVESTIGATION OF PHOTOENZYMATIC REPAIR

Compared with other enzymatic processes, photoenzymatic repair is unusual by its absolute requirement for light energy. The fact that the light-dependent reaction ("photolysis") is preceded by a light-independent reaction ("complex formation") permits, in principle, a separate investigation of each step by the

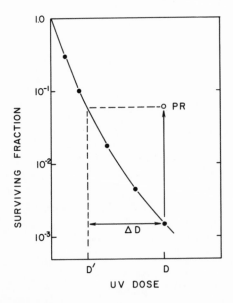

Fig. 1. *Quantitative characterization of the PR effect by the dose decrement* ΔD.

controlled use of light. Continuous illumination is often not suitable for this purpose, since PRE molecules liberated by photolysis of existing ES complexes might enter the reaction again by forming new complexes, etc. Therefore, this problem has been approached by applying intense light flashes of about 1 msec duration. The experimental work has been carried out with UV-irradiated transforming DNA from *Haemophilus influenzae* and yeast PRE *in vitro* by Helga Harm and C. S. Rupert, and with *Escherichia coli* cells and coliphages in my own laboratory. Our analysis of the photoenzymatic repair process was based on the following facts and considerations.

1. We had shown in two ways that a single flash repairs virtually all of the enzyme-substrate complexes present at that moment: (a) A flash results in maximal PR if PRE is in sufficient excess over the substrate (Harm and Rupert, 1968). (b) The PR effect is not altered by a decrease in light intensity of the flash, unless it drops to less than 0.25 times the usual intensity (Harm and Rupert, 1968; Harm et al., 1968).

2. It follows from (1) that the number of ES complexes present at the time of the flash must equal the number of pyrimidine dimers abolished by PR. The latter can be calculated from the extent of the biological effect observed (i.e. survival increase of cells or phage, or activity increase of transforming DNA). As illustrated in Fig. 1, the PR effect is expressed by the dose decrement ΔD, which is the difference between the UV dose D actually applied and a smaller UV dose D' that would have led to an equal survival without PR. ΔD expresses the amount of UV radiation whose effect is annihilated by the light flash; therefore, the number of dimers repaired should closely resemble that produced by an equal dose increment ΔD. Since the latter is known from published photochemical work for the systems we investigated,[1] the biological PR effect resulting from a flash at any time permits us to calculate the number of ES complexes present.

3. The only additional quantity needed for a more detailed analysis of PR kinetics is the number of PRE molecules involved in the reaction. We will see further below that the flash also serves as a tool for determination of this number.

It should be noticed that with *E. coli* a measurable biological PR effect resulting from a single flash can be obtained only if the cells are sufficiently sensitive to UV radiation, i.e. lacking most of their dark repair. The reason is that they contain only a low number of PRE molecules, which cannot possibly repair more than an equal number of dimers during the short time of the flash.

---

[1] The number of pyrimidine dimers formed by 1 erg mm^{-2} of 254-nm radiation is approximately 6.5 in an *E. coli* chromosome (Rupp and Howard-Flanders, 1968) and 2.2×10^9 per μg *Haemophilus* transforming DNA (for refs. see Harm and Rupert, 1968).

Therefore, strain B_{s-1} was used for most of this work. In contrast, such a restriction does not exist for the *Haemophilus* DNA system *in vitro*, where PRE can be added at sufficiently high concentration to the reaction mixture.

KINETICS OF COMPLEX FORMATION IN THE DARK

If a PRE-containing cell is UV-irradiated at time zero, the kinetics of complex formation can be easily followed by taking samples at different times and exposing them to a single flash. Figure 2 shows the increase with time in the relative number of complexes formed, expressed by $\Delta D/\Delta D_{max}$ (where ΔD_{max} is the maximal ΔD that can possibly be obtained by optimal PR treatment at a given UV dose). One sees that B_{s-1} cells, irradiated to a survival slightly below 10^{-2}, attain their equilibrium of complex formation at room temperature after about 5 min. Application of a flash at a later time permits studying a second "round" of complex formation, which involves PRE molecules liberated by flash photolysis of the complexes formed in the first round. The same approach has been used *in vitro*, where the kinetics of complex formation can of course be anything from very slow to very fast, depending on enzyme and substrate concentrations and other parameters mentioned later. Knowledge of the time required for reaching the dark equilibrium of complex formation is of basic importance for the investigation of the photolytic reaction.

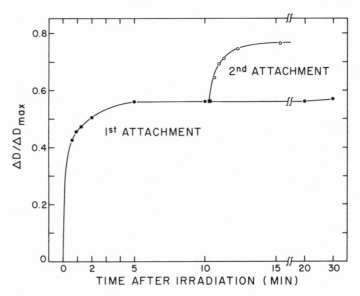

Fig. 2. *Formation of ES complexes in E. coli B_{s-1} cells as a function of time.* Cells were irradiated with 4.8 ergs mm^{-2} of 254-nm radiation, leading to a survival of about 6×10^{-3} (redrawn from Harm et al., 1968).

DETERMINATION OF THE NUMBER OF PRE MOLECULES

Information about the absolute concentration of PRE molecules in cells or in reaction mixtures *in vitro* not only is of intrinsic biological interest but also is necessary for the determination of reaction rate constants. We approached this problem by choosing conditions of substrate excess over PRE, sufficient for all enzyme molecules to be complexed. Fulfillment of this condition can be demonstrated by lack of increase in the number of ES complexes with further increase in substrate concentration. The maximal level of ΔD obtained in stationary phase B_{s-1} cells was 3 ergs mm^{-2}, corresponding to about 20 ES complexes repaired per cell, or 20 PRE molecules present (cf. Fig. 3A). Obviously an *E. coli* strain containing substantially more PRE molecules per cell could be recognized by an increased ΔD, i.e. by its greater survival increase after a single flash. This criterion was actually used for isolation of PRE-overproducing mutants (W. Harm, 1969), one of which turned out to contain about 120 PRE molecules per stationary-phase cell (cf. Fig. 3B).

Concentrations of PRE molecules have been similarly determined in reaction mixtures of UV-irradiated transforming DNA and PRE preparations (Harm and Rupert, 1968). At a given PRE concentration, an increase in concentration of DNA (irradiated with a fixed UV dose) will result in an *inversely proportional* decrease in $\Delta D/\Delta D_{max}$ if (and only if) virtually all PRE molecules are bound in ES complexes. In this case, their number can be calculated from the observed ΔD and from the known concentration of substrate molecules (for more detailed discussion see Harm et al., 1971).

REACTION RATE CONSTANT FOR COMPLEX FORMATION (k_1)

The rate constant k_1 has been determined by giving light flashes in relatively rapid sequence (every 0.5–1 sec) immediately after potential contact between PRE and substrate. Under appropriate conditions, the short time interval between flashes permits formation of only a few complexes, compared to the number of both PRE and substrate molecules present, so that the concentration of free PRE molecules, [E], is roughly constant and resembles the known concentration of total PRE molecules involved. Since the probability for dark dissociation of complexes is negligible (see next section), the decrease of lesions with time t is approximated by first-order kinetics:

$$[S]_t/[S]_0 \approx e^{-k_1 [E] t}$$

where $[S]_t/[S]_0$ equals the experimentally obtained value of $1 - (\Delta D/\Delta D_{max})$. The k_1 value thus calculated for cells of B_{s-1} (or its mutant with increased PRE content) at room temperature is approximately 10^6 liters mole^{-1} sec^{-1} (Harm,

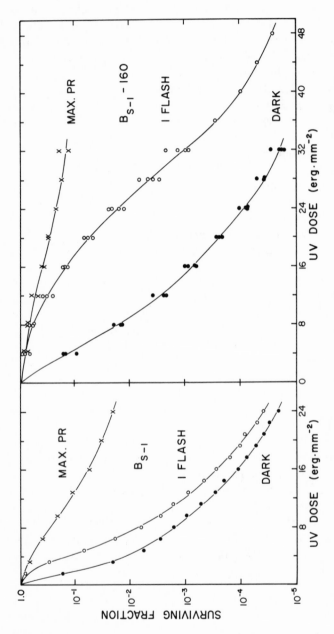

Fig. 3. Survival increase of irradiated E. coli cells resulting from PR by either a single flash (○) or continuous illumination to obtain maximal repair (X). Filled circles represent the survival without PR. Left panel: strain B_{S-1}. Right panel: strain B_{S-1}–160, which has increased PRE content (from W. Harm, 1969).

1970a). The k_1 values for the yeast enzyme *in vitro* increase with the pH from 6.0 to 7.5; they are maximal at ionic strengths between 0.17 and 0.25 (depending on the conditions), where they attain about 6×10^7 liters mole^{-1}sec^{-1} at room temperature (Harm and Rupert, 1970a). Since k_1 for the yeast enzyme *in vitro* was found to be inversely proportional to the viscosity, the lower k_1 values observed *in vivo* probably reflect the very high viscosity inside the cell.

REACTION RATE CONSTANT k_2 FOR DARK DISSOCIATION OF COMPLEXES

To measure the rate constant k_2, a great excess of competing substrate has been introduced after reaching the dark equilibrium of complex formation, so that PRE liberated by dissociation of established complexes is likely to bind with the competing DNA (whose repair is not measured by the assay). The result is a decreasing number of ES complexes, which is recorded as decreasing flash PR. For the *in vitro* system any irradiated DNA serves the purpose as competitor as long as it does not interfere with the transformation assay. In *E. coli*, equilibrium formation of ES complexes is first permitted to occur on DNA from irradiated phage T1 after infection of host cells; excess competing substrate is created by subsequent exposure of the infected cells to UV light, which increases the substrate represented by the T1 DNA only slightly. Both *in vivo* and *in vitro* the decrease of ES complexes resulting from dark dissociation corresponds to k_2 values between 10^{-2} and 10^{-3}sec^{-1} at room temperature. (Harm, 1970a; Harm and Rupert, 1970a). Similar k_2 values can be calculated from equilibria of complex formation (using the k_1 values mentioned before), and have also been obtained experimentally by applying caffeine as inhibitor of complex formation. The pH dependence of k_2 measured *in vitro* resembles that of k_1, and maximal k_2 values have been observed at about 0.25 ionic strength (Harm and Rupert, 1970a). The high k_1 values in combination with the low k_2 values observed under most conditions agree with the notion of high stability of ES complexes (Rupert, 1962b).

PHOTOLYTIC CONSTANT AND ABSOLUTE ACTION SPECTRA

The rate constant k_3 for the photolytic repair reaction can be expressed by the product $k_p I$, where I is the light intensity and k_p the "photolytic constant," which varies essentially with the wavelength. Thus k_3 is zero in the absence of photoreactivating light. A reliable measurement of k_p requires virtually all substrate to be complexed prior to the illumination; otherwise, the obtained value would be too low. This is achieved *in vitro* by supplying PRE in sufficient excess over substrate, and in the *E. coli* cell by using the PRE-overproducing mutant exposed to a low UV dose. In such cases the repair results in a decrease of ES complexes with increasing light dose L according to the function $[ES]_L/[ES]_0 = e^{-k_p L}$, from which k_p can be determined.

Action spectra are the usual choice to express the *relative* efficiency of various wavelengths in the PR reaction. In our case, the determination of k_p at different wavelengths permits construction of an action spectrum which shows, *on an absolute scale*, the efficiency with which light is used in the photolytic repair reaction. This is because any k_p value can be converted into the product of two characteristics: (1) the extent to which the complexes absorb incident photons (expressed by their molar extinction coefficient ϵ) and (2) the probability (expressed by the quantum yield ϕ) with which an absorbed photon leads to monomerization of the dimer.[2]

A plot of $\epsilon\phi$ as a function of wavelength (Fig. 4) shows that the maximal values of $\epsilon\phi$ exceed 10^4 liters mole^{-1}cm^{-1} for various biological systems. Although ϵ and ϕ cannot be determined separately by our experimental approach, we can conclude that both the absorption of light by the ES complex and the probability of an absorbed photon to accomplish repair must be high. ϵ must be at least as high as the product $\epsilon\phi$, since the quantum yield cannot exceed 1, and ϕ can hardly be lower than 10^{-1}, since this would make ϵ unreasonably high. Thus the PR reaction uses the light energy in a most economical way.

The strong absorption of the complex contrasts with the low absorption of purified free PRE from *E. coli* in the region 365–385 nm (Sutherland and Sutherland, 1972). This difference might be biologically important, assuming that repair of lesions caused by the shortest wavelengths of the solar spectrum is the essential function of this enzyme, for which it requires energy from less energetic photons of the same spectrum. Certainly, the weak absorption of free PRE reduces the likelihood of photochemical destruction at a time when it is not needed for repair. Recent investigations by Tyrrell et al. (1973) have indeed shown that PRE loses its activity upon exposure to high doses of photoreactivating light.

In contrast, the intracellular PRE is very stable in the dark. I have measured with the flash technique the activity of PRE in *E. coli* cells which were kept dark in buffer: at 5°C virtually all PRE molecules remained active through a period of over 5 weeks, at room temperature about 75% remained active, and even at 37°C after a period of 2 weeks 40% were still active. Protein extracts in aqueous solution containing yeast PRE at low degree of purification are reasonably stable for a few days when kept below 5°C, but the stability seems to decrease with the degree of purification. (As a moist cake, yeast PRE at low purification is stable for several years at −70°C).

---

[2] $\epsilon\phi$ [liters mole^{-1}cm^{-1}] $= k_p$[mm^2 erg^{-1}] $\cdot$ $(5.2 \times 10^9)/\lambda$[nm] (adapted from Rupert, 1962*b*).

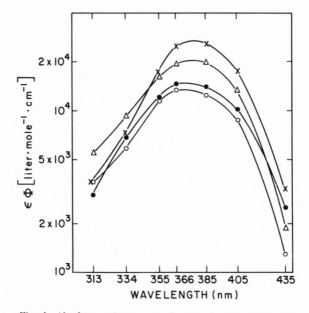

Fig. 4. *Absolute action spectra for PR of various biological systems (H. Harm, to be published).* Stationary-phase *E. coli* $B_{s-1}-160$ cells (X) (see also Harm, 1970a); stationary-phase cells of *Saccharomyces cerevisiae uvs-1-2 (Brendel et al., 1970)*, which have been mutated to increase PRE content (●) (H. Harm, unpublished); *Haemophilus influenzae* transforming DNA *in vitro* with yeast PRE either at low purification (○) (cf. Harm and Rupert, 1970b) or at high purification (△). The highly purified enzyme preparation was kindly provided by Drs. Harold Werbin and John Madden at this institution.

INFLUENCE OF TEMPERATURE ON THE REACTION RATE CONSTANTS

Both k_1 and k_2 show positive temperature dependence; Arrhenius plots of data obtained for *E. coli* cells and for the *in vitro* system under various conditions indicate activation energies for complex formation of 9–11 kcal mole^{-1}, and for dark dissociation of 4.5–6 kcal mole^{-1} (for discussion of the thermodynamics see Rupert et al., 1972). As expected, the photolytic constant k_p is virtually temperature independent over a wide range (2–40°C), but decreases drastically between 0 and −120°C (H. Harm, 1969).

INHIBITION OF PHOTOREACTIVATION BY CAFFEINE

Knowledge of the reaction rate constants and their dependencies on various parameters permits (at least qualitative) predictions for the PR kinetics to be obtained under specific conditions. An example is the inhibition of PR by caffeine (Harm, 1970*b*), which would otherwise be difficult to understand. It was observed that the PR of *E. coli* B_{s-1} cells, but not of B/r cells, is strongly inhibited by 16 mg/ml caffeine. This surprising difference is not due to differential uptake or destruction of caffeine, since PR of phage T1 is strongly inhibited by caffeine in either B_{s-1} or B/r cells. The explanation for this phenomenon is as follows.

Caffeine competes with PRE for its substrate, thereby inhibiting the formation of complexes but not their photolysis. Since B/r cells require roughly 100 times more UV radiation than B_{s-1} cells for survival decrease to the 10^{-3} to 10^{-4} level, the rate of complex formation is much greater in B/r, so that the light intensity is the limiting factor for the PR kinetics of this strain, even when complex formation is slowed down by caffeine. In contrast, the PR kinetics observed in B_{s-1} cells is limited by the rate of ES-complex formation; therefore, it is slowed down to the extent that caffeine competes with PRE for the substrate. This interpretation leads to two obvious predictions, both of which have indeed been satisfied by experimental results: (1) PR inhibition by caffeine will become observable in B/r cells at very high light intensity (when the light intensity is no longer the limiting factor); (2) PR inhibition in B_{s-1} cells will be less extensive at very low light intensity, to the extent that the latter becomes limiting for the PR kinetics of this strain.

CONCLUDING REMARKS

For convenience, some of the information and representative data discussed in this paper are summarized in Fig. 5. It reminds us that even in such a rather simple repair process as PR many experimental or environmental parameters affect the component steps in different ways, and that it is their interaction which finally determines the resulting PR kinetics. Adequate knowledge of these dependencies permits predicting, at least qualitatively, changes in the kinetics resulting from specific alterations of one or several parameters. One should be aware, however, that the given values for reaction rate constants are *weighted averages*. As described in more detail in the original papers and comprehensive review (Harm and Rupert 1970*a, b*; Harm, 1970*a, b*; Harm et al., 1971), all measurements of k_1, k_2, and k_p indicate some heterogeneity, which may reflect heterogeneity of the repairable photoproducts, of the PRE molecules, or both. Although Cook (1972) showed that the k_1 values measured for PR of only

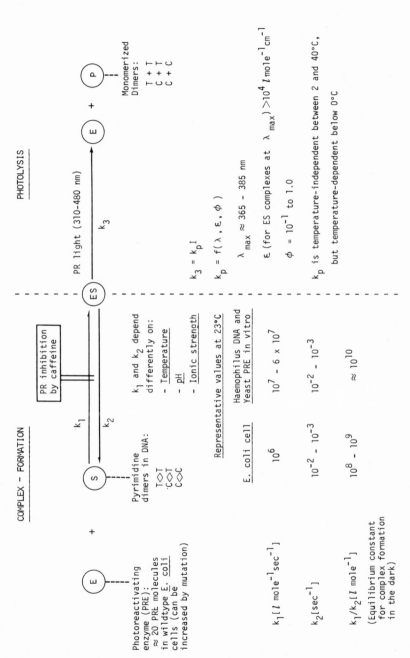

Fig. 5. Chart summarizing some of the characteristics of the photoenzymatic repair of E. coli cells and of Haemophilus transforming DNA in vitro in the presence of yeast PRE.

thymine-thymine dimers still display heterogeneity, this result does not necessarily rule out heterogeneity of substrate as a possible cause. Nucleotides adjacent to a given type of photoproduct might be a relevant part of the substrate, considering that a dimer in a sequence of less than 9 nucleotides is not sufficient for binding of PRE (Rupert, 1964; Setlow and Bollum, 1968). Obviously, in UV-irradiated *E. coli* wild-type cells, where the number of photorepairable lesions far exceeds the number of PRE molecules, any effect of substrate heterogeneity on the PR kinetics would become most apparent when the recovery approached its final level.

ACKNOWLEDGMENTS

Most of the work discussed herein was carried out in collaboration with Drs. Helga Harm and Claud S. Rupert, and was supported in various ways by the U.S. Public Health Service. Experiments on *E. coli* cells and coliphage (by W. H.) were done under Research Grant GM-12813 and a Research Career Development Award GM-34963; experiments on photoreactivation *in vitro* (by H. H. and C. S. R.) received support by Research Grants RH-00422 and GM-16547. General support was also obtained from P.H.S. Research Grant GM-13234 from the Division of General Medical Sciences; Biomedical Sciences Support Grant RR-07133 from the General Research Support Branch, Division of Research Resources, N.I.H.; and The University of Texas at Dallas Research Fund.

REFERENCES

Brendel, M., Khan, N. A. and Haynes, R. H. (1970). *Mol. Gen. Genet.* **106**, 289–295.
Cook, J. S. (1972). In *Molecular and Cellular Repair Processes* (Beers, R. F., Herriott, R. M. and Tilghman, R. C., eds.), pp. 79–94. Johns Hopkins Press, Baltimore.
Harm, H. (1969). *Mutat. Res.* **7**, 261–271.
Harm, H. and Rupert, C. S. (1968). *Mutat. Res.* **6**, 355–370.
Harm, H. and Rupert, C. S. (1970a). *Mutat. Res.* **10**, 291–306.
Harm, H. and Rupert, C. S. (1970b). *Mutat. Res.* **10**, 307–318.
Harm, W. (1969). *Mutat. Res.* **8**, 411–415.
Harm, W. (1970a). *Mutat. Res.* **10**, 277–290.
Harm, W. (1970b). *Mutat. Res.* **10**, 319–333.
Harm, W., Harm, H. and Rupert, C. S. (1968). *Mutat. Res.* **6**, 371–385.
Harm, W., Rupert, C. S. and Harm, H. (1971). In *Photophysiology* (Giese, A. C., ed.), vol. 6, pp. 279–324. Academic Press, New York.
Rupert, C. S. (1962a). *J. Gen. Physiol.* **45**, 703–724.
Rupert, C. S. (1962b). *J. Gen. Physiol.* **45**, 725–741.
Rupert, C. S. (1964). *Photochem. Photobiol.* **3**, 399–403.
Rupert, C. S., Harm, W. and Harm, H. (1972). In *Molecular and Cellular Repair Processes* (Beers, R. F., Herriott, R. M. and Tilghman, R. C., eds.), pp. 64–78. Johns Hopkins Press, Baltimore.

Rupp, W. D. and Howard-Flanders, P. (1968). *J. Mol. Biol.* **31**, 291–304.

Setlow, J. K. and Bollum, F. J. (1968). *Biochim. Biophys. Acta* **157**, 233–237.

Sutherland, J. C. and Sutherland, B. M. (1972). In *Book of Abstracts. VI Int. Congr. Photobiol.* (Bochum), No. 115.

Tyrrell, R. M., Webb, R. B. and Brown, M. S. (1973). *Photochem. Photobiol.* **18**, 249–254.

13

Purifying the *Escherichia coli* Photoreactivating Enzyme

Betsy M. Sutherland

Department of Molecular Biology and Biochemistry
University of California
Irvine, California 92664

ABSTRACT

Purification of the photoreactivating enzyme from *Escherichia coli* depends on three major factors: (1) development of a suitable assay, (2) selection of a source with high enzyme levels, and (3) development of a gentle, specific purification procedure. The preparation of purified enzyme is useful both in study of DNA repair enzymes and in the analysis of the role of cyclobutyl pyrimidine dimers in the induction of biological damage by ultraviolet light.

The *Escherichia coli* photoreactivating enzyme is important both for its role in DNA repair and for its potential use as an analytical "reagent." The latter role depends on the specific and exclusive action of the enzyme on cyclobutylpyrimidine dimers in DNA, a light-dependent reversal to the constituent monomers. If we know that UV-irradiated DNA is a substrate for enzyme X, and wish to know whether enzyme X acts on dimers, we can test the ability of the photoreactivating enzyme to remove substrate sites for enzyme X from the DNA. Thus preparation of the enzyme in at least analytical quantities would be of great potential value.

Three major hurdles in enzyme purification are selection of an assay, a suitable source, and a purification procedure. The ideal enzyme assay is sensitive (capable of measuring very low levels of enzyme activity), specific (recording as activity only that due to the enzyme being sought), insensitive to interference (by other components of the cell extracts or purification reagents), and rapid (to permit reasonably prompt purification and thus prevent degradation by the proteases of impure fractions). Table 1 shows a comparison of the major assay methods according to these criteria. It is evident that each method has its own strong points: the transformation assay (Rupert et al., 1958) indicates loss and regain of biological activity of the DNA; chromatography (Carrier and Setlow, 1971) allows the isolation, characterization, and identification of dimers; the nuclease digest–Norit adsorption method (Sutherland and Chamberlin, 1973) is reasonably rapid, specific, and resistant to interference; and the membrane filter assay (Madden et al., 1973) is particularly attractive because of its speed. For purification purposes, the most important attribute is specificity; thus the transformation and the nuclease digest methods are the best choices. The long time requirement for the transformation assay makes it somewhat inconvenient; moreover, its extreme nuclease sensitivity precludes it from use in extracts of *E. coli*, which have high nuclease levels. Therefore, the nuclease digest method is the current method of choice for purifying the *E. coli* photoreactivating enzyme. Its disadvantages, the requirements for careful quantitation and highly labeled, highly purified [^{32}P]DNA, are outweighed by its specificity, sensitivity, insensitivity to interference, and reasonable rapidity (Sutherland and Chamberlin, 1973).

The next problem is the selection of a suitable source of enzyme. For analytical purposes, *E. coli* B cells grown to stationary phase contain reasonable levels of enzyme. Since the *phr* gene is located between *gal* and λ*att*, *E. coli* cells lysogenic for a λdg will have greatly increased enzyme levels after induction (Sutherland et al., 1973a). The exact amount of enzyme in the induced λ lysogens shows considerable variation from batch to batch of cells; the factors influencing this variability are not yet understood.

Purification methods have been described both for *E. coli* B and for the induced λ lysogen (Sutherland et al., 1973b). These procedures have in common a number of preliminary "batchwise" purification steps, followed by further purification on ion-exchange or affinity columns. Our recent attempts to improve these procedures by isoelectric focusing have been unsuccessful, due to the precipitation of the enzyme at its isoionic pH. The purification must be carried out at 0–4°C, as the enzyme is thermolabile and loses activity especially rapidly when allowed to warm up when in impure preparations. In addition, better final yields are obtained when the entire preparation is performed on one day, rather than spread over several days. The enzyme is stable as a partially purified ammonium sulfate precipitate when stored in 40% glycerol at −20°C; recent experiments indicate that the pure enzyme is stable under the same conditions in the presence of 1 mg/ml bovine serum albumin (Cashion and Sutherland, unpublished results).

Table 1. Comparison of Assay Methods for the Photoreactivating Enzyme

Method	DNA type	Specificity for photoreactivation	Sensitivity (pmole DNA per assay)	Insensitivity to interference	Time for assay
Transformation	H. influenzae DNA	Yes	30–100	No	<12 hr
Formate hydrolysis, chromatography	[^{3}H]- or [^{14}C]Pyrimidine-labeled bacterial or bacteriophage DNA	No[a]	~1000	Yes	8–24 hr[b]
Nuclease digest, Norit adsorption	[^{32}P]T7 DNA	Yes[c]	30–100	Yes	2 hr
Membrane filter	[^{3}H]T7 DNA	No[d]	30–100	No[e]	~5 min

[a]Excision of dimers can cause apparent disappearance of dimers from the DNA.
[b]The assay time may be reduced by using thin-layer chromatography rather than paper.
[c]A. Braun (personal communication) has suggested that a light-dependent endonuclease cleaving immediately 5' to the dimer would give apparent photoreactivating activity in this assay.
[d]Any protein with affinity for dimers in DNA will bind UV-irradiated DNA to membrane filters.
[e]The membrane filter assay cannot be used in crude extracts because of nonspecific binding and/or nuclease interference.

The development of new assay and purification techniques for the photo-reactivating enzyme promises to open new avenues of understanding into DNA repair, both as an enzyme intriguing in itself and as a diagnostic "reagent" for determining the substrate range of other DNA repair enzymes.

REFERENCES

Carrier, W. L. and Setlow, R. B. (1971). *Methods Enzymol.* **210,** 230–237.
Madden, J. J., Werbin, H. and Denson, J. (1973). *Photochem. Photobiol.* **18,** 441–445.
Rupert, C. S., Goodgal, S. H. and Herriott, R. M. (1958). *J. Gen. Physiol.* **41,** 451–471.
Sutherland, B. M. and Chamberlin, M. J. (1973). *Anal. Biochem.* **53,** 168–176.
Sutherland, B. M., Court, D. and Chamberlin, M. J. (1973*a*). *Virology* **48,** 87–93.
Sutherland, B. M., Sutherland, J. C. and Chamberlin, M. J. (1973*b*). *J. Biol. Chem.* **284,** 4200–4205.

14

The Human Leukocyte Photoreactivating Enzyme

Betsy M. Sutherland

Department of Molecular Biology and Biochemistry
University of California
Irvine, California 92664

ABSTRACT

A photoreactivating enzyme from human leukocytes has been isolated and characterized. The enzyme requires DNA irradiated with ultraviolet light (220–300 nm) as substrate, and visible light (300–600 nm) for catalysis. In the reaction, the enzyme converts cyclobutyl pyrimidine dimers in the DNA to monomer pyrimidines.

The enzyme has an apparent monomer molecular weight of 40,000 and tends to form aggregates. The pH optimum of 7.2 and absence of a requirement for metal ions are similar to the requirements of the *Escherichia coli* enzyme; however, the ionic strength optimum of 0.05 is much lower than those for other photoreactivating enzymes.

The demonstration that human cells possess photoreactivating enzyme implies that a direct test by photoreactivation may be made of the role of pyrimidine dimers in the induction of abnormal cell growth.

Photoreactivation is unique among biological repair processes in its involvement of a single enzyme acting upon a single substrate in a light-dependent reaction (Rupert, 1960; Setlow and Setlow, 1963). Its importance in the specific repair of cyclobutylpyrimidine dimers has spurred the search for the photoreactivating enzyme (PRE) in a wide variety of organisms—but its apparent amenity to experimental manipulation (with the flip of a light switch, so to speak) may have in reality impeded our understanding of its properties.

One of the most puzzling of its properties was its distribution. In careful and extensive studies, Cook and McGrath (1967) surveyed the species distribution of the photoreactivating enzyme. Their results indicated that the enzyme occurred in all phyla and in all classes except the placental mammals. These data implied that this repair process had become superfluous to the evolving placentalia and had been lost during the development of some common ancestor. It seemed strange, however, that such a loss would occur only once during the evolution of modern plants and animals. I thus thought the enzyme might be present in placental mammals but had not been detected because of experimental difficulties.

Having decided to look for the enzyme, I next had to decide *where* to look. The data of Cook and McGrath (1967) provided a suggestion: they found different specific activities of the PRE in different tissues of the toad, with the highest in the brain and leukocyte. Since human leukocytes are easily available, I tested them for PRE *in vitro,* using the rapid, sensitive [^{32}P] DNA dimer assay (Sutherland and Chamberlin, 1973) under the conditions (0.02 M potassium phosphate buffer, pH 7.2, 0.01 M MgCl$_2$, 10^{-4} M dithiothreitol, 10^{-4} M EDTA plus NaCl to bring the ionic strength to 0.2) optimal for the *Escherichia coli* photoreactivating enzyme (Sutherland et al., 1973). The specific activity found in the leukocytes under these conditions was much too low for profitable attempts at purification. However, since it is not unlikely that enzymes with similar functions from different sources might differ in their optimal assay conditions, I determined the pH optimum (7.2), metal requirement (none), and ionic strength optimum (0.05) of the human enzyme. The last requirement is quite striking, as the enzyme shows about 5 times as much activity at an ionic strength of 0.05 as it does at 0.2. Under the optimal conditions the specific activity of the leukocyte extracts is a reasonable starting activity for purification.

The purification procedure has been described in detail (Sutherland, 1974). In brief, it consists of heparin treatment of 10 ml of freshly drawn human peripheral blood, and sedimentation of the erythrocytes by plasma gel. The leukocytes are then washed thoroughly, suspended in buffer, sonicated, and fractionated by two ammonium sulfate precipitation steps and isoelectric focusing. The resulting enzyme preparation is homogeneous by the criteria of constant specific activity across the enzyme peak obtained from isoelectric focusing and coincidence of photoreactivating enzyme activity with the protein bands obtained on polyacrylamide gels under nondenaturing conditions. The appearance of multiple PRE activities on the gels probably does not represent different PRE enzymes but aggregation of the monomers into oligomers, as is seen in glycerol gradients at low ionic strength. However, in the isoelectric focusing profile several minor peaks of enzyme activity appeared in addition to the main peak at pH 5.4. Whether these peaks represent true multiple species or artifacts of chromatography can be resolved by rechromatography of the resulting peaks: if a single peak from the first chromatographic procedure gives multiple peaks on rechromatography under the same conditions, the multiplicity of the peaks is an artifact. On the other hand, rechromatography of each of the peaks to give a single peak, occurring in the same elution position as in the first chromatographic procedure, would indicate multiple forms of the enzyme.

The tendency of the enzyme to aggregate dictates that the monomer molecular weight be determined in buffer plus 1 M NaCl, where aggregation is minimized. Gel filtration on Biogel P-100 under these conditions gives a molecular weight estimate of about 40,000, reasonably similar to that of the *E. coli* enzyme.

The enzymatic properties of the final preparation are outlined in Table 1 and have been discussed in detail (Sutherland, 1974). The reaction requires the enzyme preparation (line 1), UV-irradiated DNA (line 2), and photoreactivating light (line 3) for the activity (see line 4). Since it is also possible that a photoreversal reaction may be mediated by a small, nonprotein molecule, the effect of trypsin treatment of the enzyme was determined: the activity is labile to digestion by the protease (line 5). It is also heat labile.

In addition to these properties, a true photoreactivating enzyme must (a) use cyclobutylpyrimidine dimers as substrate, (b) cause dimers to disappear, and (c) cause monomer pyrimidines to appear. The purified human enzyme meets these criteria as follows. The most definitive determination of a requirement for dimers as substrate is the analysis of the dimer content of photoreactivated [3H] thymine-labeled DNA after TCA precipitation, hydrolysis, and chromatography (Carrier and Setlow, 1971). In this assay dimers and monomer pyrimidines are isolated and can be identified by R_fs and radioactivity; however, only dimers present in large DNA strands are detected by this procedure. In such a reaction, the human photoreactivating enzyme preparation causes the disappearance of dimers from the chromatogram only when the enzyme-DNA complex is exposed

Table 1. Enzymatic Properties of Purified Photoreactivating Enzyme
from Human Leukocytes

Enzyme	Unirradiated DNA	UV-irradiated DNA	Photoreactivating light	Photoreactivating activity
1. −	−	+	+	0
2. +	+	−	+	0
3. +	−	+	−	0
4. +	−	+	+	+
5. + (trypsin treated)	−	+	+	0

to photoreactivating light. Since the enzyme causes little or no change in the dimer content of the DNA in the dark, the possibility of dimer excision is unlikely. More solid evidence that the enzyme-catalyzed disappearance of dimers is due to photoreactivation comes from the results of the ^{32}P dimer assay (Sutherland and Chamberlin, 1973). This assay is a measure of the total dimer content of all the DNA in a reaction mixture no matter what its size. The results of such assays indicate that the dimers disappear entirely and do not appear in smaller excised fragments. Finally, a true photoreactivating enzyme should convert each dimer to two pyrimidine monomers. The demonstration of monomerization is not easy in the usual analysis of [^{3}H] thymine-labeled DNA, as the newly photoreacted monomers are lost in a sea of entirely unreacted monomers. However, an ingenious method devised by Setlow et al. (1965) allows the direct measurement of monomer products of photoreactivation. This scheme takes advantage of the heat lability to deamination of 5,6-saturated cytosines. The reaction is shown in Fig. 1. Monomer pyrimidines produced from photoreactivation are represented by U, unmonomerized dimers as ÛU or ÛT, and monomers which were never in dimers as C. The human enzyme preparation was tested for its ability to produce monomer U from [^{3}H] U-labeled dimers in just such an experiment: in a 30-min exposure to photoreactivating light, 100 μl of Fractior. II enzyme converted 45% of the ÛU and ÛT dimers to monomer U (Sutherland, 1974b). Thus the enzyme is a true photoreactivating enzyme and does convert pyrimidine dimers to their corresponding monomers in a light-dependent reaction.

Finally, it has been suggested that the enzyme in human leukocytes might actually be that of bacteria engulfed by phagocytosis by the polymorphonuclear cells and monocytes (E. Friedberg, personal communication). This possibility was tested by examining enzyme activity in bovine bone marrow: the cell extract and subsequent partially purified fractions both contained photoreactivating enzyme activity when assayed under conditions optimal for the human leukocyte enzyme. Since cells within the marrow do not phagocytose, the photoreactivating enzyme activity is due to mammalian enzyme and not one of bacterial origin.

If the photoreactivating enzyme is present in mammalian cells, why is photoreactivation not seen consistently *in vivo*? The data on photoreactivation in the cells of placental mammals are contradictory: negative results include those of Trosko et al. (1965), who observed no monomerization of dimers in Chinese hamster cells; Pfefferkorn and Coady (1968), who found that the survival of UV-irradiated pseudorabies virus grown on rabbit kidney cells was not increased by a 90-min exposure to fluorescent light; and Cleaver (1966), who followed DNA synthesis by autoradiography and survival of UV-irradiated human and mouse cells and found that a 10-min exposure to visible light increased neither cell survival nor DNA systhesis. On the other hand, Logan et al. (1959) found that the uptake of adenine by isolated calf thymus nuclei was

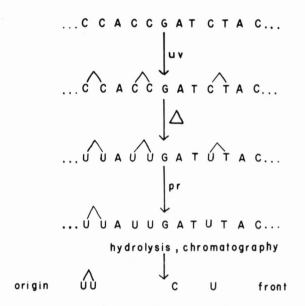

Fig. 1. Pyrimidine dimer monomerization.

photoreactivable; Rieck and Carlson (1955) reported photoreactivation of UV-induced damaged to mouse ears and killing. Pfefferkorn and Coady (1968) and Cook and Regan (cited in Cook, 1971) measured the monomerization of dimers in the [^{3}H] thymine-labeled DNA of rabbit kidney and human skin cells, respectively. Although their data show that after long exposures to photoreactivating light about 15–20% of the dimers disappeared from the DNA, this activity was much less than those in other cells tested at the same time (chick, potoroo, and wooly possum); they thus interpreted their data on photoreactivation in placental mammals as negative.

Several factors probably contribute to the difficulty of detecting photoreactivation *in vivo*. First, the enzyme may simply not be present in the cell being tested, either due to absence of the gene or to lack of its expression. Second, the enzyme may act very slowly or inefficiently on the DNA *in vivo*. For example, the apparent absence of dimer excision in the mouse (Klimek, 1966) was due to the low rate of dimer excision; at very long times, however, dimers are removed from the DNA of murine cells (R. Setlow, personal communication). It is also possible that the photoreactivating enzyme lacks access to the mammalian chromosome; Cook (1971) suggested this as an explanation of the slowness of *in vivo* photoreactivation of pyrimidine dimers in marsupial cells, even though normal levels of PRE activity could be detected in *in vitro* assays. Third, efficient excision-repair may mask the biological effects of photoreactivation: Harm et al. (1972) pointed out that the detection of photoreactivation is much easier in the excisionless *E. coli* strain B_{s-1}. Fourth, Painter (personal communication) has suggested that the photoreactivating light itself may be detrimental—or the light source may emit wavelengths not necessary for photoreactivation which are detrimental to the cell. These detrimental effects might mask any beneficial effects of dimer reversal. Finally, in some cases the photoreactivating light exposures were so brief or of such low intensities that one would not expect to see photoreactivation from any cell, whether prokaryote or eukaryote. Since some of these difficulties can be circumvented experimentally, we are currently reexamining the possibility that photoreactivation may indeed occur in mammalian cells *in vivo*.

Why is *in vivo* photoreactivation in mammals important? Since excision-repair seems capable of removing most dimers from the DNA, its contribution to cell survival may not be large. However, the photoreactivation process has much greater potential significance as an analytical tool. Wulff and Rupert (1962) showed that the substrate for the enzyme was pyrimidine dimers; moreover, Setlow and Setlow (1963) demonstrated that pyrimidine dimers were the *only* UV-induced photoproduct acted upon by the enzyme. This specific and exclusive activity allows the establishment of a criterion for the involvement of dimers in the production of biological damage by UV: if the biological effect can be reversed by true photoreactivation and the reversal is a direct effect, a major cause of the damage is the pyrimidine dimer. This approach has been used in prokaryotes and in the simple eukaryote *Paramecium* to show that dimers play a large role in production of death and mutation in these organisms.

Sunlight produces pyrimidine dimers in human cells (Trosko et al., 1970); it also causes skin cancer (Epstein, 1971). Are dimers a molecular lesion important in the induction of cancer by UV? Hart and Setlow (1973) have reported evidence that the induction of tumors by UV in the fish *Poecilia formosa* can be photoreactivated, thus implying that dimers may indeed be important in carcinogenesis. The demonstration that mammalian cells do possess photoreactivating enzyme now leads us to the opportunity for a direct evaluation of the role of dimers in human carcinogenesis.

REFERENCES

Carrier, W. L. and Setlow, R. B. (1971). *Methods Enzymol.* **210**, 230–237.

Cleaver, J. E. (1966). *Biochem. Biophys. Res. Commun.* **24**, 569–576.

Cook, J. S. (1971). In *Photophysiology* (Giese, A. C., ed.), vol. III, pp. 191–233. Academic Press, New York.

Cook, J. S. and McGrath, J. R. (1967). *Proc. Nat. Acad. Sci. U.S.A.* **58**, 1359–1365.

Epstein, J. H. (1971). In *Photophysiology* (Giese, A. C., ed.), vol. V, pp. 235–273. Academic Press, New York.

Harm, W., Rupert, C. S. and Harm, H. (1972). In *Molecular and Cellular Repair Processes* (Beers, R. F., Herriott, R. M. and Tilghman, R. C., eds.), pp. 53–63. Johns Hopkins Press, Baltimore.

Hart, R. and Setlow, R. B. (1973). In *Abstr. 1st Ann. Mtg. Amer. Soc. Photobiol. (Sarasota)*, p. 120.

Klimek, M. (1966). *Photochem. Photobiol.* **5**, 603–607.

Logan, R., Errera, M. and Ficq, A. (1959). *Biochim. Biophys. Acta* **32**, 147–155.

Pfefferkorn, E. R. and Coady, H. M. (1968). *J. Virol.* **2**, 474–479.

Rieck, A. F. and Carlson, S. (1955). *J. Cell. Comp. Physiol.* **46**, 301–305.

Rupert, C. S. (1960). *J. Gen. Physiol.* **43**, 573–595.

Rupert, C. S., Goodgal, S. H. and Herriott, R. M. (1958). *J. Gen. Physiol.* **41**, 451–471.

Setlow, J. K. and Setlow, R. B. (1963). *Nature* **197**, 560–562.

Setlow, R. B., Carrier, W. L. and Bollum, F. J. (1965). *Proc. Nat. Acad. Sci. U.S.A.* **53**, 1111–1118.

Sutherland, B. M. (1974). Nature **238**, 109–112.

Sutherland, B. M. and Chamberlin, M. J. (1973). *Anal. Biochem.* **53**, 168–176.

Sutherland, B. M., Sutherland, J. C. and Chamberlin, M. J. (1973). *J. Biol. Chem.* **284**, 4200–4205.

Sutherland, B. M., Runge, P. and Sutherland, J. C. (1974). *Biochem.* **13**, 4710–4715.

Trosko, J. E., Chu, E. H. and Carrier, W. L. (1965). *Radiat. Res.* **24**, 667–672.

Trosko, J. E., Krause, D. and Isoun, M. (1970). *Nature,* **228**, 358–359.

Wulff, D. L. and Rupert, C. S. (1962). *Biochem. Biophys. Res. Commun.* **7**, 237–240.

15

Photorepair of RNA

Milton P. Gordon

Department of Biochemistry
University of Washington
Seattle, Washington 98195

This article attempts to describe briefly the current status of our knowledge of repair mechanisms that are operative on RNA. The photorepair mechanisms active on RNA have thus far been unequivocally demonstrated only in plants (Gordon et al., 1975). The phenomenon observed is as follows: When a number of UV-irradiated RNA plant viruses and/or free viral RNAs are assayed on local-lesion hosts, an increase in the specific infectivity is observed when the assay plant is illuminated immediately after the application of the infectious material. This increase in specific infectivity is not seen when heat-inactivated or native infectious material is assayed. Preillumination of the assay host has no effect. On all plants examined, the maximum activity is obtained with black-light, which is far removed from wavelengths which are maximally effective in photosynthesis (Hidalgo-Salvatierra and McLaren, 1969; Murphy and Gordon, 1971a). Photoreactivation (PR) of UV-irradiated tobacco mosaic virus (TMV) RNA has also been observed on white leaves of a variegated mutant host, which lack chlorophyll (McLaren et al., 1970). Photoreactivated sectors ranging from 0.2 to 0.7 have been observed. The extent of repair observed depends upon the virus or viral RNA used, the wavelength used for inactivation, the wavelength used for photoreactivation, and the host plant. The *in vivo* activity appears to be enzymatic. The ability of *Xanthi* plants to photoreactivate UV-killed TMV RNA is slowly lost on dark storage. The activity can be recovered upon treatment with blue light, whereupon the kinetics of recovery resemble those of the light induction of a number of enzyme systems in plants (Murphy and Gordon, 1971a). There is a lag of about 3 hr followed by an increase in maximum activity up to 12 hr. The irradiation of assay plants with 254-nm light decreases the photoreactivated sector of subsequently applied UV-irradiated TMV RNA. Blacklight, which is most effective in photoreactivation, is also most effective in reversing the decrease of the photoreactivated sector shown by the UV-irradiated host (Murphy and Gordon, 1971b). These results can be interpreted as competition for an RNA photoreactivating enzyme between the UV-damaged RNA generated *in situ* and the exogenously applied viral RNA.

We have attempted to demonstrate the *in vitro* photoreactivation of TMV RNA using cell-free extracts of tobacco plants. Both our laboratory and that of Douglas McLaren and James Kirwan at Berkeley were able to demonstrate some *in vitro* activity, which was found to be unstable (Table 1) (Hurter et al., 1974). Our results were obtained by homogenization of leaves, spinning at $90,000g$, precipitation with $(NH_4)_2SO_4$, and gel filtration. The results are variable but indicate a significant *in vitro* reactivation. Boiling destroys activity. The second series of experiments were performed by Kirwan and McLaren. Tissues were homogenized with razor blades in Honda's medium, which contains bovine serum albumin, dextran, Ficoll, and sucrose. This medium was originally designed to avoid rupturing of organelles and thus release of RNase. Again, the results indicate light-induced photorecovery. The DNA photorepair enzymes from yeast obtained from Dr. Jane Setlow and the enzyme from pinto bean seedlings from Dr. Harold Werbin were not active on TMV RNA. There are two experimental difficulties attendant on these experiments using plant extracts. First, the results are quite variable, and, second, the presence of the plant

Table 1. *In Vitro* Photorecovery of Irradiated TMV

Procedure[a]	Number of experiments	f_p observed
A	1	0
	2	0.3
	1	0.39
	1	0.55
	2	0.6
	1	0.65
	13	0.05 (boiled or no extract)
B	1	0.14
	1	0.20
	1	0.46

[a]Procedures: (A) Leaf homogenate centrifuged at $90,000g$ for 90 min and precipitated with $(NH_4)_2SO_4$, fractionation by gel filtration on P-100 (Hurter and Gordon). (B) Leaves homogenized in Honda's medium, filtrate used (Kirwan and McLaren).

Table 2. Relationship Between Photoinduced
Lesions and Photoreactivation

Materials and conditions	Pyrimidine hydrate	Pyrimidine dimer	Pr[c]
RNA from irradiated TMV	+	−	−
TMV RNA	+	+	+
PVX RNA	+	+	+
PVX RNA irradiated in			
TMV protein[a]	+(?)	− (?)	−
TMV RNA + HCN[b]	− (?)	+ (?)	+
TMV RNA + acetone	−	+	+

[a]Assumed to be like TMV RNA in TMV protein.
[b]Extrapolated from studies using poly(rU) (Evans and McLaren, 1968).
[c]Photoreactivation.

components and other components added to the extracts causes an immediate drop in the specific infectivity of the RNA. The infectivity thereafter is stable for a number of hours.

We have shown that pyrimidine cyclobutane dimers are introduced into TMV RNA by sunlight (Huang and Gordon, 1973), and Murphy (1973) has shown that such inactivated RNA can be photoreactivated. The Murphys have made an extensive analysis of the UV sensitivities of the various components of protein-synthesizing systems in plants and have concluded that in the absence of shielding, as in epidermal cells, sunlight could cause considerable loss of activity of long-lived mRNA (30–40% per hr) (Murphy et al., 1973). The RNA-PR system would thus logically serve to repair damage to mRNA at the same time that it was occurring. Recent studies, however, have indicated that the RNA photorepair enzyme does not play a decisive role in the repair of gross UV damage, since plants stored in the dark can photoreactivate gross leaf damage in spite of a low photorepair activity toward TMV RNA (Diner and Murphy, 1973).

When RNA is irradiated, the two principal types of lesions that are produced are pyrimidine hydrates and pyrimidine cyclobutane dimers. It has been difficult to sort out the contributions of these lesions to the inactivation and the photoreactivation of plant viral RNAs. The data currently available indicate that pyrimidine dimers are both necessary and sufficient for photoreactivation to occur (Table 2). The pertinent experimental facts are as follows:

1. Free TMV RNA when irradiated in solution forms both dimers and hydrates and shows PR activity. RNA extracted from irradiated virus has no cyclobutane dimers and shows no PR activity. Potato virus X (PVX) RNA shows the same sort of behavior when irradiated free in solution or encapsulated in TMV protein (hybrid virus).

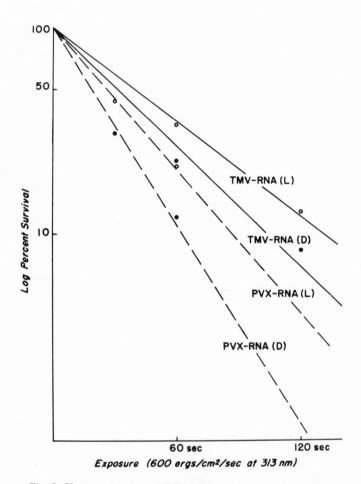

Fig. 1. Photoreactivation of TMV RNA and PVX RNA inactivated by acetone sensitization. TMV-RNA (L) and PVX-RNA (L) refer to assays in the light, and TMV-RNA (D) and PVX-RNA (D) refer to assays in the dark.

2. When poly(rU) is irradiated in the presence of HCN, dimers are still formed but hydrate formation is replaced by the addition of HCN to pyrimidines. Under identical conditions, TMV RNA can still be photoreactivated although the sector is reduced (Evans and McLaren, 1968).

3. When irradiated in the presence of acetone, TMV RNA and PVX RNA show loss of activity which can be photoreactivated. The only type of lesion that could be detected under these conditions was dimers. The oxetane content was less than 0.1 per lethal event, and chain breakage was not apparent (Fig. 1) (Huang and Gordon, 1972).

4. Finally, in PVX the RNA is not subject to the restraints exercised by TMV protein. RNA inside PVX protein can form dimers, and the intact particle or RNA extracted from irradiated intact PVX can be photoreactivated (Fig. 2) (Huang and Gordon, 1974).

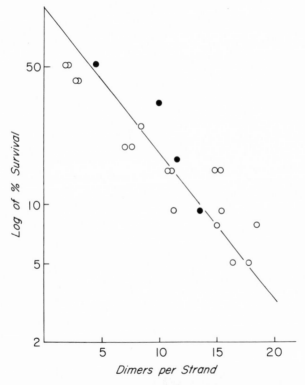

Fig. 2. Relationship between pyrimidine dimers and infectivity of UV-irradiated PVX. Closed and open circles represent experiments where charcoal or an additional paper chromatographic step were used, respectively (Huang and Gordon, 1974).

Thus, the presence of pyrimidine cyclobutane dimers is sufficient for PR to occur, but a simple relationship between the number of dimers and the photoreactivated sector does not occur. By varying the ionic strength and adding Mg^{2+}, or by running irradiations in D_2O, one can vary the ratio of dimers to hydrates but with no large effects on the photoreactivated sectors (Gordon et al., 1975; J. Kirwan, B. Singer, and A. D. McLaren, personal communication). A number of explanations such as dimer isomers are possible, but none is very convincing. There may even be position effects on lethality of lesions along an RNA molecule; i.e., dimers in certain regions of the RNA may not be lethal.

Recently in my laboratory we have undertaken a collaborative study with Drs. Fred Kramer and Don Mills in Dr. Sol Spiegelman's laboratory, using the $Q\beta$ variant MDV-1 or a mutant thereof. These self-replicating RNA molecules have completely known sequences, and we hope to have information concerning the loss of replicating activity as a function not only of the type of lesion but also the position of the lesion in the molecule.

There have been a few reports of so-called dark repair of irradiated plant viruses. It is hard to formulate a good idea as to what is occurring (Gordon et al., 1975). Recent findings that the survival of UV-irradiated double-stranded RNA of encephalomyocarditis virus is decreased on xeroderma pigmentosum fibroblasts or on normal fibroblasts in the presence of caffeine suggest that there are host-cell reactivation mechanisms other than photoreactivation active on double-stranded RNA (Závadová, 1971).

The report presented at this conference demonstrating photoreactivation of DNA extracts of human leukocytes (Sutherland, this volume) indicates that further attempts should be made to detect RNA photorepair in higher organisms.

ACKNOWLEDGMENT

This work was supported by funds from the U.S. Atomic Energy Commission [contract No. AT(45-1)2225] and the National Science Foundation (grant No. GB 24024).

REFERENCES

Diner, J. and Murphy, T. M. (1973). *Photochem. Photobiol.* 18, 429–432.
Evans, N. A. and McLaren, A. D. (1968). *Photochem. Photobiol.* 7, 661–664.
Gordon, M. P., Huang, C. W. and Hurter, J. (1975). In *Photochemistry and Photobiology of Nucleic Acids* (Wang, S. Y., ed.), Academic Press, Inc., New York (in press).
Hidalgo-Salvatierra, O. and McLaren, A. D. (1969). *Photochem. Photobiol.* 9, 417–426.
Huang, C. W. and Gordon, M. P. (1972). *Photochem. Photobiol.* 15, 493–501.
Huang, C. W. and Gordon, M. P. (1973). *Int. J. Radiat. Biol.* 23, 527–529.

Huang, C. W. and Gordon, M. P. (1974). *Photochem. Photobiol.* **19**, 269–272.

Hurter, J., Gordon, M. P., Kirwan, J. P. and McLaren, A. D. (1974). *Photochem. Photobiol.* **19**, 185–190.

McLaren, A. D., Kirwan, J., Hidalgo-Salvatierra, O., Owen, E. D. and Knight, C. A. (1970). *J. Gen. Virol.* **7**, 167–168.

Murphy, T. M. (1973). *Int. J. Radiat. Biol.* **23**, 519.

Murphy, T. M. and Gordon, M. P. (1971a). *Photochem. Photobiol.* **13**, 45–55.

Murphy, T. M. and Gordon, M. P. (1971b). *Photochem. Photobiol.* **14**, 721–731.

Murphy, T. M., Kuhn, D. N. and Murphy, J. B. (1973). *Biochemistry* **12**, 1782–1789.

Závadová, Z. (1971). *Nature New Biol.* **233**, 123.

III

Dark Repair in Bacteriophage Systems

16

Dark Repair in Bacteriophage Systems: Overview

Errol C. Friedberg

Laboratory of Experimental Oncology
Department of Pathology
Stanford University
Stanford, California 94305

Bacteriophages have proven to be extremely useful and informative biological systems with which to study various aspects of DNA repair. In these introductory comments and in the succeeding papers in this section, emphasis is placed on DNA repair in two specific bacteriophage—T4 and λ.

BACTERIOPHAGE T4

A little over 25 years ago, Salvador Luria (1947) made the observation that phage T4 is about twice as resistant to ultraviolet light at 254 nm as phages T2 and T6. At that time, he speculated that this might be due to an absence of a number of loci in T4, the presence of which in phages T2 and T6 was associated with an increased UV sensitivity.

Subsequent genetic crosses between T2 and T4 revealed that the UV sensitivity level behaved genetically as a unit character; i.e., both the T2-like and the T4-like offspring fell into two distinct classes, whose levels of UV sensitivity corresponded to those of the parental types (Luria and Dulbecco, 1949). This gene was subsequently mapped by Streisinger (1956) and termed u (presumably for ultraviolet), and in view of the interpretation made by Luria (1947), the T2 state (i.e. the most UV-sensitive state) was considered the wild-type or u^+ allele, and the T4 state the mutant or u allele. Streisinger (1956) also postulated that this gene could conceivably affect UV sensitivity of the entire genome by the inhibition of some repair mechanism by the u^+ allele, or its stimulation by the u allele.

In investigating the u gene further, Harm (1955, 1961) studied the survival of UV-irradiated u^+ phages in the presence of a functional u gene. This condition was experimentally obtained by coinfecting *Escherichia coli* with lightly UV-irradiated T2 phages (u^+) and heavily UV-irradiated T4 (u). At the dose used for the heavy irradiation, the T4 had sustained about 70 lethal hits/per phage and its survival was negligible. However, the UV sensitivity of the u gene itself was sufficiently small to ensure that many of the infected cells contained at least one undamaged copy. The results of these experiments revealed a significant increase in the survival of the lightly irradiated phage T2, approximating the survival curve for T4. This effect (termed u-gene reactivation) was not observed during infection with UV-inactivated phage T2, nor was the survival of lightly irradiated phage T4 significantly affected by infection with inactivated phage T4. Single plaques originating from the surviving u-gene-reactivated T2 were isolated, and the phage were UV-irradiated again. The phage survival was now typical of the u^+ (T2) state, thereby precluding the possibility of marker rescue during the mixed infection as an explanation of the u-gene reactivation. Harm (1961) correctly concluded that the u gene (T4 allelic state) actually coded for some function (possibly an enzyme) which influenced the intracellular environment so as to reduce the lethality of UV irradiation of the T2 phage. Since this interpretation was clearly inconsistent with the allelic designation of Streisinger (1956), Harm (1961) changed the name of the u gene to v, with the allelic state v^+ that of T4, and v as the mutant or T2 state. Subsequently, Harm (1963) mutagenized a population of T4 bacteriophage with nitrous acid and, by selecting for the phenotype of increased UV sensitivity, isolated mutants. One of these has a survival curve identical with that of phage T2 and is called T4v_1. A series of backcrosses has established that the mutation is in the v cistron itself rather than at a suppressor locus elsewhere in the genome.

The next significant observation with respect to the T4 excision-repair system came in the late 1960s, when Setlow and Carrier (1966, 1968) demonstrated that the v gene is required for excision of thymine dimers from UV-irradiated T4 DNA *in vivo*. Subsequently, Takagi's laboratory showed a dependency on the v gene for thymine dimer excision in a cell-free system (Takagi et al., 1968). Friedberg and King (1969) and Yasuda and Sekiguchi (1970a) showed that the v gene codes for an endonuclease activity expressed in phage-infected cells that degrades UV-irradiated DNA. This activity is referred to as T4 UV endonuclease (Friedberg and King, 1971) or endonuclease V of T4 (Yasuda and Sekiguchi, 1970b).

Recent studies by Sekiguchi et al. (this volume), using temperature-sensitive mutants defective in the v gene, have shown unambiguously that this gene is a structural rather than a regulator gene for the endonuclease. The enzyme has been extensively purified from extracts of T4-infected cells (Friedberg and King,

1971; Yasuda and Sekiguchi, 1970*b*). A number of its more interesting properties are summarized in the ensuing paragraphs. More detailed descriptions of individual experiments are to be found in the references cited.

1. The enzyme is induced early after infection of sensitive hosts; endonuclease activity can be detected within about 2 min after infection and continues to rise during the latent period (Friedberg and King, 1971). Recent studies by Shames et al. (1973) indicate that this enzyme, or an endonuclease very similar to it, is present in measurable amounts in disrupted purified virions. Presumably, some enzyme is trapped with the phage DNA during phage assembly. If it is active following phage infection, as the experiments of Shames et al. (1973) suggest, this provides a highly effective mechanism for initiating immediate excision-repair of UV-damaged phage DNA.

2. The purified enzyme attacks double-stranded UV-irradiated DNA, making single-stranded endonucleolytic incisions (Friedberg and King, 1971; Yasuda and Sekiguchi, 1970). Divalent cation is not required for this activity; however, both Mg^{2+} and Mn^{2+} are stimulatory (Friedberg and King, 1971; Yasuda and Sekiguchi, 1970). Other divalent cations and monovalent cations have little effect on endonuclease activity (Friedberg and King, 1971). The enzyme is insensitive to inhibition by SH inhibitors, sodium azide, 10 mM EDTA, or caffeine (Friedberg and King, 1971; Ganesan, Kahn, and Friedberg, unpublished observations). Sodium cyanide (1.0 mM) causes incomplete inhibition of enzyme acticity (Ganesan, Kahn, and Friedberg, unpublished observations). Both glucosylated and nonglucosylated DNA in either a supercoiled or a linear conformation are effective substrates following UV irradiation (Friedberg and King, 1971; Yasuda and Sekiguchi, 1970; Friedberg and Clayton, 1972).

3. When attacking duplex UV-irradiated DNA, the T4 UV endonuclease makes phosphodiester bond breaks on the 5' side of the dimers. The termini created consist of 3'-OH and 5'-P groups (Minton, Durphy, Taylor, and Friedberg, 1975).

4. The T4 UV endonuclease does not excise thymine dimers from UV-irradiated DNA (Friedberg and King, 1971).

5. An aspect of considerable interest with regard to this enzyme is a detailed understanding of its substrate specificity. In this respect, three important questions need to be addressed. (i) Are pyrimidine dimers the specific substrate sites in UV-irradiated DNA? (ii) If so, at such sites does the endonuclease make incisions exclusively in the dimer-containing strand, or can the strand opposite a dimer be attacked? (iii) Does the enzyme recognize other substrate sites in DNA?

With respect to the first question, a number of experiments indicate that pyrimidine dimers do provide substrate sites for the endonuclease.

a. By irradiation of DNA in the presence of proflavine, samples exposed to the same UV fluence but containing different dimer contents can be prepared (Setlow and Carrier, 1967). The extent of degradation of DNA with 0.5, 1.9,

3.9, and 5.4% thymine dimers by extracts containing T4 UV endonuclease shows a linear relationship to the dimer content of the DNA. On the other hand, extracts lacking endonuclease activity degrade these substrates at an equal rate (Friedberg and King, 1971).

b. UV-irradiated DNA in which pyrimidine dimers have been photoreacti- vated loses endonuclease-sensitive sites. This has been demonstrated both *in vivo* with Brij-treated UV-irradiated *Escherichia coli* (Ganesan, 1974) and *in vitro*. In the latter situation, ^{3}H-labeled T7 DNA was UV-irradiated to produce a ratio of thymine dimers/to thymine monomers of 0.14%. Incubation of part of the DNA with partially purified photoreactivating enzyme from *E. coli* resulted in a measured reduction in the thymine dimer content to 0.05%. Subsequent incuba- tion of the DNA with T4 UV endonuclease revealed a significant decrease in endonucleolytic degradation of the photoreactivated DNA compared to non- photoreactivated DNA, as estimated by sedimentation velocity in alkaline sucrose gradients.

c. There is a reasonable stoichiometric relationship between the number of endonucleolytic incisions produced and the number of pyrimidine dimers ex- pected in UV-irradiated DNA (Friedberg and King, 1971; Friedberg and Clayton, 1972). Thus, when supercoiled SV40 DNA is UV-irradiated to produce an average of one pyrimidine dimer/per molecule, 63% of the molecules are con- verted to open circular forms by incubation with purified T4 UV endonuclease— a distribution predicted by Poisson theory (Friedberg and Clayton, 1972). In evaluating these experiments, however, it should be borne in mind that currently available methods of determining the *exact* number of pyrimidine dimers in UV-irradiated DNA are of limited accuracy, and it is not possible to conclude definitively that *all* dimer sites are attacked.

Given that pyrimidine dimers are specific substrate sites in UV-irradiated DNA, it is theoretically possible that the endonuclease recognizes a particular conformational distortion in the DNA duplex produced by the dimer and can make endonucleolytic incisions on *either* strand at the dimer site. A number of experimental observations indicate that this does not occur. Rather, endonucleo- lytic incisions appear to be confined exclusively to the strand containing the dimer.

a. Electron microscopic examination of enzyme-treated UV-irradiated SV40 DNA reveals no evidence of linear forms; i.e., the endonuclease does not attack *both* strands at a dimer site (Friedberg and Clayton, 1972). Furthermore, incubation of nicked UV-irradiated T7 DNA containing 1% thymine dimers (approximately 95 thymine dimers/per molecule) with an extract containing a phage-T2-coded excision nuclease activity, results in the excision of over 90% of the thymine dimers from acid-precipitable DNA (Friedberg et al., 1974). The excision is absolutely dependent on the presence of endonuclease-induced nicks in the UV-irradiated DNA. This observation, coupled with the demonstration that only one strand in the duplex is nicked at a dimer site, suggests that the enzyme attacks the dimer-containing strand exclusively.

b. When UV-irradiated nicked DNA is denatured and incubated with bovine spleen phosphodiesterase, significant degradation of the DNA is precluded by both $5'$-P termini and adjacent pyrimidine dimers (Minton, Durphy, Taylor, and Friedberg, 1975). Even after removal of the phosphate, however, no increased rate of degradation is observed unless the dimers are monomerized with photo-reactivating enzyme. The dependency on dimer monomerization for degradation of denatured DNA by the spleen enzyme suggests that the only $5'$ sites created by the T4 UV endonuclease are at pyrimidine dimers.

c. The endonuclease does not require the duplex DNA state in order to recognize pyrimidine dimers in UV-irradiated DNA. Both alkaline-denatured UV-irradiated *E. coli* DNA and naturally occurring single-stranded UV-irradiated M-13 DNA are degraded by the endonuclease (Minton, Durphy, Taylor, and Friedberg, 1975). In the case of M-13 DNA, measurement of endonucleolytic incisions by phosphatase-dependent release of ^{32}P from labeled DNA correlates well with the number of pyrimidine dimers calculated from measurements of the thymine dimer content. Further experiments are being planned to demonstrate directly that only the strand containing the dimer is nicked by incubation with UV endonuclease.

Regarding other substrates in DNA, this question is, of course, a limitless one, since there are innumerable chemicals whose interaction with DNA might facilitate endonuclease activity. Clearly, of these, the ones of greatest interest are those known to promote lethality in DNA repair-defective organisms. We have not observed an increased sensitivity of phage T4v_1 relative to T4v^+ to any form of DNA damage other than UV irradiation (Friedberg, 1972; Mortelmans and Friedberg, 1972). Furthermore, whereas infection of UV-irradiated *E. coli* B$_{s-1}$ with inactivated (from the point of view of host-killing ability) phage T4v^+ results in a significant increase in host-cell survival (*v*-gene-mediated repair) (Friedberg, 1972; Harm, 1968), this phenomenon is not observed when the host DNA has been damaged with any of a variety of chemical agents to which it is abnormally sensitive (Friedberg, 1972). Finally, DNA treated with mitomycin C (Friedberg, 1972) or 4-nitroquinoline-*N*-oxide (Sekiguchi, personal communication) is not attacked by the endonuclease *in vitro*.

Although at the present time we would conclude that the T4 UV endo-nuclease is specific for pyrimidine dimers, a number of observations suggest alternative possibilities. Bleichrodt et al. (1972) have reported a reduced survival of phage T4v_1 compared to T4 after exposure to γ-irradiation under nitrogen. The effect of T4 UV endonuclease on γ-irradiated DNA *in vitro* is currently under investigation in our laboratory. Even more interesting is the observation by Berger and Benz (1973; and this volume) that the *v* gene is required for the repair of mismatched areas in heteroduplex DNA. *In vitro*, partially purified T4

UV endonuclease does not alter the sedimentation velocity of heteroduplex DNA. Since the x gene of T4 is also required for heteroduplex DNA repair *in vivo*, it is possible that some kind of cooperative interaction between these gene products, either with each other or with the substrate, is required for heteroduplex repair.

A number of well-characterized deoxyribonucleases with specific substrate requirements have become extremely useful tools in nucleic acid biochemistry. The specificity of the T4 UV endonuclease for pyrimidine dimers makes this enzyme a potentially powerful probe for the detection of very small numbers of pyrimidine dimers present in DNA. Indeed, the limits of this detection are a function only of the sensitivity of the method used to measure the phosphodiester bond breaks. In this regard, the endonuclease has been used, both in our laboratory and by others, for a number of photobiological studies (Ganesan, 1974; Clayton et al., 1974, and Part B of this book; Minton and Friedberg, 1974; Ganesan, 1973; Taketo et al., 1972; Duncan et al., Part B of this book). In utilizing the enzyme for such studies it is necessary to keep in mind that the specificity of the enzyme may *not* be exclusively limited to pyrimidine dimers. If the dimer-containing DNA in question has been additionally damaged some other way or contains known mismatching sequences or known peculiar conformational distortions, then, wherever feasible, control experiments utilizing DNA in which the dimers have been photoreactivated should be carried out.

The biochemical properties established thus far for the T4 UV endonuclease are totally consistent with its known role in excision-repair *in vivo*. Studies reported in this section by Sekiguchi et al. indicate that bacteriophage T4 also codes for one or possibly two exonucleases that can *excise* thymine dimers *in vitro* from UV-irradiated, specifically incised DNA. These results have been confirmed by Friedberg et al. (1974), who have found that phages T2, T3, and T6 also code for nuclease activity that can excise thymine dimers from nicked DNA. Neither laboratory has thus far been able to isolate mutants defective in dimer-excision activity.

Besides an excision-repair system, bacteriophage T4 also codes for functions required for a second mode of DNA repair. This system is defined by the genes x (Harm, 1963) and y (Boyle and Symonds, 1969), which appear to be involved in a repair pathway that bears a superficial resemblance to postreplication repair observed in bacteria (Mortelmans and Friedberg, 1972; Symonds et al., 1973). As indicated by Symonds in this section, phage T4 can also repair UV radiation damage by recombination events involving exchanges between undamaged segments of different genomes following multiple infection, a process that may be independent of x- and y-gene-mediated repair.

Finally, the role of other T4 genes in phage DNA repair has been investigated in a number of laboratores (Baldy, 1968; Ray et al., 1972; Karam and Barker, 1971; Maynard Smith and Symonds, 1973. Wallace and Melamede, 1972; Wu and Yeh, 1973). Many of these studies, however, have had to rely on the use of conditional-lethal mutants with which survival curves are only possible under conditions at least partially permissive for expression of the gene in question. Maynard Smith and Symonds (1973) have used an alternative approach involving a comparison of the functional survival of individual genes. The results of those studies are presented in detail in their publication (Maynard Smith and Symonds, 1973) and summarized in the contribution by Symonds contained in this volume.

BACTERIOPHAGE λ

Another bacteriophage that has been intensively studied with respect to the repair of UV radiation damage is phage λ. In contradistinction to phages of the T series, there is no evidence that phage λ codes for any functions that promote excision-repair. Indeed, as indicated in the accompanying review by Devoret et al. (this volume), this bacteriophage is very largely dependent on host genes for various modes of its DNA repair.

UV-irradiated phage λ can be repaired by at least three known mechanisms—host-cell reactivation, prophage reactivation, and UV reactivation. Of these, recent studies on UV reactivation are of particular interest, since they suggest that this occures by a new repair process which is "induced" in the host cell following a variety of forms of damage to the host genome, and that it is highly mutagenic. A detailed summary of this mechanism is presented by Devoret et al. (this volume).

In conclusion, it is clear that the use of bacteriophages as model systems in which to study DNA repair has been extremely informative. The studies with bacteriophage T4 have clearly defined phage-coded excision-repair functions and have provided an interesting model with which to investigate the details of the enzymology of dimer excision *in vitro*. Since excision-repair enzymes are available from a variety of other sources—i.e. *Micrococcus luteus* and *E. coli* (see this volume)—it is of interest to make detailed comparisons between them.

A number of intriguing questions still remain to be answered with respect to phage T4 repair. It is not clear, for instance, why the autonomous virulent phages T2, T4, T5, and T6 are not host-cell reactivable. This is presumably not exclusively a function of the presence of modified nucleotides in the DNA, since such modifications are not encountered in phage T5 DNA. Since both the host and phage genomes code for excision enzyme activities, the question arises, how are pyrimidine dimers normally excised from T4 DNA *in vivo*? Another aspect of T4 repair about which relatively little is known is the x-y gene pathway, and it is of great interest to know what the products of these genes are. The apparent requirement of the x gene for v-gene-mediated repair of heteroduplex DNA may facilitate purification and characterization of the x gene product. Hopefully, continued research will provide some of the answers to these questions.

ACKNOWLEDGMENTS

The author acknowledges Research Career Development Award 1 K04 CA 71005-01 from the USPHS. The author's studies were supported by grant CA-12428 from the USPHS and Contract AT(04-3)-326 with the U.S. Atomic Energy Commission.

REFERENCES

Baldy, M. (1968). *Cold Spring Harbor Symp. Quant. Biol.* 33, 333.
Benz, W. and Berger, H. (1973). *Genetics* 73, 1.
Bleichrodt, J. F., Verheij, W. S. D. and De Jong, J. (1972). *Int. J. Radiat. Biol.* 22, 325.
Boyle, J. M. and Symonds, N. (1969). *Mutat. Res.* 8, 431.
Clayton, D. A., Doda, J. and Friedberg, E. C. (1974). *Proc. Nat. Acad. Sci. U.S.A.* 71, 2777.
Friedberg, E. C. (1972). *Mutat. Res.* 15, 113.
Friedberg, E. C. and Clayton, D. A. (1972). *Nature,* 237, 99.
Friedberg, E. C. and King, J. J. (1969). *Biochem. Biophys. Res. Commun.* 37, 646.
Friedberg, E. C. and King, J. J. (1971). *J. Bacteriol.* 106, 500.
Friedberg, E. C., Minton, K., Pawl, G. and Verzola, P. (1974). *J. Virol.* 13, 953.
Ganesan, A. K. (1973). *Proc. Nat. Acad. Sci. U.S.A.* 70, 2753.
Ganesan, A. K. (1974). *J. Mol. Biol* 87, 103.
Harm, W. (1955). *Naturwissenschaften* 45, 391.
Harm, W. (1961). *J. Cell. Comp. Physiol.* 48 (Suppl. 1), 69.
Harm, W. (1963). *Virology,* 19, 66.
Harm, W. (1968). *Mutat. Res.* 6, 175.
Karam, J. D. and Barker, B. (1971). *J. Virol.* 7, 260.
Luria, S. E. (1947). *Proc. Nat. Acad. Sci. U.S.A.* 33, 253.
Luria, S. E. and Dulbecco, R. (1949). *Genetics* 34, 93.
Maynard Smith, S. and Symonds, N. (1973). *J. Mol. Biol.* 74, 33.
Minton, K., Durphy, M., Taylor, R. and Friedberg, E. C. (1975). *J. Biol. Chem.* (in press).
Minton, K. and Friedberg, E. C. (1974). *Int. J. Radiat. Biol.* 28, 81.
Mortelmans, K. and Friedberg, E. C. (1972). *J. Virol.* 10, 730.
Ray, U., Bartenstein, L. and Drake, J. W. (1972). *J. Virol.* 9, 440.
Setlow, R. B. and Carrier, W. L. (1966). *Biophys. Soc. Abstr.* 6, 68.
Setlow, R. B. and Carrier, W. L. (1967). *Nature* 213, 906.
Setlow, R. B. and Carrier, W. L. (1968). In *Replication and Recombination of Genetic Material* (Peacock, W. J. and Brock, R. D., eds.), p. 134. Australian Academy of Science, Canberra.
Shames, R. B., Lorkiewicz, Z. K. and Kozinski, A. W. (1973). *J. Virol.* 12, 1.
Streisinger, G. (1956). *Virology* 2, 1.
Symonds, N., Heindl, H. and White, P. (1973). *Mol. Gen. Genet.* 120, 253.
Takagi, T., Sekiguchi, M., Okubo, S., Nakayama, H., Shimada, H., Yasuda, S., Nishimoto, T. and Yoshihara, H. (1968). *Cold Spring Harbor Symp. Quant. Biol.* 33, 219.

Taketo, A., Yasuda, S. and Sekiguchi, M. (1972). *J. Mol. Biol.* **70**, 1.
Wallace, S. S. and Melamede, R. J. (1972). *J. Virol.* **10**, 1159.
Wu, J.-R. and Yeh, Y.-C. (1973). *J. Virol.* **12**, 758.
Yasuda, S. and Sekiguchi, M. (1970*a*). *J. Mol. Biol.* **47**, 243.
Yasuda, S. and Sekiguchi, M. (1970*b*). *Proc. Nat. Acad. Sci. U.S.A.* **67**, 1839.

17

Enzymic Mechanism of Excision-Repair in T4-Infected Cells

M. Sekiguchi, K. Shimizu, K. Sato, S. Yasuda, and S. Ohshima

Department of Biology
Faculty of Science
Kyushu University
Fukuoka 812, Japan

ABSTRACT

Excision of pyrimidine dimers from ultraviolet-irradiated DNA in a cell-free system of *Escherichia coli* infected with bacteriophage T4 consists of two different steps, one to induce a single-strand break at a point close to a pyrimidine dimer and the other to release dimer-containing nucleotide from the DNA. The enzymes responsible for these steps were isolated and the reactions were characterized; T4 endonuclease V introduces a break at the $5'$ side of a dimer and $5' \to 3'$ exonucleases, which are also induced by T4, act at the break to excise dimer-containing nucleotides. We isolated temperature-dependent v mutants, which exhibit increased sensitivity to UV at $42°C$ but not at $30°C$, and found that the mutants induce temperature-sensitive T4 endonuclease V, indicating that the v gene of T4 is indeed the structural gene for T4 endonuclease V and that the enzyme is responsible for the first step of excision-repair. A possible mechanism of excision-repair in T4-infected cells is discussed.

The first reaction in repair of ultraviolet-irradiated DNA appears to be an enzymic event which removes pyrimidine dimers from DNA. To identify enzymes responsible for this process, we have investigated excision of dimers from irradiated DNA by an extract of *Escherichia coli* infected with bacteriophage T4.

It has been shown that infection with T4 induces a rapid increase of enzyme activities concerning DNA metabolism, and thus the extract of T4-infected cells may be a good source for repair enzymes. Moreover, evidence indicates that repair of T4 is carried out by some phage-induced processes; no or only a little host-cell reactivation takes place in T4 (Ellison et al., 1960; Maynard Smith et al., 1970), and, also, some T4 mutants exhibit higher sensitivity to UV (Harm, 1963; Boyle and Symonds, 1969). The availability of such mutants should make it possible to study the role of enzymes in *in vivo* repair of DNA.

EXCISION OF PYRIMIDINE DIMERS BY T4-INDUCED ENZYMES

Incubation of UV-irradiated T4 DNA, labeled with $[^{14}C]$ thymine, with an extract of T4-infected cells in 0.01 M $MgCl_2$ –0.04 M tris-HCl, pH 7.5, results in the selective release of pyrimidine dimers from DNA; dimers are released into the acid-soluble fraction more rapidly and more extensively than is thymine (Sekiguchi et al., 1970). Analyses of the reaction mixture revealed that dimers lost from the DNA are recovered almost quantitatively in the acid-soluble fraction, indicating that dimers are not monomerized but removed from the DNA. It was also found that dimers are excised while still attached to part of the sugar-phosphate backbone and are not found in the form of free pyrimidine dimers or nucleoside dimers. Thus the excision *in vitro* is similar to the reaction observed *in vivo* in UV-resistant cells (Setlow and Carrier, 1964; Boyce and Howard-Flanders, 1964).

Subsequent experiments revealed that excision of pyrimidine dimers in T4-infected cells consists of two different steps, one to induce a single-strand break at a point close to a pyrimidine dimer and the other to release a nucleotide fragment containing the dimer from the DNA. The enzymes responsible for these steps were isolated and the reactions were characterized. Here we summarize the results of these experiments, the details of which have been or will be published elsewhere.

An enzyme which is responsible for the first step was isolated from T4-infected *E. coli* 1100 and named T4 endonuclease V (Yasuda and Sekiguchi, 1970*b*). The purification procedure includes phase partition in Dextran–polyethylene glycol and column chromatography on CM-Sephadex C-25 and on Hypatite C. The enzyme was purified about 1300-fold over the extract and is free of other endo- and exonucleases. The enzyme induces single-strand breaks in UV-irradiated DNA but does not act on native or heat-denatured DNA. The enzyme activity is dependent on the dose of UV irradiation, and the number of breaks formed is approximately equal to the number of pyrimidine dimers present in the DNA. Denatured DNA which has been exposed to UV is also attacked by the enzyme, although the rate and the extent of the reaction are greater with irradiated native DNA. The enzyme shows optimal activity at pH 7.5 and does not require added divalent ions.

When UV-irradiated T4 DNA treated with T4 endonuclease V is denatured, dephosphorylated by *E. coli* alkaline phosphatase, and then subjected to step-wise degradation by spleen phosphodiesterase, dimers are released more rapidly than thymine into the acid-soluble fraction. The degradation of the enzyme-treated, irradiated DNA by spleen phosphodiesterase requires pretreatment of the DNA with alkaline phosphatase, whereas the degradation by venom phos-phodiesterase does not. It was shown, moreover, that the 5′ termini produced by T4 endonuclease V are phosphorylated, at a relatively slow rate, by T4 poly-nucleotide kinase only after successive treatment of the DNA with yeast photo-reactivating enzyme and alkaline phosphatase. These results are compatible with the idea that T4 endonuclease V induces a break at the 5′ side of a pyrimidine dimer to yield 3′-hydroxyl and 5′-phosphoryl termini (Yasuda and Sekiguchi, 1970*b,c;* Shimizu, et al., 1971). These results are in accord with those described by Friedberg and his associates, who isolated a similar enzyme from T4-infected cells and studied its enzymic properties (Friedberg and King, 1971; Friedberg and Clayton, 1972; Clayton et al., Part B, this book).

The enzyme activity which catalyzes the second step of dimer excision was found also in an extract of T4-infected *E. coli* 1100 (Ohshima and Sekiguchi, 1972). The activity was detected by incubating T4 endonuclease V-treated, irradiated DNA in 0.04 M tris-maleate–0.01 M $MgCl_2$, pH 7.0, and then assaying the amount of pyrimidine dimers released into the acid-soluble fraction. DEAE-cellulose chromatography, as shown in Fig. 1, reveals that T4-infected cells possess three distinct dimer-excising activities, referred to as A, B, and C (Shimizu and Sekiguchi, unpublished result). Since no or only a little activity corresponding to A and B is found in normal *E. coli*, A and B appear to represent T4-induced enzyme activities. On the other hand, C is found in both normal and T4-infected cells and elutes similarly to DNA polymerase activity, which is relatively resistant to *N*-ethylmaleimide, suggesting that C is the 5′→3′ exonuclease associated with *E. coli* DNA polymerase I.

Preliminary characterization of the enzymes was performed with the DEAE fractions or more purified preparations obtained after phosphocellulose chroma-tography. A and B can release most of their nucleotide ^{32}P activity at the 5′ external termini of DNA in a short time, when only a small amount of the DNA is made acid soluble. Moreover, when DNA terminally labeled at its 3′ end with [3H]deoxythymidylate is treated with A or B, only little radioactivity is released. It therefore appears that both A and B hydrolyze a polydeoxyribo-nucleotide in a 5′→3′ direction. The two activities, however, differ in their abilities to attack internal nicks produced by two types of endonucleases. B attacks nicks produced by pancreatic DNase more efficiently than does A, whereas A and B act equally well at nicks produced in UV-irradiated DNA by T4 endonuclease V to release dimer-containing nucleotides.

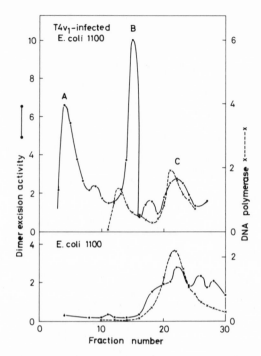

Fig. 1. *Fractionation of dimer-excising enzyme activities on DEAE-cellulose column.* An extract of normal or T4v_1-infected *E. coli* 1100 was subjected to phase partition in Dextran–polyethylene glycol and to ammonium sulfate fractionation and then applied to a DEAE-cellulose column. The column was eluted with a linear gradient of 0–0.3 M NaCl in 10 mM tris-HCl–1 mM 2-mercaptoethanol–10% ethylene glycol, pH 7.5. Dimer-excision activity (pmoles dimer made acid-soluble/30 min/50 μl of fraction) and DNA polymerase ([^{3}H]TMP incorporated (cpm $\times$ 10^{-3})/30 min/30 μl of fraction) were determined on each fraction.

The combined action of T4 endonuclease V and one of the T4 5′ → 3′ exonucleases result in the excision of dimers from circular double-stranded DNA preexposed to UV. On incubation of irradiated ϕA RF with T4 endonuclease V or 5′ → 3′ exonuclease alone, no appreciable amount of dimers or thymine is released from the DNA (Table 1). When both enzymes are present in the reaction mixture, however, selective release of dimers takes place; about 40% of the dimers originally present in the DNA become acid soluble, while only 0.3% of the thymine is released from the DNA. From the ratio of radioactivity in thymine and its dimers in the acid-soluble fraction and the thymine content of DNA, it is estimated that about six nucleotides are excised for each dimer. Thus, the RF treated by the two enzymes would possess gaps of several nucleotides in length in the molecule (Taketo et al., 1972).

INVOLVEMENT OF T4 ENZYMES IN REPAIR OF DNA

In order to see whether these enzymes function *in vivo* to repair UV-damaged DNA, we have examined the abilities of various T4 mutants to induce the enzyme activities. If the enzyme is really involved in repair of DNA, it is expected that some of the mutants might be defective in such enzyme activity.

Among the UV-sensitive mutants tested, v_1 and v_2 mutants, which were originally isolated by Harm (1963) and are sensitive to UV radiation, are unable to induce T4 endonuclease V (Friedberg and King, 1969; Yasuda and Sekiguchi, 1970a). Recently, we isolated temperature-dependent v mutants, which exhibit increased sensitivity to UV at 42°C but not at 30°C, and found that the mutants induce temperature-sensitive T4 endonuclease V, indicating that the v^+ gene of T4 is indeed the structural gene for T4 endonuclease V (Sato and Sekiguchi, 1972). It was shown, moreover, that excision of pyrimidine dimers *in vivo* as well as *in vitro* takes place in T4v^+-infected cells but not in T4v-infected cells (Setlow and Carrier, 1966; Takagi et al., 1968; Yasuda and Sekiguchi, 1970a). These results strongly suggest that T4 endonuclease V is responsible for the first step of excision-repair in T4-infected cells.

In spite of their high UV sensitivity, T4 mutants with a mutation in the x or y gene are capable of inducing both T4 endonuclease V and $5' \rightarrow 3'$ exonucleases. Since the effects of these genes and the v gene are additive, it has been suggested that the function of the x and y genes involves pathway(s) other than excision-repair (Harm, 1963; Boyle and Symonds, 1969). It was shown recently that the x mutation affects the DNA-synthesizing system of T4 (Shimizu and Sekiguchi, 1974).

Table 1. Effect of T4 Endonuclease V and $5' \rightarrow 3'$ Exonuclease on the Release of Pyrimidine Dimers from UV-Irradiated Circular Double-Strand DNA[a]

Enzyme	Dimers			Thymine		
	AS	DNA	% release	AS	DNA	% release
None	0	112	0	0	14,900	0
Endonuclease V	1	126	0.8	15	16,150	0.09
5'-Exonuclease	1	142	0.7	3	17,900	0.02
Endonuclease V plus 5'-exonuclease	45	73	38	39	15,350	0.25

[a][14 C]Thymine-labeled ϕA RF was irradiated at 600 ergs/mm^2 and incubated at 37°C for 60 min. The reaction mixture (0.3 ml) contained 0.2 μg of DNA, 12 μmoles of tris-maleate (pH 7.0), 3 μmoles of MgCl$_2$, and T4 endonuclease V (0.2 μg) and/or 5'-exonuclease (1.05 μg of activity A). After incubation, the mixture was acidified and separated into DNA and acid-soluble (AS) fractions. Each sample was hydrolyzed by heating with 6 N perchloric acid at 100°C for 3 hr, and the radioactivity of thymine and its dimers was measured after separation by Dowex-1 column chromatography.

Table 2. Properties of UV-Sensitive T4 Mutants

Strain	UV sensitivity[a]	Endonuclease V	$5' \rightarrow 3'$ exonuclease
T4D	−	++	++
F1 and 11 others (v-defective)	SS	−	++
F2 (v-leaky)	S	+ (10%)	++
F821 and 4 others (v-ts)	SS at 42°C	ts	++
F181, F195, F249	S±	++	++
v_1, v_2 (Harm)	SS	−	++
x (Harm)	S	++	++
y (Boyle and Symonds)	S	++	++
y_{100} (Smith and Symonds)	Conditional lethal	++	++

[a] SS, most sensitive; S, sensitive (intermediate between SS and T4 wild-type level); S±, slightly sensitive; −, the level of T4 wild type.

To see whether there is any UV-sensitive mutant which is defective in the second step of dimer excision, we have isolated 16 independent mutants with varied sensitivity to UV and studied their excision-repair properties (Ohshima and Sekiguchi, manuscript in preparation). As summarized in Table 2, 12 mutants were found to be defective in inducing T4 endonuclease V. One mutant, called F2, which exhibits intermediate UV sensitivity, induced a low level of endonuclease V activity. The other three produced both endonuclease V and $5' \rightarrow 3'$ exonuclease activities. It was shown, moreover, that all the amber mutants of T4 thus far examined are capable of inducing $5' \rightarrow 3'$ exonucleases. Thus, no mutant has been found to be defective in the second step of dimer excision. This may be due to the existence of more than one $5' \rightarrow 3'$ exonuclease activity in T4-infected cells.

CONCLUDING REMARKS AND DISCUSSION

Figure 2 illustrates a possible mechanism of excision-repair in T4-infected cells. As presented above, T4 endonuclease V is responsible for the first incision step, whereas T4-induced $5' \rightarrow 3'$ exonucleases may be involved in the second step of dimer excision. Although the later steps of repair have not fully been characterized, it is supposed that these steps may be catalyzed by T4-induced DNA polymerase and polynucleotide ligase, which are coded by gene 43 and gene 30 of T4, respectively (De Waard et al., 1965; Fareed and Richardson, 1967).

There is evidence that a similar mechanism functions in normal *E. coli*. We have demonstrated that T4 endonuclease V replaces the function controlled by the *uvrA, B, C,* and *D* genes of *E. coli* to reactivate irradiated infective viral DNA (Taketo et al., 1972). More recently, Braun and Grossman (this volume) reported that *E. coli* contains low but significant endonuclease activities which are specific for irradiated DNA and that one of the activities is missing in *uvrA* or *B* mutants. It is likely, therefore, that the first step of excision-repair in *E. coli* is catalyzed by an enzyme whose mode of action is similar to that of T4 endonuclease V.

It should be noted, however, that the genetic control of excision-repair in normal cells is more complex than that found in T4-infected cells; at least four genes are involved in controlling the incision or releated steps in *E. coli*. It is supposed that the repair in infected cells is carried out in the cytoplasm, where the efficiency of the reaction mostly depends on the amount and activity of enzyme involved. If we assume that the repair in normal cells takes place in some particular part of the cell. e.g. the cell membrane, it is easily understood that other factors affecting localization and transportation of enzyme and substrate are much more important.

Another notable feature which distinguishes the cellular and viral repair systems is their abilities to repair damages other than UV-induced pyrimidine

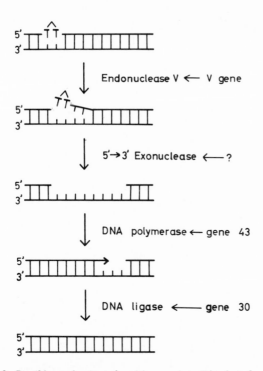

Fig. 2. Possible mechanism of excision-repair in T4-infected cells.

142   M. Sekiguchi et al.

dimers. It has been shown that *E. coli uvr* mutants are sensitive to 4-nitroquino-
line 1-oxide (4NQO), a potent carcinogen, as well as to UV, whereas T4*v*
mutants are sensitive to UV but not to 4NQO (Kondo et al., 1970; Friedberg,
1972; Ito and Sekiguchi, 1973). It is of interest to see whether this apparent
difference is due to the different specificities of T4 and *E. coli* endonucleases. It
has been shown that T4 endonuclease V does not act on DNA isolated from
4NQO-treated *uvr*B⁻ cells (Ito and Sekiguchi, 1973).

REFERENCES

Boyce, R. P. and Howard-Flanders, P. (1964). *Proc. Nat. Acad. Sci. U.S.A.* 51, 293.
Boyle, J. M. and Symonds, N. (1969). *Mutat. Res.* 8, 431.
De Waard, A., Paul, A. V. and Lehman, I. R. (1965). *Proc. Nat. Acad. Sci. U.S.A.* 54, 1241.
Ellison, S. A., Feiner, R. R. and Hill, R. F. (1960). *Virology* 11, 294.
Fareed, G. C. and Richardson, C. C. (1967). *Proc. Nat. Acad. Sci. U.S.A.* 58, 665.
Friedberg, E. C. (1972). *Mutat. Res.* 15, 113.
Friedberg, E. C. and Clayton, D. A. (1972). *Nature* 237, 99.
Friedberg, E. C. and King, J. J. (1969). *Biochem. Biophys. Res. Commun.* 37, 646.
Friedberg, E. C. and King, J. J. (1971). *J. Bacteriol.* 106, 500.
Harm, W. (1963). *Virology* 19, 66.
Ito, M. and Sekiguchi, M. (1973). *Jap. J. Genet.* 48, 421 (abstract in Japanese, details to be
    published elsewhere).
Kondo, S., Ichikawa, H., Iwo, K. and Kato, T. (1970). *Genetics* 66, 187.
Maynard Smith, S., Symonds, N. and White, P. (1970). *J. Mol. Biol.* 54, 391.
Ohshima, S. and Sekiguchi, M. (1972). *Biochem. Biophys. Res. Commun.* 47, 1126.
Sato, K. and Sekiguchi, M. (1972). *Jap. J. Genet.* 47, 371 (abstract in Japanese, details to be
    published elsewhere).
Sekiguchi, M., Yasuda, S., Okubo, S., Nakayama, H., Shimada, K. and Takagi, Y. (1970). *J.
    Mol. Biol.* 47, 231.
Setlow, R. B. and Carrier, W. L. (1964). *Proc. Nat. Acad. Sci. U.S.A.* 51, 226.
Setlow, R. B. and Carrier, W. L. (1966). *Biophys. Soc. Abstr.* 6, 68.
Shimizu, K. and Sekiguchi, M. (1974). *J. Virol.* 13, 1.
Shimizu, K., Yasuda, S. and Sekiguchi, M. (1971). *Seikagaku,* 43, 495 (abstract in Japanese,
    details to be published elsewhere).
Takagi, Y., Sekiguchi, M., Okubo, S., Nakayama, H., Shimada, K., Yasuda, S., Nishimoto, T.
    and Yoshihara, H. (1968). *Cold Spring Harbor Symp. Quant. Biol.* 33, 219.
Taketo, A., Yasuda, S. and Sekiguchi, M. (1972). *J. Mol. Biol.* 70, 1.
Yasuda, S. and Sekiguchi, M. (1970*a*). *J. Mol. Biol.* 47, 243.
Yasuda, S. and Sekiguchi, M. (1970*b*). *Proc. Nat. Acad. Sci. U.S.A.* 67, 1839.
Yasuda, S. and Sekiguchi, M. (1970*c*). *Seikagaku* 42, 696 (abstract in Japanese, details to be
    published elsewhere).

18

The Repair of Ultraviolet Damage by Phage T4: The Role of the Early Phage Genes

Neville Symonds

University of Sussex
School of Biological Sciences
Falmer, Brighton, Sussex

ABSTRACT

The characteristic behavior of phage T4 after UV irradiation is discussed in terms of the involvement of the early T4 genes in excision and postreplication repair.

Genetic studies with phage T4 have led to the identification of two repair processes for reversing UV damage. The first is defined in terms of the v gene originally isolated by Harm (1963) and subsequently shown to code for an endonuclease controlling the initial step in excision-repair (Yasuda and Seki-guchi, 1970; Friedberg and King, 1971). The second is defined by the genes x and y (Harm, 1963; Boyle and Symonds, 1969), which are involved in a repair pathway that bears some analogy to postreplication repair in bacteria (Symonds et al., 1973). Mutations in either of the genes v, x, or y increase the UV sensitivity of wild-type T4 approximately twofold. The effects of the x and y mutations on UV sensitivity are not additive (which is the evidence that these genes, which are genetically unlinked, are in the same repair pathway), but the double mutants vx and vy are both nearly 4 times as sensitive as wild type. This suggests that the two repair processes largely act independently, and for convenience we shall refer to them as vR and yR.

A puzzling feature of the radiobiology of T4 is the extreme sensitivity of phage particles to ultraviolet light as compared to the sensitivity of bacteria. Although

T4 controls two supposedly efficient repair processes, the ratio of thymine dimers to lethal hits for wild-type phage is less than 10, while the comparable figure for *Escherichia coli* K12 is around 1000. This disparity reflects a basic difference between the availability of the various enzymes and substrates necessary for repair in the two situations. When a bacterial culture is irradiated, the cells already contain the essential nucleases, polymerases, DNA precursors, etc., which make up the repair pathways. However, when a virulent phage like T4 is irradiated extracellularly and then adsorbed to sensitive cells, the numerous early proteins necessary for phage DNA synthesis still have to be synthesized. As UV lesions cause functional damage which inhibits transcription, any UV damage within genes which are part of a repair pathway cannot successfully be repaired by that route, as an essential gene product will be lacking in the infected cells. In other words, the reason for the abnormal UV sensitivity of extracellularly irradiated T4 is that a considerable fraction of the radiation damage falls within essential early genes whose products are required for repair.[1]

This interpretation of the abnormal UV sensitivity of T4 forms the basis for a classification of the early genes with regard to their involvement in vR (or yR) by comparing the *functional survival* of individual genes. If the functional survival of a particular gene is identical as measured in experiments using v^+ and v phage, then that gene is classified as being in the vR pathway; and similarly, if functional survival is unaffected by the state of the y allele, then that gene is considered part of the pathway for yR (Maynard Smith and Symonds, 1973).

In Table 1 the results of experiments of this kind are presented for a series of early T4 genes, using the technique for measuring functional survival pioneered by Krieg (1959). It can be seen that the state of the y allele does not appreciably affect the functional survival of any of the early genes tested. This implies that they are all part of the pathway for yR; or, put another way, normal phage DNA synthesis is a prerequisite for yR. The effect of the v allele is different, as it influences the functional survival of some early genes but not others. Only some early genes are, therefore, involved in vR, so the DNA synthesis occurring after excision is fundamentally different from that occurring during normal phage replication. The result that the genes 32 and 43 are not involved in vR is of particular interest in that it is generally accepted that the gene-43 product is the polymerase responsible for the semiconservative replication of T4 phage DNA and that the unwinding protein coded by gene 32 cooperates specifically with the polymerase in this synthesis. If these two proteins are not necessary for the repair synthesis which occurs after excision, then another polymerase must be involved. It seems that this is the bacterial polymerase, *pol*1, as phage survival,

---

[1] A fairer comparison is between the UV sensitivities of bacteria and of infected cells irradiated *after* the early genes have had time to be expressed. Under these conditions the infected cells are more than 20 times as UV-resistant as free phage particles (Luria and Latarjet, 1947), suggesting that the efficiencies of repair of T4 and K12 are not markedly different.

Table 1. Influence of the v and y Alleles on the
Functional Survival of T4 Early Genes

Gene	Relative sensitivity of functional survival		Gene product (where known)
	$\dfrac{vy^+}{v^+y^+}$	$\dfrac{vy}{vy^+}$	
1	1.10	1.00	Hydroxymethyl-dCMP-kinase
30	0.95	1.03	Polynucleotide ligase
32	1.53	1.20	Unwinding (Alberts) protein
41	2.20	0.95	
42	1.01	1.09	dCMP hydroxymethylase
43	1.66	1.05	DNA polymerase
44	1.50	0.87	
45	0.96	1.11	
56	1.06	1.07	dCTPase

multiplicity reactivation, and functional survival of T4v^+ phage are all appreciably more sensitive to UV irradiation in $pol1$ than in $pol1^+$ strains, while the behavior of T4v phage is identical in the two bacterial strains (Maynard Smith et al., 1970).

Another facet of the radiobiology of T4 in which the early genes play a central role is the phenomenon of multiplicity reactivation (MR). This was probably the first of all repair processes to be recognized (Luria, 1947) and refers to the finding that the rate of inactivation of cells multiply infected with irradiated T4 phage is much less than would be predicted from considerations of the survival of the phage particles individually (Fig. 1a). The MR curve shows an extended shoulder followed by a linear portion whose slope is about four times less than that of the phage survival curve. Although the state of the v allele influences the final slope of the MR curve, the phenomenon is essentially similar in v^+- and v-infected cells and is much more easily discussed and investigated in the latter case, when there are no complications arising from excision-repair.

Current ideas about MR date back to the work of Barricelli (1956). The survival of cells which have been multiply infected with UV-irradiated T4v depends both on complementation (the cells must contain a complete set of undamaged early genes so that phage DNA synthesis is possible) and on the efficiency of repair (which is influenced particularly by the x and y genes).

There are a number of questions to which one would like answers in order to have further insight into the mechanism of MR. These include: (1) What is the ratio of the fraction of infected cells which make phage DNA after a particular dose of ultraviolet to the fraction which produces viable phage particles? (2) How do the rates of DNA synthesis and phage production in cells undergoing

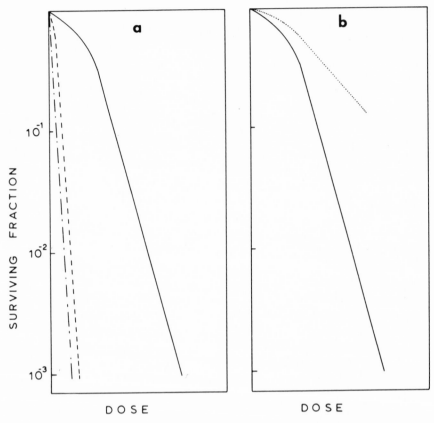

Fig. 1. (a) Phage survival (– · –) and MR (——) curves for T4v phage. The curve (– – –)
represents the theoretical survival of the multiply infected cells (multiplicity = 5) if there
were no cooperation between the infecting phages. (b) A comparison between the fraction of
multiply infected cells making phage DNA (······) and the fraction producing viable
phage (——). These experiments were performed by Sheila Maynard Smith. The DNA
determinations were made from radioautographs of cells which had been grown after
infection in a medium containing tritium.

MR compare with those in normal T4-infected cells? (3) Does genetic recombination play any role in MR? (4) Do infected cells which make DNA but not
infectious phage snythesize late phage proteins? (5) Is there any evidence for the
induction of specific repair enzymes during MR?

Some information on these points is now available. Radioautographs of
infected cells undergoing MR have recently been studied in order to answer the
first question, with the results shown in Fig. 1b. It can be seen that considerably
more infected cells produce phage DNA than support the growth of viable
phage, so, as might be expected, the repair is far from being 100% efficient. The

kinetics of DNA synthesis and phage production has also been studied with infected cells where the survival was about 1% (i.e., where 20% support DNA synthesis). Compared to cells infected with unirradiated phage there is a small delay in the onset of DNA synthesis (of about 4 min) but a considerable diminution in rate. Phage synthesis is substantially delayed; the first phage particles are not produced until nearly 35 min after infection, as compared to 12 min in an unirradiated control.

As far as the third question is concerned, it is known that all T4 variants (including x and y) which display a reduced potential for recombination also are more sensitive to ultraviolet both in terms of survival and in MR experiments. Therefore, there must be at least a considerable overlap in the pathways of genetic recombination and postreplication repair. It is worth noting, however, that multiply infected T4v cells still have the option to repair UV damage by either of two routes. The first is that usually pictured in postreplication repair with bacteria (and after single infection with phage), whereby DNA synthesis which generates gaps is followed by an interaction between daughter genomes. The second is genetic recombination between the damaged parental genomes, which can directly obviate the UV damage.

Some preliminary experiments relating to last two points have been performed by looking at the patterns of phage proteins synthesized at different times after infection, using SDS slab gels. It appears that nonviable infected cells which synthesize DNA do turn on late phage genes; also there is some evidence for the synthesis of increased amounts of some early proteins soon after infection, but the nature of these proteins is as yet unknown.

REFERENCES

Barricelli, A. (1956). *Acta Biotheoret.* 11, 107–120.
Boyle, J. B. and Symonds, N. (1969). *Mutat. Res.* 8, 431–439.
Friedberg, E. C. and King, J. J. (1971). *J. Bacteriol.* 106, 500–507.
Harm, W. (1963). *Virology* 19, 66–71.
Krieg, D. (1959). *Virology* 8, 80–98.
Luria, S. (1947). *Proc. Nat. Acad. Sci. U.S.A.* 33, 253–264.
Luria, S. and Latarjet, R. (1947). *J. Bacteriol.* 53, 149–163.
Maynard Smith, S. and Symonds, N. (1973). *J. Mol. Biol.* 74, 33–44.
Maynard Smith, S., Symonds, N. and White, P. (1970). *J. Mol. Biol.* 54, 391–393.
Symonds, N., Heindl, H. and White, P. (1973). *Mol. Gen. Genet.* 120, 253–259.
Yasuda, S. and Sekiguchi, M. (1970). *Proc. Nat. Acad. Sci. U.S.A.* 67, 1839–1845.

19

Repair of Heteroduplex DNA in Bacteriophage T4

Hillard Berger and Wendy C. Benz[1]

Department of Biology
The Johns Hopkins University
Baltimore, Maryland 21218

ABSTRACT

Evidence is presented that T4 phage posesses a system which repairs "loops" in heteroduplex DNA. The T4 v gene (endonuclease V) is required for repair.

Repair of noncomplementary bases in heteroduplex DNA has been proposed as a mechanism to explain gene conversion in eukaryotes (Holliday, 1964) and may explain aberrant recombination frequencies (Tessman, 1965) and high negative interference observed over short intervals in phage crosses (Chase and Doermann, 1958; Amati and Meselson, 1965).

In this report we present evidence that T4 possesses an intracellular system which removes segments of excess DNA (loops) from heteroduplex DNA. The heteroduplex molecules are produced *in vivo* by recombination subsequent to mixed infection with wild-type and deletion-mutant phage or *in vitro* by reannealing of the separated DNA strands (Guha and Sybalski, 1968) of wild-type and deletion-mutant phage. One prediction of this repair system is that alleles associated with the segment of excess DNA should be selectively lost. Thus, in mixed infections, the frequency of wild-type alleles would be expected to decrease in the viable phage output. As shown in Table 1 and by Benz and Berger (1973), selective loss of wild-type alleles is clearly seen in the phage

---

[1] Current address: Harvard University, Cambridge, Massachusetts.

Table 1. Allele Loss Following Deletion-Mutation–
Wild-Type and Addition-Mutation–Wild-Type
Mixed Infection[a]

Mutant	Type	Allele loss coefficient
r61	deletion	0.64
rb17	deletion	0.69
rdb52	deletion	0.68
rj101	deletion	0.87
rdb145	deletion	0.82
ejr13	addition (5 bases)	1.29
ej28	addition (4 bases)	1.13
ejr14	addition (2 bases)	1.20
ejd6	addition (1 base)	1.18
ejd5	deletion (2 bases)	0.84

[a]Mixed infections were carried out with limited DNA synthesis using the conditions of Sechaud et al. (1965). Wild-
type and rII phage were differentiated by plaque morphology. Wild-type and e phage were differentiated by plaque
morphology on lysozyme-supplemented agar plates (Okada
et al., 1966). Allele loss coefficients are the ratio of output
to input wild-type allele frequencies, normalized for the
wild-type advantage observed in mixed infection with wild-
type and point mutants (Benz and Berger, 1973).

output following mixed infection with rII deletion mutants and wild type. As
also shown in Table 1, we observe loss of alleles associated with small regions of
excess DNA in e-gene frameshift mutants. Allele loss is not observed after mixed
infection with rII point mutants and wild type; in fact, there is a small but
consistent wild-type advantage (Benz and Berger, 1973) apparently associated
with mixed growth of rII and wild-type phage.

Table 2 shows that selective loss of wild-type alleles does not occur when the
phage carry a mutation in the v gene. This observation indicates that the
T4-induced UV endonuclease, endonuclease V (Friedberg, this volume; Seki-

Table 2. Effect of the v Gene on Allele Loss in
Wild-Type–Deletion Mixed Infections[a]

Deletion	Genotype	Allele loss coefficient
r61	v^+	0.65
rdb52	v^+	0.68
r61	v^-	0.98
rdb52	v^-	0.99

[a]Crosses were carried out as described in Table 1.

Table 3. Increased Heterozygosity for a Five-Base
Deletion Marker Under v^- Conditions[a]

Cross	Genotype	Number of Hets	Total progeny examined	% Hets
$ejr13 \times$ w.t.	v^+	39	2237	1.7
$ejr13 \times$ w.t.	v^-	111	2308	4.8

[a]Crosses were carried out with limited DNA synthesis according
to the methods of Sechaud et al. (1965). Individual progeny
plaques from uncrowded plates were picked, resuspended, and
plated on citrate agar (Okada et al. 1966) supplemented with
500 µg of egg-white lysozyme. Wild type and e mutants have
easily distinguishable morphologies on these plates. All plaques
which exhibited greater than 10:1 segregation frequency were
scored as Hets.

guchi et al., this volume), is also involved in repair of loops in heteroduplex
DNA. It is likely that the mechanism of "loop repair" is analogous to the
excision-repair of pyrimidine dimers.

One genetic consequence of loop repair is the elimination of regions of
heterozygosity for unmatched bases. This may partially account for the low
frequency of "internally heterozygous" phage observed among the progeny of

Table 4. Transformation with Heteroduplex DNA
Following Treatment with Cell-Free Extracts[a]

Extract	Heteroduplex DNA transformation frequency ($\times 10^{-9}$)	Homoduplex DNA transformation frequency ($\times 10^{-9}$)
v^+	2	20
v^-	13	47
None	28	18

[a]DNA strands of T4B wild type and $r1589$ were separated according to the methods
of Guha and Sybalski (1968). Strands were reannealed in 0.3 M NaCl, 0.01 M tris,
pH 7.3, for 6–7 hr at 66°C. Reannealed strands were dialyzed against 0.02 M NaCl
at 4°C and stored frozen. Cell-free extracts were prepared using *Escherichia coli*
D110. Cells were infected with T4B $r1589$-v^+ or T4B $r1589$-v^- at a m.o.i. of 5 and
harvested by centrifugation at 13 min postinfection. The infected cells were resuspended on 1/200 volume of 0.05 M tris, pH 7.5, 0.001 M mercaptoethanol and
disrupted by sonication. The supernatant after 10,000g centrifugation for 15 min
was used to treat heteroduplex DNA. In these experiments DNA was treated for 60
min at 37°C with 200 µg of extract per 5 µg of DNA. The reaction was stopped by
placing DNA in a boiling H_2O bath for 5 min, chilling in ice-cold H_2O, and diluting
to approximately 0.6 µg of DNA per ml with 0.01 sodium phosphate buffer, pH 7.0.
0.5-ml aliquots were distributed to tubes with freshly made spheroplasts of *Aero-
bacter aerogenes*, and transformation was carried out with $r1589$ helper phage as
described by Wais and Goldberg (1969). Total phage yields and transformants were
determined by plating on *E. coli* B and *E. coli* K (λh). The results given are the
average of 3–8 individual experiments.

deletion–wild-type and addition–wild-type mixed infections (Nomura and Benzer, 1961; Sechaud et al., 1965). As shown in Table 3, the frequency of heterozygotes for a five-base addition site increases when the phage lack endonuclease V. In our previous studies, heterozygote frequencies for extended rII deletion did not increase in the absence of endonuclease V activity (Benz and Berger, 1973). It is likely that DNA containing large "loops" is not incorporated into viable phage when repair activity is absent.

We have constructed rII deletion–wild-type heteroduplex DNA in order to study loop repair *in vitro*. This DNA was exposed to cell-free extracts of phage-infected cells and subsequently tested for the amount of remaining rII⁺ DNA, utilizing the transformation assay of Wais and Goldberg (1969). In agreement with our *in vivo* findings, exposure of DNA to cell-free extracts containing endonuclease V (v^+) resulted in selective loss of the rII⁺ alleles (Table 4). It is likely that the allele loss results from the concerted action of a system

Table 5. Heterozygote Formation for Point Mutant Markers in the Presence and Absence of v Gene Function[a]

Point mutant	Type	v genotype	Number of Hets	Hets
*bc*11	Transition	v^+	126	4.7
*bc*11		v^-	85	5.5
*uv*375	Transition	v^+	190	8.2
*uv*375		v^-	157	7.8
*em*84	Transition	v^+	87	10.5
*em*84		v^-	111	8.2
*c*204	Transition	v^+	145	6.1
*c*204		v^-	112	6.1
*nt*88	Transition	v^+	100	5.6
*nt*88		v^-	106	6.4
*sn*103	Transition	v^+	120	6.2
*sn*103		v^-	80	6.9
*sm*94	Transversion	v^+	116	10.5
*sm*94		v^-	181	7.8
*uv*74	Transversion	v^+	78	6.7
*uv*74		v^-	57	6.8

[a]Crosses were carried out according to the procedure of Sechaud et al. (1965). Hets were determined by their "mottled" appearance on S/6 indicator bacteria.

Table 6. Reduction of High Negative Interference
in the Presence or Absence of v-Gene Function When
the Central Marker Is a Deletion[a]

r70   r59   r2–20

rdb52

	Cross	v genotype	Coefficient of coincidence
(1)	+ rdb52 + X r70 + r2–20	v^+	8.2
(2)	+ rdb52 + X r70 + r2–20	v^-	12.4
(3)	+ r59 + X r70 + r2–20	v^+	36.1
(4)	+ r59 + X r70 + r2–20	v^-	39.7

[a]Crosses were carried out as described in Berger and Warren (1969). The rdb52 mutation is a relatively small rII deletion, and r59 is a point mutation covered by the rdb52 lesion. The flanking markers are both rII point mutants spanning a distance of approximately 1 map unit.

which requires endonuclease V as an early step in the repair process. Additional activities are probably responsible for the degradation of the segment of excess bases. Preliminary experiments (Berger, unpublished) indicate that purified endonuclease V does not result in allele loss. However, utilizing the purified enzyme, it may be possible to identify and isolate additional activities involved in "loop" repair.

Mismatched bases in DNA (arising, for example, by mutation or by recombination in point-mutant–wild-type crosses) do not appear to be repaired by the v-gene system. No selective allele loss can be demonstrated for a large number of rII point mutants (indicating either a lack of repair or random repair of either strand). As shown in Table 5, the frequency of heterozygotes is independent of v-gene function for a sampling of rII point mutants. If v-gene activity were responsible for heterozygous to homozygous conversions for single-base mismatches, heterozygote frequencies would be expected to increase in the absence of this function.

We have also examined the effects of v-gene-mediated repair upon recombination in three-factor crosses in which the central marker is a deletion or a point mutant. Under v^+ conditions a deletion as central marker reduces the level of high negative interference as measured by the frequency of wild-type recombinants (Berger and Warren, 1969) (Table 6). Crosses with v^- phage do not alter this pattern (Table 6), suggesting that loop repair is not responsible for the decreases observed. A more plausible picture is that wild-type alleles in "loop" segments cannot be used to "convert" the complementary DNA strand. It is possible, however, that loop repair does contribute to the frequency of triple-mutant double recombinants produced in these crosses.

ACKNOWLEDGMENTS

This is Contribution Number 830 from the Department of Biology of The Johns Hopkins University and was supported in part by National Institutes of Health research grant AI 08161.

REFERENCES

Amati, P. and Meselson, M. (1965). *Genetics* **51**, 369.
Benz, W. and Berger, H. (1973). *Genetics* **73**, 1.
Berger, H. and Warren, A. J. (1969). *Genetics* **63**, 1.
Chase, M. and Doermann, A. H. (1958). *Genetics* **43**, 332.
Guha, A. and Sybalski, W. (1968). *Virology* **34**, 608.
Holliday, R. (1964). *Genet. Res.* **5**, 282.
Nomura, M. and Benzer, S. (1961). *J. Mol. Biol.* **3**, 684.
Okada, Y. E. and Terzaghi, E., Streisinger, G., Emrich, J., Inouye, M. and Tsugita, A. (1966). *Proc. Nat. Acad. Sci. U.S.A.* **57**, 1692.
Sechaud, J., Streisinger, G., Emrich, J., Newton, J., Lanford, H., Reinhold, H. and Stahl, M. (1965). *Proc. Nat. Acad. Sci. U.S.A.* **54**, 1333.
Tessman, T. (1965). *Genetics* **51**, 63.
Wais, A. C. and Goldberg, E. B. (1969). *Virology* **39**, 153.

20

Recovery of Phage λ from Ultraviolet Damage

Raymond Devoret, Manuel Blanco,[1] Jacqueline George,[2] and Miroslav Radman[3]

Laboratoire d'Enzymologie, C.N.R.S.
F 91190 Gif sur Yvette, France

ABSTRACT

Recovery of phage λ from ultraviolet damage can occur, in the dark, through three types of repair processes as defined by microbiological tests: (1) host-cell reactivation, (2) prophage reactivation, and (3) UV reactivation. This paper reviews the properties of the three repair processes, analyzes their dependence on the functioning of bacterial and phage genes, and discusses their relationship.

Progress in the understanding of the molecular mechanisms underlying the three repair processes has been relatively slow, particularly for UV reactivation. It has been shown that host-cell reactivation is due to pyrimidine dimer excision and that prophage reactivation is due to genetic recombination (prereplicative).

We provide evidence showing that neither of these mechanisms accounts for UV reactivation of phage λ. Furthermore, UV reactivation differs from the other repair processes in that it is inducible and error-prone. Whether UV-damaged bacterial DNA is subject to a similar repair process is still an open question.

---

[1] Present address: Instituto de Investigaciones Citologicas, Amadeo de Saboya 4, Valencia 7, Spain.
[2] Present address: Unité de Génétique Cellulaire, Institut de Biologie Moléculaire, 2 Place Jussieu, 75005 Paris, France.
[3] Present address: Biophysique et Radiobiologie, Université Libre de Bruxelles, Rue des Chevaux 67, 1640 Rhode St. Genese, Belgium.

A large part of the research carried out in our laboratory has been devoted to elucidating the mechanisms of the different cellular processes which tend to restore bacterial or phage DNA damaged by irradiation. The present article intends to review the different processes whereby a temperate phage like phage λ can be repaired in its host, *Escherichia coli* K12.

It has been known for 20 years that phage λ can recover from UV damage through three types of repair processes, which are, as defined by microbiological tests, *host-cell reactivation* (Garen and Zinder, 1955; Harm, 1963*a*), *prophage reactivation* (Jacob and Wollman, 1953), and *UV reactivation* (Weigle, 1953). Yet, progress in the understanding of their molecular mechanisms has been relatively slow, particularly for the last-mentioned one. In this short review we would like to (1) recall the known properties of the three repair processes, (2) analyze their dependence on the functioning of bacterial and phage genes, and (3) discuss their relationship in the same host cell.

DEFINITION OF REPAIR; EVALUATION OF ITS EFFICIENCY

In recent years there has been a tendency to admit that, if a genetic or physiological change in a host cell produces a different survival curve of the UV-damaged phage, such a result would likely indicate the existence of a repair process in the wild-type host. It has proved to be legitimate and also fruitful in the study of the ways whereby UV-damaged phage λ can be reactivated.

Consequently, when a repair process has been shown to operate on the UV-damaged phage (e.g. pyrimidine dimer excision), it is particularly useful to be able to estimate the number of repaired lesions by comparison of the proper survival curves. As the evaluation of repair efficiency (E) is considerably at variance in many publications, it must be pointed out that the most direct way of calculating it is the following (Jagger, 1960; see Fig. 1*A*):

$$E = (\log S_2 - \log S_1)/(\log S_2 - \log S_0) \tag{1}$$

The repair efficiency can also be derived from the value of the dose-reduction factor (DRF):

$$E = 1 - DRF = 1 - D_1/D_2) \tag{2}$$

when the curves are such (Fig. 1*B*) that curve 1 can be superimposed on curve 2 after expanding the scale of the abscissa.

For instance, calculation of the efficiency of UV reactivation using formula (1) shows that, if the *reactivation factor,* as defined first by Kellenberger and Weigle (1958), is 10 at a phage survival of 10^{-2} or 100 at a phage survival of 10^{-4}, this means that the repair efficiency is the same in both cases; 50% of the lethal damage is removed. Therefore, one should normalize all the results of UV-reactivation experiments by expressing them as repair efficiency values. This

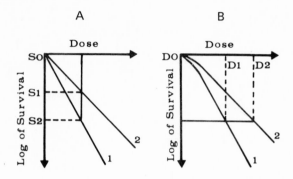

Fig. 1. (A) Dose-survival curves for calculating repair efficiency by Eq. (1); (B) curves for calculating repair efficiency from the dose-reduction factor by Eq. (2).

method is certainly valid when the host bacteria have been exposed to 500 ergs/mm^2 or less, such UV doses leaving unimpaired the capacity of the host cell to reproduce λ phage.

HOST-CELL REACTIVATION

Host-cell reactivation has experimentally been defined by the fact that UV-damaged phage λ has a different plating efficiency depending on the genetic background of the host cell in which it multiplies (Harm, 1963a).

Repair Efficiency. This repair process has an efficiency of about 85% (Harm, 1963a), so that it is by far the most efficient of all repair processes. Furthermore, the repair is accurate, and no phage mutagenesis results from the action of the repair enzymes.

Role of Bacterial Genes. In *E. coli*, host-cell reactivation depends on the functioning of at least four genes–*uvrA, uvrB, uvrC* (Howard-Flanders et al., 1966; van de Putte et al., 1965), and *uvrD* (Ogawa et al., 1968). The genes *uvrA* and *uvrB* seem to be involved in the first biochemical step in the process of excision of pyrimidine dimers. They code for a UV endonuclease (see Braun et al., this volume). The *uvrC* gene product seems to antagonize the action of DNA ligase (Seeberg and Rupp, Part B of this book).

Ubiquity of Host-Cell Reactivation. Host-cell reactivation of temperate and semitemperate phages has been shown to exist in all the various bacterial species where it has been sought. Moreover, host-cell reactivation occurs in mammalian cells. UV-damaged herpes virus is reactivated in normal human fibroblasts, but fibroblasts from humans suffering from xeroderma pigmentosum are unable to carry out host-cell reactivation of herpes virus. The more severe the human cases, the less are the fibroblasts able to effect repair of the UV-damaged virus (Bootsma et al., 1971).

Molecular Mechanism. Host-cell reactivation of phage λ has been demonstrated to be due to pyrimidine dimer excision (Boyle and Setlow, 1970), which is carried out by *cell enzymes only.* This is in contrast to phage T4, whose DNA codes for enzymes able to perform the first step in the excision process (Harm, 1963b; Setlow and Carrier, 1966; Friedberg, this volume). Therefore, recent biochemical findings show the validity of the term used to designate host-cell repair of phage λ.

In spite of the recent progress on the knowledge of pyrimidine dimer excision-repair, many questions concerning its mechanism still remain unanswered. At the genetic level, we do not yet know precisely the respective roles of genes *uvrA* and *uvrB*, the function of *uvrD*, or why *uvrD* is dominant over *uvrD$^+$*. At the biochemical level, the role of DNA polymerase I in host-cell reactivation is still to be clarified (Glickman, 1974, and this volume). Furthermore, there are no precise data as to the velocity of excision and of local DNA resynthesis as compared to the timing and speed of phage DNA replication itself. Evidently, the repair process to be efficient must take place before the beginning of phage DNA replication (Radman et al., 1970).

PROPHAGE REACTIVATION

Survival of ultraviolet-irradiated phage λ is higher in a host carrying a homologous heteroimmune prophage than in a host nonlysogenic or lysogenic with a nonhomologous prophage. This phenomenon was discovered by Jacob and Wollman (1953) and has been known since then as prophage reactivation. This type of repair, which also occurs in phage P2 (Chase, 1964) and P22 (Yamamoto, 1967), is due to recombination between the homologous DNAs of the UV-damaged phage and the intact resident prophage (Chase, 1964; Devoret and Coquerelle, 1966; Yamamoto, 1967; Hart and Ellison, 1970; George and Devoret, 1971).

Efficiency of Prophage Reactivation. About 50% of the lethal UV lesions are repaired by this process; this value is constant and independent of whether or not the monolysogenic host is able to perform pyrimidine dimer excision (Devoret, unpublished results; Hart and Ellison, 1970; George and Devoret, 1971). This means, therefore, that repair by prophage reactivation operates either on the remaining DNA lesions left by the pyrimidine dimer excision process or on the primary DNA lesions in the case of excision-repair-deficient hosts. At any rate, the absolute number of lesions dealt with by prophage reactivation is the same. One can state that the efficiency of repair by prophage reactivation is rather small as compared to pyrimidine dimer excision; in a uvr^+ host, in which 85% of the lesions are repaired by pyrimidine dimer excision, half of the remaining 15%, i.e. 7.5% of the total damage, can be removed by prophage reactivation. What is the timing of repair of UV-damaged phage λ in a wild-type host? It might be that pyrimidine dimer excision occurs first, followed by prophage reactivation.

   Role of Bacterial and Phage Genes Involved in Recombination. If repair of UV-irradiated phage λ by prophage reactivation is due to recombination, one expects that this process may not take place in some recombination-deficient mutants either of the phage—*red* and *int* (Echols and Gingery, 1968; Signer and Weil, 1968; Gottesman and Yarmolinsky, 1968)—or of the host cell—*recA, recB,* and *recC* (Howard-Flanders and Boyce, 1966).

   Blanco and Devoret (1973) have investigated the relative contributions of genes *recA, red,* and *int* to ensure efficient prophage reactivation. They showed that the extent of reactivation in $recA^+$ hosts is 53%, as compared to about 13% in a *recA* mutant. Therefore, prophage reactivation of a UV-damaged phage depends mostly but not only on the presence in the cell of the $recA^+$ product. Other bacterial *rec* gene functions such as *recC* are dispensable, as shown by Hart and Ellison (1970).

   Experiments with *red* and *int* mutants of phage λ tested the dependence on the phage recombination systems of prophage reactivation obtained in a *recA* host. It can be concluded that the *red* system of the UV-irradiated phage is responsible for the small amount of reactivation found; in contrast, there is no *int*-promoted prophage reactivation (Blanco and Devoret, 1973).

   Influence of DNA Homology. The influence of the extent of homology between phage and prophage on the efficiency of prophage reactivation was tested by using *recA* bacteria carrying two prophages either in *cis* i.e. in tandem on the bacterial chromosome, or in *trans,* on the chromosome and on an episome. In the two double-lysogenic strains, prophage reactivation can be as high as 26% (Table 1). Under these conditions the efficiency of the red^+ gene product is about 60% that of the bacterial $recA^+$ gene.

   Prophage reactivation is indeed very sensitive to qualitative differences in DNA homology. In effect, repair of UV-damaged phage λ is rather low when the host carries prophage 434; there is no repair at all when UV-irradiated phage λ has to pair with prophage 80 (Blanco and George, unpublished). This reflects the structural differences found between the DNAs of the lambdoid phages. The preceding results taken all together indicate that prophage reactivation is closely dependent on genetic recombination.

   Absence of Mutagenesis of UV-Damaged Phage λ in Prophage Reactivation. Experiments were carried out by Blanco and Devoret (1973) in order to determine whether repair by prophage reactivation led to mutagenesis of the rescued phage. Clear plaque mutations were scored. As shown in Table 2,

Table 1. Prophage Reactivation in *recA* Single and Double Lysogens and
in *lexA1* Single Lysogens

UV dose to phage[a] (ergs/ mm^2)	Heteroimmune superinfecting phage		Host[b]	Reactivation in (%)[c]
	int	*red*		
2000	+	+	*recA* (λhimm434)	13 (4)
2000	+	+	*recA* (λhimm434)^2	30 (4)
2000	6	3	*recA* (λhimm434)	<0
2000	6	3	*recA* (λhimm434)^2	<0
2000	+	+	*recA*(λhimm434)/F$_1$-*gal*$^+$(λhimm434)	26 (3)
2000	+	+	*lexA*$^-$(λhind$^-$)	45 (4)
2000	102	3	*lexA*$^-$(λhind$^-$)	37 (3)
2000	+	+	*rec*$^+$ *lex*$^+$ (λhind$^-$)	53 (4)
2000	+	+	*rec*$^+$ *lex*$^+$ (λhind$^-$)^2	64 (3)
2000	102	3	*rec*$^+$ *lex*$^+$ (λhind$^-$)	39 (3)

[a]Phage survival was 10^{-2} in a *rec*$^+$ *lex*$^+$ nonlysogenic host.
[b]The *rec*$^+$ *lex*$^+$ strain was AB1157; the *recA* strain was AB2463 *recA13*; the *lexA*$^-$ strain was AB2494 *lex-1*. In a strain which carries two prophages the name of the prophage is followed by exponent 2.
[c]Percent of reactivation is calculated as indicated in the text. Values are the average of a number of experiments shown in parentheses.

prophage reactivation does not cause mutagenesis of the UV-damaged phage. Absence of correlation between UV mutagenesis and recombination of phage λ has also been reported by Miura and Tomizawa (1970).

In conclusion, prophage reactivation consists essentially of a repair process which tends to bypass by a recombination process the pyrimidine dimers left unexcised in the UV-damaged phage.

MODES OF RECOMBINATION REPAIR

Two types of recombination repair will be discussed here. Recombination repair of *one DNA homoduplex* containing pyrimidine dimers might occur, according to Howard-Flanders et al. (1968), in two steps: First, DNA replication proceeds; this process leaves gaps opposite the dimers in the newly formed DNA strand. Then, in this newly formed four-stranded structure, sister-strand exchanges, triggered by the free ends of the gaps, restore the original DNA structure. This repair process has been termed *postreplicative recombination repair.* Considerable experimental evidence has been accumulated in favor of this mechanism of repair (Rupp et al., 1971; Ganesan, 1974, and this volume).

Repair by recombination of *two DNA duplexes* was shown to be involved in prophage reactivation (Jacob and Wollman, 1953; Yamamoto, 1967) and in similar repair processes such as *multiplicity reactivation* (Luria, 1947) and *cross-reactivation* or *marker rescue* (Doermann, 1961). Although these three processes are similar, they differ in that, in multiplicity reactivation, all phage DNA molecules are irradiated, whereas in prophage reactivation the phage molecule alone is damaged. Hence prophage reactivation can be equated neither with multiplicity reactivation nor with cross-reactivation, in which process only the rescued marker comes from irradiated DNA. In all these recombination processes, in contrast to postreplication repair, there is in principle no need for any extensive DNA synthesis, since all the DNA strands required in the recombination process are present in the cell at its inception. Therefore, one might support the view that all the repair processes based on recombination between two DNA duplexes are mostly *prereplicative recombination repair*.

Table 2. UV Reactivation of Phage λh⁺imm434 in λhcIind⁻ Lysogenic Hosts: Extent of Reactivation, Production of λhimm434 Recombinants, and Phage Mutagenesis

Bacterial strains[a]	UV dose (ergs/ mm²) to		Frequency of λhimm434 recombinants[c]	Reactivation (%)[d]	Frequency of clear mutations (× 10⁻⁴)
	Phage[b]	Bacteria			
lex⁺ λ⁻	800	0	–	0	1.4
	800	500	–	63	34.1
lex⁺ (λhind⁻)	800	0	28 (1.0)	53	1.8
	800	500	21 (0.75)	69	25.6
	0	0	0.5 (1.0)	–	4.8
	0	500	1.7 (2.9)	–	1.7
lexA⁻ λ⁻	800	0	–	0	<1
	800	100	–	–30[e]	<1
lexA⁻ (λhind⁻)	800	0	26 (1.0)	59	<1
	800	100	28 (1.1)	40[f]	<1
	0	0	0.9 (1.0)	–	–
	0	100	1.0 (1.1)	–	–

[a]The lex⁺ strain was AB1157; the lexA⁻ strain was AB2494 lex-1.
[b]Phage survival is about 10⁻¹ in a nonlysogenic host.
[c]The frequency of recombinants is expressed in percent; the numbers in parentheses represent the ratio of recombinants in irradiated bacteria over that in nonirradiated hosts.
[d]Reactivation is calculated as indicated in the text. Repair in the unirradiated nonlysogenic host is taken as background repair and is therefore assigned the value 0, although all strains are uvr⁺ and fully subject to excision-repair.
[e]This negative value indicates that repair of phage in the nonlysogenic host decreases with UV irradiation of the host.
[f]Reactivation is calculated taking as a control the phage survival on the UV-irradiated host.

However, in recent years, it has been shown by many authors (see review by Radding, 1973) that recombination is always associated with DNA synthesis; extensive DNA synthesis can occur in recombining phage λ (Stahl et al., 1972). Consequently, in order to avoid the use of confusing terms in the future, one should be careful to distinguish postreplicative recombination in a DNA homo-duplex from that involving at least two DNA duplexes.

Our hypothesis that $recA^+$-promoted prophage reactivation might be a pre-replicative recombination process is supported by the following arguments:

1. Recombination appears to be triggered by the presence of pyrimidine dimers in the phage DNA, and it is known that UV-induced recombination occurs before DNA replication (Baker et al., 1971; Jacob and Wollman, 1955). Similarly, as shown by Howard-Flanders and Lin (1973), marker rescue between phage and prophage can occur when phage DNA replication is blocked by immunity.

2. When the phage is UV-irradiated, a delay in the appearance of mature intracellular phage is observed. This might be a consequence of the inhibition of DNA replication and would, consequently, facilitate the pairing and recombination of chromosomes before replication.

As to red^+-promoted prophage reactivation in a $recA^-$ host, phage DNA replication might be necessary in order to allow the red^+ function to operate. This would account for the increased efficiency of red^+-promoted prophage reactivation in double lysogens (Blanco and Devoret, 1973); in such a case, two rounds of recombination triggered by replication would enhance the efficiency of prophage reactivation. Experiments are in progress to test the exact role of DNA replication in prophage reactivation (Blanco, personal communication).

UV REACTIVATION

The survival of ultraviolet-irradiated phage λ is increased when the host cell has been exposed to a low UV dose before infection. This phenomenon, known as UV reactivation (Weigle, 1953), is accompanied by a high rate of mutation (UV mutagenesis) of the reactivated phage.

Repair Efficiency. Repair efficiency can be as high as 50%. However, as for prophage reactivation, UV reactivation acts on DNA lesions remaining after pyrimidine dimer excision, so that, in a cell where excision-repair takes place, UV reactivation accounts for the repair of about 8% of the total lesions. Although this type of repair has a rather low comparative efficiency, its biologi-cal significance is of high interest, as will be seen below.

UV Reactivation is a Host-Cell Process. UV reactivation of phage λ depends only on host-cell processes. No phage genes seem to be required; $\lambda cI857bio10$, which is deleted from *int* into *cIII*, is reactivated normally (Devoret, unpub-lished). This fact excludes the possibility that phage genes such as *redA*, *redB*, and *gam,* which promote or allow phage λ recombination, are involved in UV reactivation. These results are in sheer contrast with those obtained for prophage reactivation, in which the *red* gene functions can operate.

A few bacterial mutations such as *recA, lex-1* (Howard-Flanders and Boyce, 1966), *zab* (Castellazzi et al., 1972*b*), or *lex-30* (Blanco et al., 1974) are known to prevent UV reactivation. The results of Castellazzi et al. (1972*b*) and Blanco et al. (1974) taken together lead to the suggestion that *lex* mutations belong to two groups, *lexA* and *lexB,* respectively located near *malB* and *recA.* In contrast, *recB* and *recC* bacteria are able to carry out UV reactivation of phage λ (Elsevier and Dove, 1969).

Interestingly enough, a bacterial mutation called *tif-1,* located near *recA,* confers on the cells when they are shifted from 32° to 41°C the ability to reactivate UV-damaged phage λ (Castellazzi et al., 1972*a*). Reactivation of UV-damaged phage λ by temperature shift of cells carrying mutations *tif-1* and *sfiA* is even more spectacular, for the *sfiA* mutation suppresses the cellular lethal effects of *tif-1* and leaves undiminished the cell capacity to reproduce λ (George, Castellazzi, and Buttin, to be published).

HYPOTHESES ON THE MECHANISM OF UV REACTIVATION

To account for the mechanism of UV reactivation, two hypotheses have been offered which postulate either an enhanced efficiency of host-cell reactivation of the irradiated phage genome after UV irradiation of the host cell (Harm, 1963*b*) or the existence of some kind of recombinational repair between homologous parts of the UV-irradiated host and phage genomes (Garen and Zinder, 1955; Kellenberger and Weigle, 1958; Weigle, 1966). Extending the latter hypothesis, Hart and Ellison (1970) have postulated that UV reactivation is analogous to prophage reactivation.

Does UV Reactivation Result from an Enhancement of Excision-Repair? The question of the involvement of pyrimidine dimer excision of UV reactivation has been a controversial matter in the past (Harm, 1966; Ogawa and Tomizawa, 1973). Yet, it has been shown that UV reactivation is not the result of an enhanced dimer excision in Uvr⁺ cells (Boyle and Setlow, 1970), since it takes place in Uvr⁻ bacteria, in which there is no pyrimidine dimer excision (Defais et al., 1971; Radman and Devoret, 1971).

Does UV Reactivation Result from Recombination? In order to analyze whether or not UV reactivation was dependent on recombination, Blanco and Devoret (1973) studied UV reactivation under conditions of maximum homology of the infecting phage to the host cell. Since recombination is highly dependent upon the degree of homology exhibited by the prospective recombining parts, absence of influence of homology upon UV reactivation would indicate that recombination is not involved in this repair process.

For these reasons, UV reactivation of phage λ*imm434* in nonlysogenic host cells was compared to that obtained in hosts lysogenic for $\lambda hcIind^-$. The presence of such a prophage in the host cell had the following three advantages: (1) it provided as much homology as can be attained between the infecting phage and the host cell; (2) it permitted one to check the rescue, by the developing phage, of the h marker carried by the prophage (considered to be a valid index for recombination); and (3) it allowed one to UV-irradiate the host cells, which could, nevertheless, still carry the noninducible prophage.

The data reported in Table 2 show that UV reactivation was slightly increased when the host cell carried a prophage homologous to the infecting phage. However, the relative frequency of recombinants between phage and prophage was decreased; this was due to the fact that, in spite of an enhanced phage survival, the number of recombinants remained constant; i.e., no new recombinants resulted from UV exposure of the lysogenic host. This indicates that there is no correlation between the efficiency of restoration of the UV-damaged phage and the frequency of recombinant formation in UV reactivation. The fact that UV reactivation takes place in the UV-damaged lysogenic host is demonstrated by the level of clear plaque mutations of the repaired phage (Table 2, lines 2 and 4). These results rule out the hypothesis that UV reactivation occurs as a result of recombination between phage and bacterial chromosomes.

Is Postreplicative Recombination Repair of Phage λ Involved in UV Reactivation? Even though this hypothesis has not yet been offered in the literature, we would like to discuss it as a possible occurrence. As postreplicative recombination repair operates on UV-damaged chromosomal DNA (Rupp et al., 1971), one can imagine that such a process might also act on UV-damaged phage λ DNA during the process of UV reactivation. Past experimental evidence discards this possibility. It has been established that not only phage S13 can be UV-reactivated (Tessman and Ozaki, 1960) but also phage ϕR (Ono and Shimazu, 1966), which is a single-stranded phage analogous to ϕX (Burton and Yagi, 1968). Since postreplicative recombination repair can only operate on double-stranded damaged DNA, the abovementioned results seem to exclude its involvement in UV reactivation.

EVIDENCE FOR A NEW REPAIR MECHANISM INVOLVED IN UV REACTIVATION

UV reactivation as well as UV mutagenesis of phage λ requires for its occurrence the bacterial *recA, lexA,* and *lexB* (*zab*) gene functions (Miura and Tomizawa, 1968; Defais et al., 1971; Blanco et al., 1974). This requirement is shared by UV induction of prophage λ (Brooks and Clark, 1967; Hertman and Luria, 1967; Donch et al., 1970; Devoret and Blanco, 1970; Castellazzi et al., 1972b), as well as by UV mutagenesis of *E. coli* K12 (Witkin, 1967, 1969a,b). This has led Defais et al. (1971) to suggest that common pathways might be involved in the abovementioned phenomena.

Since lysogenic induction in a lysogenic recipient cell can be obtained "indirectly" by conjugation with a UV-irradiated F⁺ (Borek and Ryan, 1958; Devoret and George, 1967) or F-prime donor (George and Devoret, 1971), if the above hypothesis is correct, one should be able to produce UV reactivation of phage λ as well as its mutagenesis in a recipient host cell mated with a UV-damaged F-prime donor. It has been shown that this is indeed the case (George et al., 1974).

Indirect UV Reactivation of Phage λ. The number of plaques formed by UV-irradiated phage λ on an F⁻ recipient host cell crossed with an F-*lac*⁺ or Hfr donor increases when the donor has been exposed to UV light before mating (Fig. 2). However, the reactivation factor under these conditions is 3–5, lower than that found (about 10) on directly UV-irradiated host cells (*direct UV reactivation*). Such a difference can almost completely be accounted for by the variable efficiency of DNA transfer to the recipient.

The phage reactivation process described is not an enhanced *host-cell reactivation,* since it is not abolished in an F⁻ recipient λ host deficient in pyrimidine dimer excision-repair (Table 3, experiments A₁ and A₂). This finding parallels what is found in direct UV reactivation (Defais et al., 1971; Radman and Devoret, 1971). Yet, bacteria carrying *uvrB5,* used in this series of experiments, appear to be less proficient in UV reactivation of λ than some other *uvrA, uvrB,* or *uvrC* bacteria (Radman and Devoret, 1971; Defais et al., 1971).

Furthermore, a high rate of mutation of λ to clear plaque mutants is observed when UV-irradiated phage λ is reactivated in a recipient crossed with a

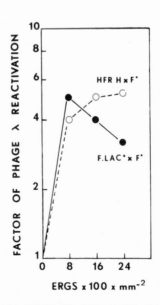

Fig. 2. Indirect UV reactivation determined on GY 688 mated with UV-irradiated Hfr GY 1151 (o – – – o) and with F-lac⁺ GY 854 (●——●). The titer of phage λ used was 3 × 10⁶/ml, its survival was precisely 10⁻² in this experiment. The percentages of recipients having acquired lac⁺ were 59% (cross GY 854 × GY 688) and 17% (cross GY 1151 × GY 688). The reactivation factor is plotted in the ordinate against the UV dose given to the donor in the abscissa. The maximal value of E (repair efficiency) was 0.35.

Table 3. Indirect UV Reactivation[a]

Experiment	Donor	λ host	pfu/ml λ	pfu/ml λ*	λ survival (%)	Reactivation factor	E
A_1	GY 854 F-lac^+	GY 688	10^6	3.9×10^3	0.4	11.7	0.44
	GY 854 F-lac^+*	GY 688	10^6	4.7×10^4	4.7	18.4	0.49
	None	GY 688	10^6	2.6×10^3	0.26		
	None	GY 688*	10^6	4.8×10^4	4.8		
A_2	GY 854 F-lac^+	GY 744 $uvrB$	10^6	1.4×10^3	0.14	3.8	0.20
	GY 854 F-lac^+**	GY 744 $uvrB$	10^6	5.3×10^3	0.53	3.1	0.16
	None	CY 744 $uvrB$	10^6	8×10^2	0.08		
	None	GY 744 $uvrB$*	10^6	2.5×10^3	0.25		
B_1	GY 4010 F-lac^+	GY 688	9×10^5	10^4	1.1	3.3	0.27
	GY 4010 F-lac^+**	GY 688	8×10^5	3×10^4	3.7	8.5	0.42
	None	GY 688	8×10^5	5×10^3	0.6		
	None	GY 688*	7×10^5	3.6×10^4	5.1		
B_2	GY 4006 F⁻	GY 4007 F-lac^+	8×10^5	6×10^3	0.75	1	0
	GY 4006 F⁻*	GY 4007 F-lac^+	8×10^5	6×10^3	0.75	16	0.50
	None	GY 4007 F-lac^+	8×10^5	3.5×10^3	0.4		
	None	GY 4007 F-lac^+**	5×10^5	3.2×10^4	6.4		
C_1	GY 746 Hfr	GY 688	5.2×10^6	8.7×10^4	1.7	4.4	0.37
	GY 746 Hfr*	GY 688	6.5×10^6	4.9×10^5	7.5	2.6	0.21
	None	CY 688	8×10^6	8×10^4	1		
	None	GY 688*	6.5×10^6	1.7×10^5	2.6		
C_2	GY 746 Hfr	GY 743 (PICM)⁺	4×10^6	9.2×10^4	2.3	2.3	0.23
	GY 746 Hfr*	GY 743 (PICM)⁺	7×10^6	3.8×10^5	5.4	5	0.32
	None	GY 743 (PICM)⁺	7.9×10^6	9.4×10^4	1.2		
	None	CY 743 (PICM)⁺*	5.9×10^6	3.5×10^5	5.9		

[a]The sign * in this and the subsequent table indicates that bacteria or phage have been UV-irradiated. The percentages of recipients having acquired lac^+ were 72% in A_1 and 82% in A_2; 80% in B_1, and 100% in B_2; 27% in C_1 and less than 10^{-4} in the restrictive cross C_2. To avoid its restriction in GY 743, λ.PICM (λ grown on a PICM lysogen) was used in experiments C_1 and C_2.

UV-damaged HfrH donor (Table 3). The rate of mutation is of the same magnitude as that found in direct UV reactivation.

Since UV reactivation can be characterized as (1) a repair process that is highly mutagenic for the recovered phage, and (2) taking place in uvr^- hosts, we propose to call this repair process *indirect UV reactivation* by analogy to *indirect UV* induction of prophage λ (Devoret and George, 1967).

Indirect UV Reactivation Is Induced by Transferred UV-Damaged DNA. To prove this point and eliminate the possibility that indirect UV reactivation could be either due to some component released into the medium by the irradiated donor or promoted solely by cell-wall contact, we sought to demonstrate that indirect UV reactivation is determined by the direction of conjugal transfer. Indeed, only in those λ hosts that were recipients of UV-damaged transferred DNA did UV reactivation of λ take place (Table 3, experiments B_1 and B_2).

It has been demonstrated that conjugal transfer of UV-damaged episomal DNA (George and Devoret, 1971; Monk, 1969) as well as Hfr DNA (George, unpublished) is not diminished after UV irradiation of the donor bacteria, at least in the range of UV doses used here. This, taken together with the control experiments described just above, suggests that indirect UV reactivation is induced in a nonirradiated host cell by the introduction of UV-damaged bacterial DNA.

Indirect UV Reactivation Versus Indirect UV Induction. It was previously established (Devoret and George, 1967; George and Devoret, 1971; Rosner et al., 1968) that indirect UV induction of prophage λ results from the transfer of an entire UV-damaged *replicon*. Restriction in the recipient lysogen of the incoming DNA abolishes prophage induction (George, 1966). In contrast to indirect prophage induction, indirect UV reactivation can be promoted with equal efficiency by either F-prime or Hfr donors, as shown in Fig. 2 and Tables 3 and 4.

Moreover, restriction of the incoming UV-damaged donor DNA in the recipient by a factor of 10^4 only slightly decreases the efficiency of indirect UV reactivation of unrestricted phage λ infecting the same recipient (Table 3, experiments C_1 and C_2). If degradation of the transferred DNA very likely takes place, this does not prevent the appearance of the phenomenon.

Is UV Reactivation an Inducible Repair Process? The aforementioned data show that the introduction of UV-damaged DNA into a host cell leads to the reactivation of UV-irradiated phage λ. This new kind of phage reactivation has the same essential features as direct UV reactivation: it is associated with the mutagenesis of the restored phage and it can be produced in a strain that is deficient in pyrimidine dimer excision. Introduction by conjugation of a small piece of UV-damaged DNA such as an F-*lac*$^+$ factor has evidently a notable effect in inducing an error-prone repair process.

As demonstrated in the preceding sections, repair processes such as pyrimidine dimer excision, recombination between phage and host DNAs, and

Table 4. Comparison of Indirect Reactivation and Induction in F-*lac* and Hfr Hosts[a]

Donor	Indirect UV-reactivation in GY 688 F⁻				Indirect prophage induction in GY 701 F⁻(λ)⁺ LamR	
	Infective centers (pfu/ml)		λ survival (%)	λ Reactivation factor	Free phages (pfu/ml)	Efficiency of indirect induction
	λ	λ*				
GY 854 F-*lac*	8×10^5	3.5×10^3	0.4		4.3×10^6 (A)	B/A : 30
				5		
GY 854 F-*lac**	8×10^5	1.6×10^4	2		1.3×10^8 (B)	
GY 1151 Hfr	8×10^5	4.1×10^3	0.5		3.2×10^6 (a)	b/a : 2.5
				3.4		
GY 1151 Hfr*	7×10^5	1.2×10^4	1.7		8×10^6 (b)	

[a]The host for indirect UV reactivation of phage λ was GY 688, and the recipient lysogen used for indirect UV induction was GY 701. Indirect UV reactivation was estimated as described in the text. The technique for obtaining and estimating indirect UV induction has been fully described by Devoret and George (1967). The percentages of recipients which had received at 20 min *lac*⁺ were 55% (cross GY 854 × GY 688), 28% (cross GY 854 × GY 701), 19% (cross GY 1151 × GY 688), 4.6% (cross GY 1151 × GY 701).

postreplicative recombination repair cannot account for the mechanism of UV reactivation. Consequently, we think that the phenomenon of indirect UV reactivation provides evidence in favor of the hypothesis that there exists a *third mechanism* of dark repair of UV damage different from pyrimidine dimer excision and from pre- or postreplicative recombination.

By analogy with indirect UV induction of prophage λ, we have postulated (George et al., 1974) that the newly identified repair process is *induced* by the presence in the recipient host of UV-damaged DNA. However, there is no direct evidence that *induction* is indeed involved. It might also be that introduction of UV-irradiated DNA *activates* some repair enzymes in the recipient host cell.

UV Reactivation of Phage λ and SOS Repair. The SOS repair hypothesis (Radman, 1973, and this volume) states that an inducible repair process acts on both UV-damaged phage and bacterial DNA in such a way that a common biochemical pathway relates UV mutagenesis, UV lysogenic induction, UV reactivation, and UV filamentation. It is based on such main evidence as the pleiotropic effects of *recA* and *lex* mutations on the four abovementioned phenomena, the UV-like effect of *tif* expression, and the existence of indirect UV reactivation of λ. Although theoretically attractive, this general hypothesis needs more experimental support.

There are already some recent experimental findings that lead us to discuss this general hypothesis. In effect, for instance, treatments that in lysogens promote lysogenic induction, such as UV irradiation, thymineless death, or thermal shift of *tif* bacteria, also promote UV reactivation of phage λ in the nonlysogenic host (Hart and Ellison, 1970; Castellazzi et al., 1972a). In contrast, conditions that promote indirect UV reactivation are not *sufficient* to promote indirect lysogenic induction of prophage λ (George et al., 1974). Even though both phenomena are triggered by pyrimidine dimers carried by the donor DNA transferred to the recipient, the phenomenon of lysogenic induction, in order to be produced, requires that the lesions be on a transmitted *replicon* (Devoret and George, 1967; George and Devoret, 1971; Rosner et al., 1968). The relation between the conditions of UV lysogenic induction and of UV reactivation is not a two-way relationship.

In view of the importance for the theory of SOS repair of the function of gene *tif,* and as regards its possible involvement in an inducible repair process, one must point out that, in lysogenic induction, the expression of *tif* has been shown to be constitutive (Shinagawa and Itoh, 1974). Consequently, one may question whether, in *tif* mutants, repair and mutagenesis in phage λ would necessarily be inducible processes (George, experiments in progress).

It has been shown by Witkin (1974) that mutagenesis is stimulated by *tif* expression in bacteria exposed to low UV doses, and that this stimulation is inhibited by chloramphenicol. However, attempts to produce mutagenesis in slightly UV-irradiated bacteria mated with a UV-damaged ColV sex factor have failed so far (Green, Rothwell, and Bridges, personal communication).

CONCLUSION

Studies on UV reactivation of phage λ have provided evidence in favor of the hypothesis that there exists in bacteria an inducible error-prone repair process different from other known repair mechanisms. This type of repair acts on phage DNA. To decide whether or not it operates also on bacterial DNA awaits further experimental evidence.

ACKNOWLEDGMENTS

The excellent technical assistance of Mrs. M. Pierre has been highly appreciated as well as discussions with A. Bailone. This work was aided by funds from the Fondation pour la Recherche Médicale Française. Manuel Blanco and Miroslav Radman were, respectively, supported by a Euratom Fellowship and by a "Joliot-Curie" postdoctoral fellowship.

REFERENCES

Baker, R., Doniger, J. and Tessman, I. (1971). *Nature New Biol.* **236**, 23–25.
Blanco, M. and Devoret, R. (1973). *Mutat. Res.* **17**, 293–305.
Blanco, M., Levine, A. and Devoret, R. (1975). This volume, part B.
Bootsma, D., de Weerd-Kastelein, E. A., Veldhuisen, G. and Keijzer, W. (1971). *DNA Repair Mechanisms.* F. K. Schattauer Verlag, Stuttgart.
Borek, E. and Ryan, A. (1958). *Proc. Nat. Acad. Sci. U.S.A.* **44**, 374–377.
Boyle, J. M. and Setlow, R. B. (1970). *J. Mol. Biol.* **51**, 131-144.
Brooks, K. and Clark, A. J. (1967). *J. Virol.* **1**, 283–293.
Burton, A. J. and Yagi, S. (1968). *J. Mol. Biol.* **34**, 481–486.
Castellazzi, M., George, J. and Buttin, G. (1972*a*). *Mol. Gen. Genet.* **119**, 139–152.
Castellazzi, M., George, J. and Buttin, G. (1972*b*). *Mol. Gen. Genet.* **119**, 153–174.
Chase, M. C. (1964). Ph. D. Thesis, University of Southern California, Los Angeles, California.
Defais, M., Fauquet, P., Radman, M. and Errera, M. (1971). *Virology* **43**, 495–503.
Devoret, R. and Blanco, M. (1970). *Mol. Gen. Genet.* **107**, 272–280.
Devoret, R. and Coquerelle, T. (1966). *Genetical Aspects of Radiosensitivity: Mechanisms of Repair,* pp. 89–95. International Atomic Energy Agency, Vienna.
Devoret, R. and George, J. (1967). *Mutat. Res.* **7**, 713–734.
Doermann, A. H. (1961). *J. Cell. Comp. Physiol.* **58** (suppl. 1), 79–93.
Donch, J., Greenberg, J. and Green, M. H. L. (1970). *Genet. Res.* **15**, 87–97.
Echols, H. and Gingery, R. (1968). *J. Mol. Biol.* **34**, 239–249.
Elsevier, S. and Dove, W. F. (1969). In *Phage Meetings Cold Spring Harbor* (abstract).
Ganesan, A. K. (1974). *J. Mol. Biol.* **87**, 103–119.
Garen, A. and Zinder, N. D. (1955). *Virology* **1**, 347–376.
George, J. (1966). *C. R. Acad. Sci. Paris* **262**, 1805–1808.
George, J. and Devoret, R. (1971). *Mol. Gen. Genet.* **111**, 103–119.
George, J., Devoret, R. and Radman, M. (1974). *Proc. Nat. Acad. Sci. U.S.A.* **71**, 144–147.
Glickman, B. W. (1974). *Biochim. Biophys. Acta* **335**, 115–122.
Gottesman, M. E. and Yarmolinsky, M. B. (1968). *Cold Spring Harbor Symp. Quant. Biol.* **33**, 735–747.

Harm, W. (1963a). Z. Vererbungsl. 94, 67–79.

Harm, W. (1963b). Virology 19, 66–71.

Harm, W. (1966). Virology 29, 494.

Hart, M. G. R. and Ellison, J. (1970). J. Gen. Virol. 8, 197–208.

Hertman, I. and Luria, S. E. (1967). J. Mol. Biol. 23, 117–133.

Howard-Flanders, P. and Boyce, R. P. (1966). Radiat. Res. 6 (suppl.) 156–184.

Howard-Flanders, P. and Lin, P. F. (1973). Genetics (suppl.) 73, 85–90.

Howard-Flanders, P., Boyce, R. P. and Theriot, L. (1966). Genetics 53, 1119–1136.

Howard-Flanders P., Rupp, W. D., Wilkins, B. M. and Cole, R. S. (1968). Cold Spring Harbor Symp. Quant. Biol. 33, 195–205.

Jacob, F. and Wollman, E. L. (1953). Cold Spring Harbor Symp. Quant. Biol. 18, 101–121.

Jacob, F. and Wollman, E. L. (1955). Ann. Inst. Pasteur 88, 724–749.

Jagger, J. (1960). In Radiation Protection and Recovery (Hollaender, A., ed.), pp. 352–377. Pengamon, New York.

Kellenberger, G. and Weigle, J. J. (1958). Biochim. Biophys. Acta 30, 112–124.

Luria, S. E. (1947). Proc. Nat. Acad. Sci. U.S.A. 33, 253–264.

Miura, A. and Tomizawa, J. (1968). Mol. Gen. Genet. 103, 1–10.

Miura, A. and Tomizawa, J. (1970). Proc. Nat. Acad. Sci. U.S.A. 67, 1722–1726.

Monk, M. (1969). Mol. Gen. Genet. 106, 14–24.

Ogawa, H. and Tomizawa, J. (1973). J. Mol. Biol. 73, 397–406.

Ogawa, H., Shimada, K. and Tomizawa, J. (1968). Mol. Gen. Genet. 101, 227–244.

Ono, J. and Shimazu, Y. (1966). Virology 29, 295.

Radding, C. M. (1973). Ann. Rev. Genet. 7, 87–109.

Radman, M. (1974). In Molecular and Environmental Aspect of Mutagenesis (Drakash, L., Sherman, F., Miller, M., Lawrence, C. W., Taber, H. W., eds.), pp. 128–142. C. C. Thomas, Springfield, Illinois.

Radman, M. and Devoret R. (1971). Virology 43, 504–506.

Radman, M., Cordone, L., Krsmanovic-Simic, D. and Errera, M. (1970). J. Mol. Biol. 49, 203–212.

Rosner, J. L., Kass L. R. and Yarmolinsky, M. B. (1968). Cold Spring Harbor Symp. Quant. Biol. 33, 785–789.

Rupp, D. W., Wilde, C. E., Reno, D. L. and Howard-Flanders, P. (1971). J. Mol. Biol. 61, 25–44.

Setlow, R. B. and Carrier, W. L. (1966). Biophys. Soc. Abstr. 6, 68.

Shinagawa, H. and Itoh, T. (1974). Mol. Gen. Genet. 126, 103–110.

Signer, E. R. and Weil, J. (1968). J. Mol. Biol. 34, 261–271.

Stahl, F. W., McMilin, K. D., Stahl M. M. and Nozu, Y. (1972). Proc Nat. Acad. Sci. U.S.A. 69, 3598–3601.

Tessman, E. S. and Ozaki, T. (1960). Virology 12, 431–439.

van de Putte, P., van Dillewijn, J., van Sluis, C. A. & Rörsch, A. (1965). Mutat. Res. 2, 97–110.

Weigle, J. J. (1953). Proc. Nat. Acad. Sci. U.S.A. 39, 628–636.

Weigle, J. J. (1966). In Phage and the Origins of Molecular Biology (Cairns, J., Stent, G. S. and Watson, J. D., eds.), pp. 226–235. Cold Spring Harbor Laboratory of Quantitative Biology, New York.

Witkin, E. M. (1967). Brookhaven Symp. Biol. 20, 17–55.

Witkin, E. M. (1969a). Mutat. Res. 8, 9–14.

Witkin, E. M. (1969b). Ann. Rev. Microbiol. 23, 487–513.

Witkin, E. M. (1974). Proc. Nat. Acad. Sci. U.S.A. 71, 1930–1934.

Yamamoto, N. (1967). Biochem. Biophys. Res. Commun. 27, 263–269.

IV

Enzymology of Excision-Repair in Bacteria

21

Enzymology of Excision-Repair in Bacteria: Overview[1]

Lawrence Grossman

Graduate Department of Biochemistry
Brandeis University
Waltham, Massachusetts 02154

The incision step in the repair process is, perhaps, rate limiting, controlling the remainder of events leading to the ultimate removal of photoproducts and the reinsertion of nucleotides for the restoration of biological activity and strand continuity of irradiated DNA.

INCISION—CORRENDONUCLEASE-CATALYZED REACTIONS

Correndonuclease II Activities on Diadduct Damage

There are two generic classes of damage to DNA in which either mono- or diadduct products are formed (Regan and Setlow, 1973). Monoadducts involve a single base modification, base change, or the formation of a non-complementary base pair. Diadduct damage is that which involves more than one nitrogenous base, held through intrastrand or interstrand bonds.

---

[1] Definitions: exonuclease—a phosphodiesterase which requires a terminus for hydrolysis; such enzymes can exhibit specificities for terminal phosphodiester bonds or those that are internal; endonuclease—a phosphodiesterase which does not require a terminus; correndonuclease—an endonuclease whose specificity eventually leads to correctional mechanisms and which is limited to a DNA whose structure is modified such that its nitrogenous bases are no longer complementary, are removed from the DNA, or can interact with bases on the same strand (intrastrand dimers) or with bases on the opposite strands (interstrand dimers, cross-links).

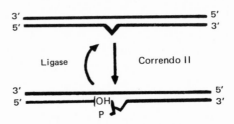

Nucleases recognizing the distortion arising from diadduct formation, such as pyrimidine-pyrimidine dimers, psoralen cross-links, mitomycin-induced cross-links, and nitrous-acid-induced cross-links, will be referred to as correndonuclease II.

The *Micrococcus luteus* correndonuclease II has received the most attention, and more recently the correndonuclease II from *Escherichia coli* has been isolated and studied. This class of enzyme specifically acts on the strand containing pyrimidine dimers, and from reports by Patrick and his collaborators at this meeting, the correndonuclease II of *M. luteus* is capable of acting on both the cytosine and thymine species of pyrimidine dimers. The pyrimidine adducts $Thy(\alpha\text{-}5)II_2$ Thy, those non-hydrogen-bonded regions formed by point mutation heteroduplexes, apparently are not acted upon by such enzymes.

The correndonucleases II isolated from *M. luteus* and *E. coli* have been shown to act specifically on the strand damaged by pyrimidine dimers. The data reported by M. Radman for the *E. coli* endonuclease III, and in collaboration with Andrew Braun for the *E. coli (uvrA uvrB)* correndonuclease II were obtained through the use of heteroduplexes in which selectively labeled UV-damaged strands were resolved and annealed with differentially labeled undamaged strands. Carrier and Setlow (1970) showed by annealing ^{32}P-labeled irradiated denatured DNA with tritiated unirradiated single-stranded DNA that the *M. luteus* correndonuclease II selectively acts on the damaged strands.

Site Specificity

The correndonuclease II enzymes from bacteria act 5′ at, or no more than one nucleotide from, pyrimidine dimers. This was also observed by Braun with the *E. coli* enzyme, and previously reported in detail by Kushner et al. (1971) for the correndonuclease II of *M. luteus*. The fact that all of these enzymes do break on the 5′ side to the damage imposes restrictions on the types of enzymes which may act subsequently to excision of damage.

Specificity of the Break

Specificity of the break is on the a ($|_a P^b|$) side of the phosphodiester bond, resulting in the formation of hydroxyl groups and 5′-phosphoryl groups which become associated with the damaged strand. It is not apparent, at this point, whether there are more than one species of correndonuclease II enzymes in microorganisms, but apparently there are enzymes which will hydrolyze, resulting in the appearance of a $3'$-phosphoryl group and a $5'$-hydroxyl group. The enzymes discussed during this session specifically break so that a $3'$-hydroxyl group is formed, providing a nucleophilic site for the initiation of polymerization. Exposure of the $5'$ terminus provides susceptibility to action by the polymerase-associated $5' \rightarrow 3'$ exonuclease or independent exonucleases. That a $5'$-phosphoryl is formed as a result of incision by correndonuclease II also provides a polynucleotide ligase site. The significance of this is discussed below.

The biological identity of the *M. luteus* correndonuclease II was inferred through the use of mutants lacking correndonuclease II activity (Mahler et al., 1971) and identified as the *uvrA* and *uvrB* gene products of *E. coli* by Braun and his coworkers (Braun and Grossman, 1974; Braun et al., this volume). Furthermore, the diadduct damage specificity of this enzyme is substantiated by the requirements of the *uvrA* and *uvrB* genes for the removal of psoralen cross-links (Kohn et al., 1961; Cole and Sinden, Part B of this book).

In organisms possessing both photolyase and excision capabilities, there is an apparent competition which is controlled by their respective K_ms and dissociation constants. Thus, the photoreactivating enzyme, if binding to the pyrimidine site, will preclude incision or vice versa. Likewise, phosphodiester bond cleavage associated with the action of the correndonuclease II renders the site unavailable for photoreactivation, even though the photolyase enzymes can bind to such sites (Patrick and Harm, 1973).

The use of correndonucleases for identifying pyrimidine dimers has provided a fairly sensitive and specific tool for measuring the removal of pyrimidine dimers *in vivo*. The chromatographic assays that have routinely been employed for measuring dimer removal do not have the sensitivity of this procedure. Ganesan (this volume) describes the use of this enzyme in conjunction with permeabilized cell systems, in order to follow the removal of pyrimidine dimers *in vivo*.

Correndonuclease I (Adduct Damage Incision)

At these meetings, the repair of alkylation damage, the formation of apurinic sites arising from alkylation, and thymine modification arising from γ radiation have been discussed. The 5,6-dihydroxydihydrothymine derivatives arising from γ-radiation provide substrate sites for endonucleases other than the correndonuclease II in which phosphodiester bond breaks are found in proximity to such nucleotide damage (Hariharan and Cerutti, 1974; Hariharan et al., this volume; Strniste and Wallace, Part B of this book). The *uvrA uvrB* gene product, though not participating directly seems to enhance the incision of phosphodiester bonds near γ-radiation products (Hariharan and Cerutti, 1974). From discussions at these meetings, it would appear that the correndonuclease I breaks 5′ to the damage, although it is not certain whether a 3′-phosphoryl or a 3′-hydroxyl group arises from such hydrolysis reactions.

Reversibility of the Incision Step by Correndonuclease II

The reversibility of the incision step by correndonuclease II catalyzed by polynucleotide ligase involves the 5′-phosphorylated terminus adjacent to the dimer with the juxtaposed 3′-OH strand. This reversibility by ligase provides a potential for an abortive repair step, which is operative in repair pathways under certain circumstances. The indications of Seeberg and Rupp (this volume) with the double mutant *uvrC ligts* (conditionally lethal ligase mutant) would tend to functionally localize the *uvrC* gene product at some point immediately after incision and before the excision step. The *uvrC* gene product is presumed to control the activity of ligase at this enzymatic locus.

Dark repair in prokaryotic systems is specifically inhibited by caffeine and proflavin, which are intercalating agents. It is assumed, as a consequence of this DNA-inhibitor interaction, that the specificity in inhibition of the repair process occurs at the substrate level. Braun et al. (this volume) show that caffeine is a competitive inhibitor of the enzyme, and from preincubation studies they conclude that this purine analog most likely acts by binding at the substrate site of the *E. coli* correndonuclease II. Proflavin, on the other hand, acts as a noncompetitive inhibitor, with the site of correndonuclease II inhibition probably at the DNA substrate level, reducing its availability to the enzyme.

EXCISION

The removal of the damaged nucleotides from DNA can proceed by two alternate and, perhaps, interdependent routes (see scheme, p. 179).

Polymerase-Associated Excision System

In addition to its polymerizing properties, *E. coli* polymerase I has associated exonucleases which are capable of two sorts of repair processes. *A 5′→3′-associated double-strand-specific exonuclease* is capable of removing pyrimidine dimers in its path concomitant with polymerization. Furthermore, when noncomplementary nucleotides are incorrectly inserted during polymerization, *the 3′→5′ single-strand-specific exonuclease* is capable of digesting in a direction opposite of that of polymerization to the point of the first stable hydrogen bond (Brutlag and Kornberg, 1972; Muzyczka et al., 1972). Hamilton et al. (this

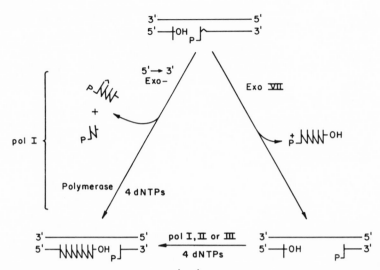

volume) rule out the associated $3'\rightarrow5'$ exonuclease activity in the excision process. Therefore, the $5'\rightarrow3'$ exonucleolytic activity of DNA polymerase I enzymes suitably assumes the excision role in this repair pathway. The ability of this $5'\rightarrow3'$ exonuclease activity to hydrolyze nonterminal phosphodiester bonds, bypassing photoproducts, renders this activity suitable for repair of this type. Most of the evidence, obtained from *M. luteus* and *E. coli* correndonuclease II experiments, is that the phosphodiester bond incised is no more than one nucleotide away from the pyrimidine dimer. Therefore, the $5'\rightarrow3'$ exonucleolytic activity is capable of circumventing such damage during polymerization.

The rate and extent of pyrimidine removal depend upon the simultaneous polymerization to provide nucleotides reasonably juxtaposed to the $5'$ termini destined for removal during the nick-translation process (Cozzarelli et al., 1969; Friedberg, this volume; Lehmann, Part B of this book). The assignment of the $5'\rightarrow3'$ exonuclease of polymerase I in the excision-repair process *in vivo* is attained through the polymerase mutants of Glickman (this volume) and Uyemura et al. (this volume), in which the phenotypic aberrancy specifically affects the $5'\rightarrow3'$ exonuclease function. The $5'\rightarrow3'$ mutants described by Uyemura et al. are conditional lethals, and at restrictive temperatures there is a loss of viability. An accumulation of 10S replication intermediates (Okazaki fragments) probably contributes to the lethality at restrictive temperatures. The mutants described by Glickman lack $5'\rightarrow3'$ activity and exhibit UV sensitivity; however, pyrimidine-dimer excision is reported to be normal. The contribution, therefore, of the $5'\rightarrow3'$ activity is difficult to assess, since strand-displacement reactions associated with DNA polymerase activity may operate. The pyrimidine dimers located on such a displaced strand may be excised by an unrelated process. An additional pathway for dimer excision in this mutant could occur via unassociated exonuclease-catalyzed excision, discussed in the next section.

Unassociated Exonucleases

There are correctional exonucleases (correxonucleases) capable of excising pyrimidine dimers which are not associated with polymerases. These enzymes, which have properties in common, will be referred to as correxonucleases for the sake of uniformity in identification. One such enzyme from *E. coli* has been reported at this meeting by Chase and is similar in many ways to the enzyme from *M. luteus* described by Kaplan et al. (1971). These enzymes are insensitive to EDTA and initiate hydrolysis from both 5′ and 3′ termini of single-stranded DNA. That these enzymes can initiate hydrolysis from a 5′ terminus at internal phosphodiester bonds, releasing oligonucleotides, makes them uniquely suitable for removing pyrimidine dimers from incised DNA. These enzymes have an absolute specificity for single-stranded DNA, unlike the 5′→3′ exonucleolytic activity of polymerase I. It is feasible, therefore, that these unassociated exonucleases may act at loci different from the 5′→3′-polymerase-associated exonuclease. The conformational changes arising from incision may be varied as a consequence of the structural environment in and around the photoproducts, thereby necessitating different exonucleolytic specificities. Double mutants controlling the 5′→3′ activity of either polymerase I or exonuclease VII (*E. coli* correxonuclease) would be useful. A comparison of the extent of excision of dimers controlled by each gene product would, as a result, be feasible. Until such double mutants can be constructed, the exact biological role of these enzymes remains unclear.

REINSERTION

Hamilton went into detail concerning the control *in vitro* of both excision of pyrimidine dimers and reinsertion of nucleotides catalyzed by the polymerase I enzyme of *M. luteus*. This polymerase, for all intents and purposes, is biochemically and immunologically identical to the polymerase I of *E. coli*. Polymerase I, unlike polymerases II and III and phage-induced DNA polymerases, by virtue of specifically binding at "nicks," satisfies an important and specific requirement for the excision process. It would appear, therefore, that the nick-binding and translating properties of these enzymes provide an advantage for the purposes of the concomitant mechanisms of reinsertion and excision. Nick-translation conditions result in a one-to-one stoichiometry between the removal of photoproducts with accompanying nucleotides and the reinsertion of nucleotides. Furthermore, polymerase I is capable of repairing gaps. Incubation of this polymerase at $10°C$ and optimal Mn^{2+} levels leads to the stoichiometric excision-reinsertion property which, furthermore, can be controlled by the presence of polynucleotide ligase. Under conditions of strand displacement (at $37°C$), the presence of ligase appears to restrict the stoichiometric ratio to one nucleotide excised per nucleotide polymerized. Therefore, *in vitro* the polymerase−correndonuclease II combination accompanied by ligase is sufficient for the complete repair of single-strand breaks associated with incision and restoration of biological properties of UV-irradiated transforming DNA. It appears from Hamilton's presentation that at doses of 50% inactivation or less, there is a quantitative restoration of biological activity. At higher doses, the maximum repair capability of this enzyme system declines considerably. This is attributed to the formation of double-strand breaks effected during repair by virtue of the unidirectionality of the repair process. This unidirectionality of repair on strands of opposite polarity leads to effective double-strand breaks during the excision process.

The value of such an *in vitro* system lies in its adaptability to cofactor examination for the correlation of the differences during *in vitro* and *in vivo* repair. It was reported (Hamilton et al., this volume) that the use of polyamine levels comparable to those found in both pro- and eukaryotic systems is sufficient to hold the molecules together so that such effective double-strand breaks are prevented. It is not certain whether the presence of spermine or spermidine enhances photoproduct removal and restoration of biological activity to irradiated transforming DNA at high doses.

In toluenized bacteria, polymerase substitution and consequent involvement in reinsertion may occur at gaps to which the latter polymerases bind specifically (Masker et al., this volume). If the specificity of these polymerases is to be taken at face value, it would duplicate independent-exonuclease-catalyzed excision-repair *in vivo*.

LIGASE STEP

The possible participation of polynucleotide ligase in the early steps in the excision pathway is augmented by its obvious involvement in the final steps of the preparation of fully active DNA. The specificity of this enzyme, requiring a $5'$-phosphoryl group next to a juxtaposed $3'$-hydroxyl group, is capable of end-to-end joining (Sgaramella et al., 1970). Furthermore, DNA containing noncomplementary nucleotides at its $3'$ end is also a suitable substrate for the ligase (Tsiapalis and Narang, 1970). Conformational changes associated with some pyrimidine-dimer-containing regions are obviously not too extreme for this enzyme to function. The ligase, therefore, is capable of assuming, in addition to sealing the final phosphodiester bond, a level of control at earlier steps in the excision-repair process.

REFERENCES

Braun, A. G. and Grossman, L. (1974). *Proc. Nat. Acad. Sci. U.S.A.* **71**, 1838 (1974).
Brutlag, D. and Kornberg, A. (1971). *J. Biol. Chem.* **247**, 241.

Carrier, W. L. and Setlow, R. B. (1970). *J. Bacteriol.* **102**, 178.

Cozzarelli, N. R., Kelly, R. B. and Kornberg, A. (1969). *J. Mol. Biol.* **45**, 513.

Hariharan, P. V. and Cerutti, P. A. (1974). *Proc. Nat. Acad. Sci. U.S.A.* **71**, 3532.

Kaplan, J. C., Kushner, S. R. and Grossman, L. (1971). *Biochemistry* **10**, 3315.

Kohn, K., Steigbigel, N. and Spears, C. (1961). *Proc. Nat. Acad. Sci. U.S.A.* **53**, 154.

Kushner, S. R., Kaplan, J. C., Ono, H. and Grossman, L. (1971). *Biochemistry* **10**, 3325.

Mahler, I., Kushner, S. R. and Grossman, L. (1971). *Nature New Biol.* **234**, 47.

Muzyczka, N., Poland, R. I. and Bessman, M. J. (1972). *J. Biol. Chem.* **247**, 7116.

Patrick, M. H. and Harm, W. (1973). *Photochem. Photobiol.* **18**, 371.

Regan, J. D. and Setlow, R. B. (1973). In *Chemical Mutagens* (Hollaender, A., ed.), vol. 3, p. 151. Plenum, New York.

Sgaramella, V., van de Sande, J. H. and Khorana, H. G. (1970). *Proc. Nat. Acad. Sci. U.S.A.* **67**, 1468.

Tsiapalis, C. M. and Narang, S. A. (1970). *Biochem. Biophys. Res. Commun.* **39**, 631.

22

The *Escherichia coli* UV Endonuclease (Correndonuclease II)

Andrew Braun, Peggy Hopper, and Lawrence Grossman

Graduate Department of Biochemistry
Brandeis University
Waltham, Massachusetts 02154

ABSTRACT

An endonuclease from *Escherichia coli* which acts specifically upon UV-irradiated DNA (correndonuclease II) and is absent from the *uvrA* and *uvrB* mutants has been isolated and partially characterized. The enzyme is present in normal amounts in the *uvrC* mutant. It elutes from phosphocellulose at about 0.25 M potassium phosphate (pH 7.5) and passes through dialysis tubing. The enzyme binds tightly to UV-irradiated DNA but does not bind to unirradiated DNA. The enzyme incises irradiated DNA to the $5'$ side of a pyrimidine dimer and leaves a $5'$-phosphoryl terminus which can be resealed with polynucleotide ligase. The K_m of the enzyme is about 1.5×10^{-8} M dimers. Endonucleolytic activity of the enzyme is inhibited by caffeine with a K_I of about 10mM.

While the enzyme involved in the initial step of pyrimidine-dimer excision-repair has been isolated from *Micrococcus luteus* (Kaplan et al., 1969) the *Escherichia coli* infected by bacteriophage T4 (Friedberg and King, 1969), this enzymatic activity has not been isolated from uninfected *E. coli*. Endonucleolytic activity acting preferentially against UV-irradiated lambda DNA has been observed by Takagi et al. (1969) in extracts of *E. coli*. However, little progress in the isolation of the *repair* enzyme has been made, since the activity was observed in *hcr⁻* cells, as well as the wild type. Since this ubiquitous activity was present in very low quantities, sedimentation of treated DNA in alkaline sucrose was required to detect the presence of the enzyme. A standard biochemical assay, such as the "BAP" assay used by Kaplan et al. (1969) and by Yasuda and Sekiguchi (1970) to detect the *M. luteus* and T4 enzymes, is not sufficiently sensitive for routine assays of the *E. coli* activity in crude extracts. A more sensitive and rapid assay for endonucleolytic activity has been developed by Center and Richardson (1970). Using this assay and a filter binding assay similar to that of Riggs et al. (1970), we have detected at least two UV-specific endonucleolytic activities in extracts of *E. coli*. One of these activities appears to be absent in the excision-defective *uvrA* and *uvrB* mutants.

Crude extracts of *E. coli* contain substantial amounts of endonucleolytic activity against unirradiated RFI DNA. Even higher levels of activity against UV-irradiated DNA can be demonstrated, as shown in Figure 1. The difference between endonucleolytic activity on irradiated and unirradiated DNA is a measure of the endonucleolytic activity specific for UV-damaged DNA. Figure 1 shows that this UV-specific activity is roughly the same in extracts of *hcr*$^+$ (left) and *hcr*$^-$ (right) strains. Indeed, wild-type levels of UV-specific endonucleolytic activity have been observed in crude extracts of the *uvrA, uvrB,* and *uvrC* excision-defective mutants of *E. coli*.

Since excision appears to be no more than a three-step process (Grossman,

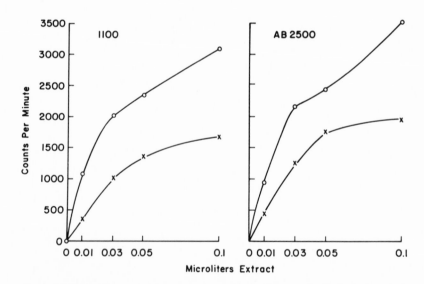

Fig. 1. *Endonuclease assay of crude extracts of a hcr*$^+$ *(left) and a hcr*$^-$ *(right) strain of* E. coli. Extracts were prepared by suspending frozen cells in a phosphate buffer, subjecting the suspension to sonic oscillation, and removing cell debris by centrifugation. Various amount of extract were assayed for endonucleolytic activity against unirradiated RFI DNA (X) and irradiated DNA (o). Reprinted from Braun and Grossman (1974).

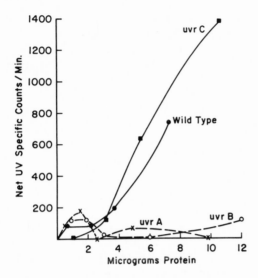

Fig. 2. *Endonuclease assay of material from crude extracts forced through dialysis tubing.* Protein concentration was estimated spectrophotometrically. The curves show the net correndonuclease II activity for various amounts of material. Reprinted from Braun and Grossman (1974).

this volume), it had been expected that at least one of the *uvr* loci would lack the correndonuclease II. A resolution of this apparent paradox is provided by the appearance of at least two separable UV-specific incision activities from extracts of *E. coli* when eluted from phosphocellulose. These two activities, designated as peak I (eluting with about 0.25 M potassium phosphate, pH 7.5) and peak II (0.35 M), are quite similar in biochemical characteristics. However, a feature of peak I which allows it to be distinguished from the other UV-specific activity is the fact that the peak I activity passes through dialysis tubing, while that from peak II will not pass through. By forcing material from crude extracts through dialysis tubing and assaying for UV-specific endonucleolytic activity, it was possible to determine specifically whether peak I activity was present in the extract. Figure 2 shows that peak I activity was observed in material derived from wild-type and *uvrC* extracts, while no activity was detected in extracts derived from the *uvrA* and *uvrB* mutants. Presumably, the peak II activity is present in all excision-defective mutants of *E. coli*. Its function, and the reason for its absence of activity against UV-irradiated DNA *in vivo,* remain a mystery.

An alternate method of distinguishing peaks I and II is based on a binding assay technique similar to that of Riggs et al. (1970). A peak of binding activity to UV-irradiated DNA was observed to elute from phosphocellulose coincidentally with the peak I endonucleolytic activity, while no UV-specific binding activity was associated with peak II. Similar assays of analytical-scale phospho-cellulose chromatograms of crude extracts of the *uvrA, uvrB,* and *uvrC* mutants are shown in Fig. 3. Here a distinct peak of binding activity is seen in the chromatogram of the *uvrC* mutant. No similar peak of activity is observed in the chromatograms of the *uvrA* and *uvrB* mutants.

The experiment depicted in Figure 4 indicates that the peak I activity exposes a 5′-phosphoryl terminus and that no nucleotides are excised. In this experiment, irradiated ^{32}P-labeled *E. coli* DNA has been incised with the peak I enzyme and subsequently treated with various quantities of *E. coli* polynucleo-tide ligase. After treatment, the number of exposed ^{32}P termini is measured by the standard BAP assay (Kaplan et al., 1969). Incision by the peak I UV endonuclease substantially increases the number of exposed phosphoryl termini in UV-irradiated DNA, while little increase is observed when unirradiated DNA is treated with the enzyme. Figure 4 shows that the incision is sealed when irradiated DNA is treated with *E. coli* polynucleotide ligase. *E. coli* poly-nucleotide ligase seals only at juxtaposed termini ending with 5′-phosphoryl and 3′-hydroxyl groups (Becker et al., 1967) and does not act when a gap occurs between the termini (Gefter, et al., 1967). Hence, we conclude that the incision

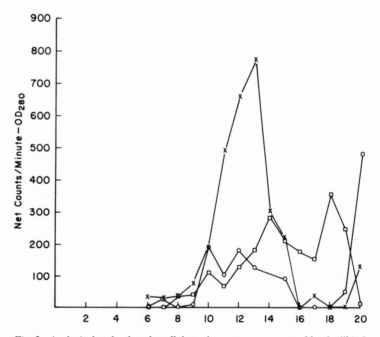

Fig. 3. Analytical-scale phosphocellulose chromatograms assayed by the "bind-ing assay." ×, derived from *uvrC* strain; ○, derived from *uvrA* strain; □, derived from *uvrB* strain. Technical details in Braun and Grossman (1974).

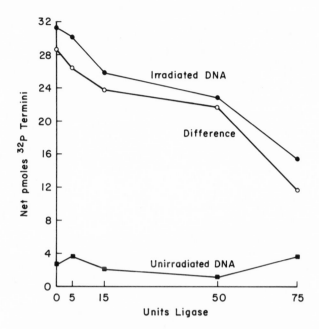

Fig. 4. *Sealing of the correndonuclease II incision by polynucleo-tide ligase.* Irradiated and unirradiated *E. coli* DNA labeled with ^{32}P was treated with the DNA-cellulose fraction of the peak I correndonuclease II, and the reaction was terminated by heating to 68°C. Various amounts of the *E. coli* polynucleotide ligase were added, and the mixture was incubated for a further 30 min at 37°C. Terminal phosphates were assayed by the standard BAP assay (Kaplan et al., 1969). •, Terminal phosphates after treatment of irradiated DNA; ■, unirradiated DNA; ○, difference between irradiated and unirradiated DNA.

(a) is at a phosphodiester bond and (b) leaves a 5'-phosphoryl terminus, and (c) that no nucleotides are excised by the enzyme.

The data presented in Table 1 indicate that the incision produced by the peak I correndonuclease II is adjacent to the pyrimidine dimer. In these experiments, ^{3}H-labeled *E. coli* DNA was irradiated, incised, and then treated with the exonuclease VII of *E. coli* (kindly provided by Dr. J. Chase). Table 1 shows that pyrimidine dimers were excised preferentially from the irradiated DNA after this sequence of steps. Since exonuclease VII (correndonuclease) acts in both directions on single-stranded DNA (Chase and Richardson, this volume), these results indicate that the incision is in the vicinity of a pyrimidine dimer and that after incision the dimer-containing region has a single-stranded character.

Other experiments, not shown here, indicate that the incision is 5' to the pyrimidine dimer. In these experiments, incised, denatured, alkaline-phosphatase-treated DNA was found to be refractory to digestion by calf spleen phosphodiesterase. Since this enzyme acts exonucleolytically in the 5'-to-3' direction and is inhibited by photoproducts, we have concluded that the incision is 5' to the photoproduct.

Figure 5 shows a Lineweaver-Burk plot of the "nicking" activity of the enzyme in the presence and in the absence of caffeine. The plot is consistent with a hypothesis that caffeine is a competitive inhibitor of the UV endonuclease with a K_I of about 10mM. Caffeine at this concentration has been shown to inhibit excision of pyrimidine dimers *in vivo* (Shimada and Takagi, 1967; Sideropoulous and Shankel, 1968). Figure 5 also indicates that the K_m of the enzyme is about 1.5×10^{-8} M (pyrimidine dimers).

ACKNOWLEDGMENTS

The authors wish to express appreciation to Dr. Jack Chase for his gift of *E. coli* exonuclease VII, to Marguerite Cahoon for the pyrimidine dimer assays, and to Dr. Inga Mahler for general encouragement.

This is Publication No. 947 of the Graduate Department of Biochemistry, Brandeis University. The work reported in this paper was supported by the following grants: NP–8D of the American Cancer Society; AT(11–1)3232–2 of

Table 1. Dimer Excision from UV-Irradiated *E. coli* DNA Treated with Exonuclease VII[a]

	% net acid-soluble counts	% dimers (acid insoluble)	Fraction dimers excised
Experiment 1			
Irradiated DNA	0.8	3.04	–
+ Correndonuclease II	3.1	3.03	–
+ Exonuclease VII	2.4	3.31	–
+ Correndo II + Exo VII	5.9	2.49	25%
Experiment 2			
Irradiated DNA	0.2	2.68	–
+ Correndonuclease II	1.6	2.47	–
+ Exonuclease VII (3)	1.0	3.17	–
+ Correndo II + Exo VII (2)	4.8	2.08	34%

[a]Irradiated [^{3}H]DNA (0.4 µg) was treated at 37°C with 1 unit of DNA-cellulose fraction of peak I UV endonuclease for 90 min, heated to 68°C for 10 min, and then treated with exonuclease VII (20 units) for 60 min at 37°C. Pyrimidine dimer assays were carried out by paper chromatography, as discussed by Hamilton (1974). For unknown reasons, a higher number of counts were observed in the dimer region after exonuclease VII treatment. Several duplicate samples in experiment 2 were averaged. The numbers in parentheses indicate the numbers of samples averaged.

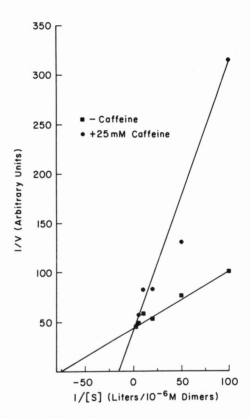

Fig. 5. Lineweaver-Burk plot of correndonuclease II activity as measured by the BAP assay (Kaplan et al., 1969). Since the DNA-cellulose fraction of the peak I enzyme is not homogeneous, it was not possible to express velocity in terms of the molar enzyme concentration. Dimer concentration was estimated in separate experiments by the number of terminal phosphates exposed by the enzyme when the irradiated DNA substrate was totally incised. •, Enzyme alone; ■, in presence of 25 mM caffeine.

the Atomic Energy Commission; GM 15881–14 of the National Institutes of Health and GB 29172X of the National Science Foundation. One of the authors (A. B.) gratefully acknowledges the support of a Damon Runyon Memorial Fund Cancer Research postdoctoral fellowship.

REFERENCES

Becker, A., Lyn, G., Gefter, M. and Hurwitz, J. (1967). *Proc. Nat. Acad. Sci. U.S.A.* **58**, 1996–2003.
Braun, A. and Grossman, L. (1974). *Proc. Nat. Acad. Sci. U.S.A.* **71**, 1838–1842.
Center, M. S. and Richardson, C. C. (1970). *J. Biol. Chem.* **245**, 6285–6291.
Friedberg, E. C. and King, J. J. (1969). *Biochem. Biophys. Res. Commun.* **37**, 646–651.
Gefter, M. L., Becker, A. and Hurwitz, J. (1967). *Proc. Nat. Acad. Sci. U.S.A.* **58**, 240–247.
Hamilton, L., Mahler, I. and Grossman, L. (1974). *Biochemistry* **13**, 1886–1896.
Kaplan, J. C., Kushner, S. R. and Grossman, L. (1969). *Proc. Nat. Acad. Sci. U.S.A.* **63**, 144–151.
Riggs, A. D., Suzuki, H. and Bourgeois, S. (1970). *J. Mol. Biol.* **48**, 67–83.

Shimada, K. and Takagi, Y. (1967). *Biochim. Biophys. Acta* **145**, 763–770.

Sideropoulous, A. S. and Shankel, D. M. (1968). *J. Bacteriol.* **96**, 198–204.

Takagi, Y., Sekiguchi, M., Okubo, H., Nakayama, H., Shimada, K., Yasuda, S., Michimoto, T. and Yoshihara, H. (1969). *Cold Spring Harbor Symp. Quant. Biol.* **33**, 219–227.

Yasuda, S. and Sekiguchi, M. (1970). *Proc. Nat. Acad. Sci. U.S.A.* **67**, 1839–1845.

23

Endonuclease II of *Escherichia coli*

David A. Goldthwait, Dollie Kirtikar, Sheik M. Hadi,[1]
and Errol C. Friedberg[2]

Biochemistry Department
Case Western Reserve University
Cleveland, Ohio 44106

ABSTRACT

Endonuclease II has been purified from *Escherichia coli* and characterized. The enzyme degrades both alkylated and depurinated DNA, as well as γ-irradiated DNA. It also has been shown to attack DNA containing O^6-methylguanine following treatment of the DNA with the carcinogen methylnitrosurea.

Enzyme fractions from *Bacillus subtilis* and *Micrococcus lysodeikticus* which would attack DNA alkylated with methyl methanesulfonate (MMS) were reported by Strauss and Robbins (1968). Because of an interest in genetic recombination and the possibility that there was an enzyme which recognized an area where double-stranded DNA was converted to single-stranded DNA, we proceeded to search for an enzyme which would recognize distortions in DNA (Friedberg and Goldthwait, 1968). The easiest means of creating distortions chemically was with MMS, and an initial 330-fold purification of Endonuclease II of *E. coli* was accomplished (Friedberg and Goldthwait, 1969) using as an assay alkylated DNA entrapped in a polyacrylamide gel (Melgar and Goldthwait, 1968).

---

[1] Present address: Department of Chemistry, Aligarh Muslim University, Aligarh, U.P., India.
[2] Present address: Department of Pathology, Stanford University Medical Center, Stanford, California 94305.

Endonuclease II, with an S value of 3.6, was shown to degrade alkylated DNA. The enzyme had no divalent metal requirement. It was found to be active on the alkylated DNA substrate in the presence of various chelating agents such as 8-hydroxyquinoline and 1,10-phenanthroline. The activity of the enzyme was stimulated to twice the normal level by Mg^{2+} at an optimum concentration of 1.3×10^{-2} M. EDTA did inhibit the enzyme activity 71% at 10^{-4} M and 90% at 10^{-3} M, while tRNA had no inhibitory effect. The enzyme was also sensitive to p-chloromercurisulfonate. Nonrandom enzyme breaks in alkylated DNA were suggested by the observation that one double-strand break was made for every four single-strand breaks. At that time it was also observed that the enzyme made a limited number of single-strand breaks in native T4 and T7 DNA (Friedberg et al., 1969).

Degradation of partially depurinated DNA by the enzyme was observed with a 1600-fold purified enzyme (Hadi and Goldthwait, 1971). DNA, either free or entrapped in polyacrylamide gel, was heated at pH 3.5 for 6 min at temperatures varying from 30 to 80°C in order to produce substrate. The DNA heated at 60°C or above was totally converted to single-stranded DNA (as determined by its density in CsCl) and yet was still a substrate for the enzyme. Whether single-stranded depurinated DNA is the substrate or whether microareas of double-stranded DNA are the substrate is not known. Depurinated DNA is unstable and undergoes an alkali-catalyzed β-elimination reaction which results in phosphodiester bond cleavage. It was found that this alkali-catalyzed degradation could be prevented completely by reaction of the partially depurinated DNA with $NaBH_4$ to convert the aldehyde of the deoxyribose to an alcohol. With depurinated reduced DNA, it was then possible to observe that again the enzyme produced about one double-strand break per four single-strand breaks, suggesting a nonrandom depurination. Evidence was provided to suggest that the enzyme was capable of hydrolyzing the phosphodiester bond at every depurinated site. Stabilization of depurinated DNA in alkali was brought about not only by reaction with $NaBH_4$ but also by reaction with hydroxylamine (NH_2OH), which produces an oxime derivative at the C-1 position (Hadi and Goldthwait, 1971). The activity of a similar enzyme from E. coli on depurinated DNA was observed by Paquette et al. (1972).

Evidence for the limited degradation of native T7 DNA (2.4 breaks per single strand at 37°C) by the 1600-fold purified enzyme has been provided (Hadi et al., 1973). The amount of enzyme required to make single-strand breaks in native DNA was approximately 1000 times as great as that required to degrade depurinated reduced DNA. However, if the temperature was increased from 37 to 45°C, much less enzyme was required and more single-strand breaks were made. This suggested that the enzyme recognized (A + T)-rich areas in the DNA which could undergo increased "breathing" at the higher temperature. The activity on native DNA did not require divalent metal, was stimulated by Mg^{2+}, was inhibited by EDTA, and was not affected by tRNA. For these reasons, the activity on native DNA was considered to be due to the same enzyme which degraded alkylated DNA. Poly d(AT), either untreated or treated with alkylating agent, was degraded by a fraction of the 1600-fold purified enzyme isolated by isoelectric focusing and thus freed of a minimal exonuclease contamination. Poly(dG·dC), alone or reacted with half sulfur mustard, was not a substrate for

Table 1. The Percent of the Total Counts Present on a
Thin-Layer Chromatogram Which Were in
Different Bases as a Function of Time[a]

		Time (min)				
	Enzyme	0	15	30	45	60
Origin	–	85.6	81.0	75.5	70	68.2
	+	86.0	75.1	70.0	61	53.6
I	–	0.30	0.28	0.95	0.98	1.25
	+	0.14	0.40	1.08	1.22	1.92
7-MeG	–	7.03	8.90	9.45	14.00	15.9
	+	7.16	8.20	9.00	13.95	13.70
l-MeA	–	0.09	0.45	0.85	0.98	0.68
	+	0.23	0.62	1.38	1.83	2.37
7-MeA	–	1.70	2.36	3.50	3.16	3.28
	+	1.80	3.78	4.00	4.68	4.65
3-MeA	–	2.80	3.56	4.87	6.00	5.16
	+	2.88	7.20	7.46	10.60	13.40
II	–	–	0.59	0.59	0.65	0.66
	+	–	0.65	1.08	–	2.02
O^6-MeG	–	1.48	2.10	2.38	3.23	3.05
	+	1.58	2.90	4.40	4.94	5.04

[a]The incubation mixtures of 0.1 ml contained 20 nmoles of [^{3}H]MNU-T4 DNA (sp. act. 2500 ^{3}H cpm per nmole nucleotide), 1×10^{-4} M β-mercaptoethanol, 1×10^{-4} M 8-hydroxyquinoline, 5×10^{-2} M tris-HCl, pH 8.0, and 0.2 unit of enzyme where indicated. The reactions were terminated with EDTA after incubation for 0, 15, 30, 45, or 60 min at 37°C. The contents were concentrated 3- to 4-fold under vacuum at 20°C. Aliquots of 20 μl containing approximately 5000 cpm plus alkylated bases were subjected to thin-layer chromatography using 2-methyl-l-propanol–0.8 M boric acid–14.8 M NH_4OH (100:14:0.4 by volume) and Eastman chromatogram sheets (No. 6065). I was a UV spot with R_{f_A} 0.2–0.3. II was a UV spot with R_{f_A} 0.98. These have not been identified.

the enzyme. Single-stranded T4 DNA denatured by alkali was degraded to pieces of approximately 25S, but heat-denatured T4 DNA which already was approximately 25S was not degraded further by the enzyme. These results suggested that degradation occurred at a limited number of sites in a single strand, which could form hydrogen bonds with other sites in the strand.

Analysis with AT^{32}P and polynucleotide kinase of the action of endonuclease II on native DNA (Hadi et al., 1973) indicated that the enzyme produced 5'-phosphates and 3'-hydroxyls. However, when depurinated reduced DNA was treated with endonuclease II and then tested with polynucleotide kinase without and with alkaline phosphatase, no significant increase in ^{32}P incorporation could be detected above the control value. After the demonstration that this substrate was not an inhibitor of the kinase, it was concluded that the kinase could not react with the endonuclease II product of the depurinated reduced DNA, because the kinase had no specificity for a depurinated sugar residue. This would place the hydrolysis of the phosphodiester bond by endonuclease II on the 5' side of the depurinated reduced residue, a situation comparable to that observed with the UV-specific endonuclease (Kushner et al., 1971).

Since Endonuclease II of *E. coli* had the specificity to degrade both alkylated DNA and depurinated DNA, it was hypothesized that depurination was an intermediate step in the enzymatic degradation of alkylated DNA. To test this hypothesis, unlabeled DNA was treated with ^{3}H-methyl-labeled dimethylsulfate, the labeled DNA was incubated with the enzyme, and methyl-labeled free bases liberated by endonuclease II were then sought by both column and paper chromatography. In these experiments, evidence for the enzymatic release of 3-methyladenine, but not of 7-methylguanine was obtained (Kirtikar and Goldthwait, 1974). When DNA was treated with ^{3}H-methyl-labeled methylnitrosourea, then incubated with and without enzyme, and the reaction mixtures were examined by thin-layer chromatography, evidence was obtained for the enzymatic release of O^6-methylguanine, 1-, 3-, and 7-methyladenine, but not 7-methylguanine (Table 1). Data on the stoichiometry of enzymatic release of the bases O^6-methylguanine and 3-methyladenine and their disappearance from the DNA have also been presented (Kirtikar and Goldthwait, 1974). On the assumption that O^6-methylguanine is the mutagenic and carcinogenic base (Loveless, 1969; Gerchman and Ludlum, 1973), this represents the first purified enzyme known to recognize a specific chemical carcinogen on a DNA molecule. The *in vitro* results obtained with endonuclease II parallel in base specificity the results obtained *in vivo* with *E. coli* (Lawley and Orr, 1970).

Finally, endonuclease II recognizes some form of damage in DNA caused by γ-irradiation. With increasing doses of irradiation up to 22.5 krads, there is an increase in enzyme-sensitive sites with a correspondingly smaller increase in single-strand breaks. Above this dose, the single-strand breaks increase more rapidly than the number of enzyme-sensitive sites. The enzyme-sensitive sites

Table 2. The Effect of Preincubation at 37°C for 4 hr and of NaBH₄ on the Nonenzymatic and Enzymatic Single-Strand Breaks in Irradiated DNA[a]

Sample	Preincubation (4 hr, 37°C)	NaBH₄ before preincubation	NaBH₄ before incubation	Single-strand breaks		
				No enzyme	Plus enzyme	Enzyme-induced
a	–	–	–	4	21	17
b	–	–	+	2	22	20
c	+	–	–	18	110	92
d	+	+	–	3	88	85

[a] [³H]DNA in 0.1 M KPO₄–0.005 M tris-HCl, pH 7.0, was irradiated at 0°C under N₂ with 22 krads. The pH of samples (a) and (c) was adjusted to 8.5. To samples (b) and (d) several additions of NaBH₄, to a final concentration of 0.25 M were made during a 1-hr period. The pH was then adjusted to 8.5. Samples (a) and (b) were held at 0°C while samples (c) and (d) were incubated at 37°C for 4 hr. All samples were then incubated with and without enzyme for 60 min. NaOH was added to a concentration of 0.1 M and the incubation at 37°C was continued for 10 min prior to layering of aliquots on sucrose gradients for centrifugation and analysis.

produced by the irradiation are not depurinated or depyrimidinated sites, because after the incubation with and without enzyme, the reaction mixture is incubated in 0.1 N NaOH at 37°C to break all phosphodiester bonds at such sites. It was noted that on storage of irradiated DNA at 4°C for up to 30 days there was an increase in both single-strand breaks and enzyme-sensitive sites. Preincubation of irradiated DNA at 37°C for 4 hr produced the same phenomenon (Table 2). Preincubation in O_2 resulted in a marked increase in single-strand breaks compared with a comparable preincubation in nitrogen, and $NaBH_4$ treatment of the DNA after irradiation and prior to the preincubation prevented the appearance of most of the non-enzyme-induced single-strand breaks but did not affect significantly the appearance of enzyme-sensitive sites. The nature of these effects is under investigation. Similar enzyme-sensitive sites produced by γ irradiation and recognized by an extract of *M. luteus* have been reported (Setlow and Carrier, 1973).

ACKNOWLEDGMENTS

This work was supported by grants from the National Institutes of Health (CA-11322), The Health Fund of Greater Cleveland, and the Cuyahoga Unit of the American Cancer Society. D. A. G. was the recipient of a National Institutes of Health Research Career Award Fellowship K6-6M-21444.

REFERENCES

Friedberg, E. C. and Goldthwait D. A. (1968). *Cold Spring Harbor Symp. Quant. Biol.* 33, 271–275.
Friedberg, E. C. and Goldthwait, D. A. (1969). *Proc. Nat. Acad. Sci. U.S.A.* 62, 934–940.
Friedberg, E. C., Hadi, S. M. and Goldthwait, D. A. (1969). *J. Biol. Chem.* 244, 5879–5889.
Gerchman, L. L. and Ludlum, D. B. (1973). *Biochim. Biophys. Acta,* 308, 310–316.
Hadi, S. M. and Goldthwait, D. A. (1971). *Biochemistry* 10, 4986–4994.
Hadi, S. M., Kirtikar, D. and Goldthwait, D. A. (1973). *Biochemistry* 12, 2747–2754.
Kirtikar, D. and Goldthwait, D. A. (1974). *Proc. Nat. Acad. Sci. U.S.A.* 71, 2022–2026.
Kushner, S. R., Kaplan, J. C., Ono, H. and Grossman, L. (1971). *Biochemistry* 10, 3325–3334.
Lawley, P. D. and Orr, D. J. (1970). *Chem.-Biol. Interact.* 2, 154–157.
Loveless, A. (1969). *Nature,* 223, 206–207.
Melgar, E. and Goldthwait, D. A. (1968). *J. Biol. Chem.* 243, 4401–4408.
Paquette, Y., Crine, P. and Verly, W. G. (1972). *Can. J. Biochem.* 50, 1199–1209.
Setlow, R. B. and Carrier, W. L. (1973). *Nature New Biol.* 241, 170–173.
Strauss, B. and Robbins, M. (1968). *Biochim. Biophys. Acta* 161, 68–75.

24

Endonuclease III: An Endonuclease from *Escherichia coli* That Introduces Single Polynucleotide Chain Scissions in Ultraviolet-Irradiated DNA[1]

Miroslav Radman[2]

The Biological Laboratories
Harvard University
Cambridge, Massachusetts 02138

ABSTRACT

An endonuclease that makes single polynucleotide chain scissions in UV-irradiated DNA has been purified from *Escherichia coli*. The activity has the following properties: (1) unirradiated DNA is attacked very little if at all; (2) single-stranded DNA is not attacked, whether irradiated or not; (3) there is no requirement for divalent cations, and the activity is not affected by the addition of EDTA; (4) the pH optimum is approximately 7; (5) the activity is inhibited by 1 M NaCl, single-stranded DNA, transfer RNA, and unirradiated double-stranded DNA; (6) the sedimentation coefficient, $S_{20, w}$, is approximately 2.6; (7) it is a basic protein. The enzyme is tentatively named *E. coli* endonuclease III. The physiological function of the endonuclease has not yet been established.

This study represents an initial attempt to isolate and characterize enzymes involved in correction of mismatched base pairs in DNA. Such local irregularities

---

[1] A full report will be published elsewhere (Radman, 1975).
[2] Present address: Laboratoire de Biophysique et Radiobiologie, Université Libre de Bruxelles, Rue des Chevaux 67, B 1640 Rhode St. Genèse, Belgium

in the double-stranded structure of DNA can arise either as a result of the reaction of mutagenic chemicals or radiation with DNA or as a consequence of the formation of hybird DNA during genetic recombination (for review see Radding, 1973). Mismatches, i.e. noncomplementary base pairs, will occur during recombination whenever the hybird region includes portions of the parental molecules containing different genetic markers. In such instances, genetic markers which are single-base-pair substitutions will produce single-base-pair mismatches, whereas insertion or deletion will produce extensive regions of mispairing, i.e. loops. Genetic evidence exists for the correction in *Escherichia coli* of both single-base-pair mismatches (Baas and Jansz, 1972; Wildenberg and Meselson, 1975; White and Fox, 1974) and loops (Berger and Benz, this volume).

Using labeled heteroduplex λ DNA with three single-base-pair mismatches as substrate, we have been unable to detect in *E. coli* extracts any major endonucleolytic activity specific for heteroduplex DNA. In an alternative approach, we have used UV-irradiated DNA in our assay system in order to purify endonucleases specific for damaged DNA and to test their activity on mismatched DNA.

Two different assays for measuring the extent of single-strand breakage in the DNA were used: (1) velocity sedimentation analysis of λ DNA in alkaline sucrose gradients and (2) measurement of the proportion of intact covalently closed circular DNA molecules (RFI) of phage ϕX174 by the reversibility of denaturation of this particular DNA. A typical reaction mixture (0.25 ml) contained (per ml) 50 μM Tris·HCl—Tris base buffer (pH 7.1), 5 μM EDTA, 5 μM 2-mercaptoethanol, 100 μg bovine serum albumin or gelatin, the enzyme sample to be assayed, and either approximately 10^{11} labeled λ DNA molecules or about 0.1 μg labeled ϕX174 RFI DNA. The purification scheme includes DEAE-cellulose and phosphocellulose column chromatography, Sephadex gel filtration, velocity sedimentation in a glycerol gradient, and preparative isoelectric focusing in polyacrylamide gels. All methodological details were described elsewhere (Radman, 1975).

Of at least two, and possibly three, different "UV endonucleases," one, which we tentatively call *endonuclease III*, has been considerably purified through the above-mentioned steps. The principal property of the enzyme, the exclusive single-strand breakage of UV-irradiated DNA only, is illustrated in the experiment shown in Fig. 1.

Some properties of the endonuclease III have been studied and are summarized in the abstract above. The full paper (Radman, 1975) discusses endonuclease III in relation to all other known *E. coli* endonucleases, and we suggest that endonuclease III is different from endonuclease I, endonuclease II, restriction endonuclease R-K, *recBC* nuclease, and *E. coli* UV endonuclease. This latter enzyme has been studied by Braun and Grossman (1974, and this volume) and shown to be missing in the *uvrA* and *uvrB* mutants, which are deficient in excision-repair of pyrimidine dimers.

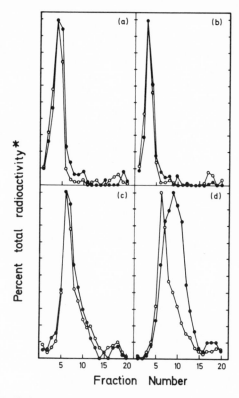

Fig. 1. Neutral and alkaline sedimentation analysis of UV-irradiated and nonirradiated λ *DNA treated with endonuclease III.* Two standard reaction mixtures (0.4 ml each) contained approximately 2 μg each of UV-irradiated ^{32}P-labeled and nonirradiated ^{3}H-labeled λ DNA. One mixture was incubated without enzyme and to the second mixture approximately 1 μg protein from phosphocellulose fraction was added. Incubation was at 37°C for 60 min. After addition of EDTA, the two reaction mixtures were divided into two equal parts for sedimentation analysis at neutral and alkaline pH. Neutral exponential gradients of 6–25% sucrose contained 0.01 M tris buffer, pH 8.0, 0.001 M EDTA, 0.9 M NaCl, and 1% Duponol. Alkaline exponential gradients of 6–25% sucrose contained 0.005 M EDTA and 0.3 M NaOH. Centrifugation was performed in an IEC, SB405 rotor at 55,000 rpm, at 20°C, for 60 min in the case of neutral gradients, and for 130 min in the case of alkaline gradients. Fractions were collected from the bottom onto filter papers. Radioactivity of dried papers was determined by liquid scintillation counting in Omnifluor-toluene scintillation fluid. Gradients were routinely plotted using a computer program in which peaks were normalized to the same level, and the curves were plotted as percent total radioactivity of each isotope in one gradient. Peak values are *(a)* ^{32}P = 27.2%, ^{3}H = 30.3%, *(b)* ^{32}P = 33.5%, ^{3}H = 30.7%; *(c)* ^{32}P = 21.9%, ^{3}H = 23.1%; *(d)* ^{32}P = 14.7%, ^{3}H = 23.2%. Total radioactivity per gradient was approximately 1000 cpm of each isotope. *(a)* Neutral gradient, no enzyme added; *(b)* neutral gradient, with enzyme; *(c)* alkaline gradient, no enzyme added; *(d)* alkaline gradient, with enzyme. o——o, ^{3}H-labeled, nonirradiated λ DNA; ●——●, ^{32}P-labeled, UV-irradiated λ DNA.

The investigation of the types of distortion in the DNA structure which are recognized by endonuclease III and the screening of mutants deficient in its activity should furnish more information about the biological role of endonuclease III.

ACKNOWLEDGMENTS

I thank Matt Meselson for the advice and encouragement and hospitality in his laboratory. I have enjoyed excellent assistance of S. Bott, J. Landers, and R. Burger. This work was done while I was holding an EMBO postdoctoral fellowship and a fellowship from the Jane Coffin Childs Memorial Fund for Medical Research with a grant-in-aid. This work was supported by an NSF grant to M. Meselson.

REFERENCES

Baas, P. D. and Jansz, H. S. (1972). *J. Mol. Biol.* **63**, 557–568.
Braun, A. and Grossman, L. (1974). *Proc. Nat. Acad. Sci. U.S.A.* **71**, 1838–1842.
Radding, C. M. (1973). *Ann. Rev. Genet.* **1**, 87–111.
Radman, M. (1975). *J. Biol. Chem.* (in press).
White, R. and Fox, M. S. (1974). *Proc. Nat. Acad. Sci. U.S.A.* **71**, 1544–1548.
Wildenberg, J. and Meselson, M. (1975). *Proc. Nat. Acad. Sci. U.S.A.* (in press).

25

An *Escherichia coli* Endonuclease
Which Acts on X-Irradiated DNA

Gary F. Strniste and Susan S. Wallace

Department of Biological Sciences
H. H. Lehman College
City University of New York
Bronx, New York 10468

ABSTRACT

An endonuclease activity which acts on X-irradiated DNA has been found in extracts of *Escherichia coli*. For every single-strand break induced directly in the DNA by X rays, there exists approximately one additional X-ray damage sensitive to this endonuclease. Similar activities have been demonstrated in *E. coli* B/r and B_{s-1}, which differ in their repair capabilities.

Ionizing radiation can induce a variety of lesions in the genetic material of a cell, including single- and double-strand breaks as well as damages to the sugar and base structures. The physical reconstitution of radiation-induced single-strand breaks has been shown to occur in many organisms. In *Escherichia coli* this process of repair is mediated by at least two distinct mechanisms: rapid repair, involving DNA polymerase I, and a slower, presumably recombination repair involving the *recA* gene product (Town et al., 1971). Furthermore, it has been reported that, in the radioresistant organism *Micrococcus radiodurans,* radiation-induced double-strand breaks may be repaired (Dean et al., 1966; Kitayama and Matsuyama, 1968). Recently, Hariharan and Cerutti (1972) have described a radiolysis product of thymine which appears in the acid-soluble fraction during postirradiation incubation of *M. radiodurans*. Hariharan and Cerutti (this volume) have presented evidence that extracts of *E. coli* can remove this hydrolysis product of thymine from γ-irradiated poly[d(A-T)]. These findings indicate the presence of an excision-type repair mechanism in bacteria which could act on damaged bases. Additional evidence to support this hypothesis has been provided by Setlow and colleagues, who have shown that extracts of *Micrococcus luteus* contain endonuclease activity which reacts with DNA γ-irradiated *in vivo* and *in vitro* (Paterson and Setlow, 1972; Setlow and Carrier, 1973). In this report we show that extracts of *E. coli* contain endonuclease activity which is specific for X-irradiated DNA.

Samples of ^{3}H-labeled ϕX174 RF (replicative form) DNA were X-irradiated for a series of doses and then incubated with or without the addition of *E. coli* B/r extract. Since RFI DNA is converted into the slower-sedimenting RFII DNA by the addition of one or more single-strand breaks into either strand of the DNA molecule (Burton and Sinsheimer, 1965), quantification of single-strand breaks induced directly by X-rays or indirectly by the action of the endonuclease can be readily achieved by monitoring with sucrose gradient sedimentation the loss of RFI material and subsequent appearance of RFII material. This conversion of RFI to RFII as a function of X-ray dose and subsequent extract treatment is shown in Fig. 1. The D_{37} for X-ray-induced single-strand breaks (dose required to convert 63% RFI to RFII) under our experimental conditions is 11 krads. Furthermore, it is observed that incubation of the irradiated RF DNA prior to centrifugation with the extract results in a substantial increase in the conversion of RFI to RFII (D_{37} equals 6.0±0.5 krads). It can be calculated from our data that for every single-strand break induced directly by the X-rays, 0.7–1.0 site is induced which is sensitive to the endonuclease activity in the B/r extract. Similar activities were found in extracts isolated from the radiosensitive strain B_{s-1} (data not shown). It should be noted that treatment of unirradiated RF DNA with extract containing twice the protein amount noted in Fig. 1 did not result in any transfer of RFI to RFII.

The enzymatic conversion of X-irradiated RFI to RFII has been effectively competed by extraneous X-irradiated calf thymus DNA but not by calf thymus DNA irradiated with ultraviolet light at fluences giving comparable numbers of lesions (Strniste and Wallace, 1975). Only at very high fluences of ultraviolet light can competition be observed. These studies as well as those by Paterson and Setlow (1972) and Wilkins (1973a,b) with the *M. luteus* enzyme, indicate that bacteria may possess an excision-repair mechanism for ionizing-radiation damages distinct, at least with respect to the first step, from UV excision-repair. Endonucleases which are active on depurinated sites (Verly and Paquette, 1972) and on alkylated sites (Friedburg and Goldthwait, 1968) in DNA have also been isolated. It has been reported that the latter enzyme has activity on X-irradiated calf thymus DNA (Goldthwait et al., this volume).

Recently we have examined extracts of *E. coli* AB3027 (kindly supplied by P. Howard-Flanders) for endonuclease activity which reacts with X-irradiated RF DNA. This strain, which lacks endonuclease II, has the same X-ray-specific

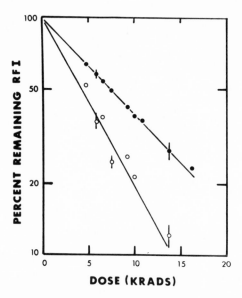

Fig. 1. Conversion of φX174 RFI to RFII as a function of X-ray dose and incubation with E. coli B/r extract. Solutions of RF DNA (6 μg/ml) in 10 mM tris buffer (pH 8.0) and 1 mM EDTA were irradiated with X-rays for a series of doses. Reaction mixtures (0.100 ml) containing 0.05 M phosphate buffer (pH 6.5), 0.5 mM EDTA, 2.5 mM 2-mercaptoethanol, 5 μg calf thymus DNA, 0.150 μg irradiated ³H-labeled RF DNA (1.7 × 10⁴ cpm/μg) were incubated with or without the addition of 10.2 μg (protein) extract at 37°C for 10 min. The reactions were terminated by the addition of 0.033 ml 4 M KCl and rapid freezing. Thawed reaction samples were layered onto preformed 6.5–20.0% sucrose gradients containing 1 M KCl, 10mM tris buffer (pH 7.5), and 1mM EDTA, and the gradients were centrifuged in a Beckman SW50L rotor at 50,000 rpm at 4°C for 3 1/3 hr. Gradients were then fractionated and assayed for ³H cpm in the RFI and RFII peak positions. •, Experiments in which the irradiated RF DNA was incubated in the absence of extract; ○, experiments in which the reactions included the addition of the extract. Details of the isolation of the extract are described elsewhere (Strniste and Wallace, 1975).

endonuclease activity as the parent strain. Thus it appears that the endonuclease activity measured in these studies is not endonuclease II.

ACKNOWLEDGMENTS

This work was supported by a PHS Grant from the National Cancer Institute and a CUNY Research Grant.

REFERENCES

Burton, A. and Sinsheimer, R. L. (1965). J. Mol. Biol. 14, 327–347.
Dean, C. J., Feldschreiber, P. and Lett, J. T. (1966). Nature 209, 49–52.

Friedberg, E. C. and Goldthwait, D. A. (1968). *Cold Spring Harbor Symp. Quant. Biol.* **33**, 271–275.

Hariharan, P. V. and Cerutti, P. A. (1972). *J. Mol. Biol.* **66**, 65–81.

Kitayama, S. and Matsuyama, A. (1968). *Biochem. Biophys. Res. Commun.* **33**, 418–422.

Paterson, M. C. and Setlow, R. B. (1972). *Proc. Nat. Acad. Sci. U.S.A.* **69**, 2927–2931.

Setlow, R. B. and Carrier, W. L. (1973). *Nature New Biol.* **241**, 170–172.

Strniste, G. F. and Wallace, S. S. (1975). *Proc. Nat. Acad. Sci. U.S.A.* (in press).

Town, C. D., Smith, K. C. and Kaplan, H. S. (1971). *Science* **172**, 851–853.

Verly, W. G. and Paquette, Y. (1972). *Can. J. Biochem.* **50**, 217–224.

Wilkins, R. J. (1973*a*). *Nature New Biol.* **244**, 269–271.

Wilkins, R. J. (1973*b*). *Biochim. Biophys. Acta* **312**, 33–37.

26

Substrate Specificity of *Micrococcus luteus* UV Endonuclease and Its Overlap with DNA Photolyase Activity

M. H. Patrick

Institute for Molecular Biology
The University of Texas at Dallas
Dallas, Texas

ABSTRACT

The action of an endonuclease from *Micrococcus luteus* that operates on UV damage in DNA overlaps with that of DNA photolyase from yeast: homo- and heterocyclobutane dipyrimidines in DNA are substrates for both enzymes, but pyrimidine adducts or the "spore photoproduct" in DNA are not. As expected from this overlap, the action of the two enzymes is mutually interfering: single-strand nicks introduced by the endonuclease effectively preclude photoreactivation; conversely, formation of a photolyase-cyclobutane dipyrimidine complex can prevent nicking by the endonuclease.

Preparations of UV-endonuclease activity from *Micrococcus luteus* have been shown to act on both Thy- and Cyt-containing Pyr<>Pyr but not on Pyr adduct or the "spore product," Thy(α-5)H$_2$Thy, and thus has the same photoproduct substrate specificity as DNA photolyase. These conclusions are based on the following evidence (Patrick and Harm, 1973, and references cited therein):

    1. There is no demonstrable UV-endonuclease activity toward irradiated native T7 DNA (254 nm, 25°C, 0.1 M sodium phosphate buffer, pH 7) which has been maximally photoreactivated by incubation with DNA

photolyase from yeast and exposed to blacklight. For the fluences used in these experiments ($\sim$ 100 Jm^{-2}), there are roughly 60 Pyr<>Pyr per molecule in the ratio of 2 Thy<>Thy:1 Cyt<>Pyr (Unrau et al., 1973) and approximately 20 Pyr adducts (M. H. Patrick, manuscript in preparation). Under the *in vitro* photoreactivation conditions used, more than 90% of all Pyr<>Pyr are reversed *in situ*, leaving only nonphotoreactivable Pyr adducts (Patrick, 1970).

2. Irradiated poly(dI·dC) (750 J/m^2, 280 nm, 25°C) is a substrate for the *M. luteus* UV endonuclease, as assayed by Sepharose chromatography. Chromatograms of acid-hydrolyzed samples showed Cyt<>Cyt as the only photoproduct formed.

3. The approximate yields of photoproducts in *Escherichia coli* DNA irradiated at −76°C in 15% ethylene glycol (4000 J/m^2, 254 nm) are Pyr<>Pyr/Pyr $\cong$ 0.015, Pyr adduct/Pyr $\cong$ 0.015, and Thy(α-5)hThy/Thy $\cong$ 0.03. The corresponding yields at 25°C in 0.1 M phosphate buffer (pH 7) are Pyr $\cong$ 0.06, Pyr adduct/Pyr $\cong$ 0.01, and no Thy(α-5)hThy/Thy. Thus, at −76°C, the major product is Thy(α-5)H_2Thy, and the absolute yield of Pyr adduct is increased; after maximum photoreactivation, the only products remaining are Pyr adducts and Thy(α-5)H_2Thy. When the extent of strand nicking by the *M. luteus* UV endonuclease is compared for T7 DNA (assayed by alkaline sucrose gradient sedimentation) or *Haemophilus influenzae* DNA (assayed by bacterial transformation) irradiated at 25°C or −76°C, we find that the extent of UV-endonuclease activity on frozen irradiated DNA is reduced roughly in relation to the decrease in Pyr<>Pyr yield, and that after maximum photoreactivation there is no observable endonuclease activity in DNA irradiated at either temperature.

The lack of UV-endonuclease activity toward Pyr adducts is interesting in view of other recently completed experiments, in which I could find no evidence for a lethal role of Pyr adducts (as assayed by host-cell reactivation of irradiated phage, infectious phage, or bacterial transforming DNA) (M. H. Patrick, manuscript in preparation). On the other hand, Varghese and Day (1970) observed Pyr adduct in fragments excised from the DNA of UV-irradiated *Micrococcus radiodurans*. If Pyr adducts are generally nonlethal products and not recognized by the incision step of an excision-resynthesis repair scheme, these results suggest that clustering of photoproducts in DNA may not be limited to Pyr<>Pyr.

If *H. influenzae* transforming DNA is exposed to relatively low fluences ($\sim$100 J/m^2, 254 nm, 25°C, 0.1 M sodium phosphate, pH 7) and incubated with the *M. luteus* UV endonuclease, the ability of this DNA to be enzymatically photoreactivated decreases with increasing incubation time. In one similar experiment using [³H] Thy-labeled *E. coli* DNA, we found that while ≥90% of Pyr<>Pyr usually are reversed photoenzymatically, incubation with UV endo-nuclease rendered only ∿30% of the dimers susceptible to DNA photolyase (Patrick and Harm, 1973). Conversely, stoichiometric binding of DNA photo-lyase to Pyr<>Pyr in DNA completely prevents UV-endonuclease action.

This inhibition of photoenzymatic repair is probably due to the proximity of the nick to the dimer, since the same number of non-dimer-specific nicks (e.g., induced by pancreatic DNase or X rays) does not alter the photoreactivation ability. Thus, in addition to the restriction of nucleotide chain length imposed on the function of the DNA photolyase, there is the additional restriction of the position of the dimer on a polynucleotide strand.

By competitive inhibition studies and the use of the flash technique for studying the dark-dependent events in photoenzymatic repair (Harm, this vol-ume), we have shown that the

$$E + S \underset{k_2}{\overset{k_1}{\rightleftharpoons}} ES$$

step in the complex of DNA and photolyase-nicked, irradiated DNA is the same as for nonnicked DNA. The light-dependent step, however, fails to occur. These results are not necessarily at odds with the observation by Kushner et al. (1971), who find that dimers in DNA treated with UV endonuclease can be eliminated by prolonged photoreactivation treatment, since our results do not exclude a low rate (e.g., ≤10% of normal) for the photolytic step.

Our findings are also consistent with the observation that with increasing time allowed for dark-repair processes *in vivo*, a progressive loss in photoreactiv-ability occurs in cells proficient in excision-resynthesis dark repair but not in UV-sensitive cells lacking that proficiency (Harm, 1968). Thus, if irradiated cells are held in buffer before undergoing photoreactivation, the photoreactivated sector decreases to zero after 24 hr, and, concomitantly with the increase in cell survival elicited by the liquid holding period, incision proceeds without a significant amount of excision (Patrick and Harm, 1971). This suggests that the incision step of the ERR process and photoreactivation are mutually exclusive events *in vivo* as well as *in vitro*.

Finally, we have been able to estimate rate constants associated with binding and dissociation of the *M. luteus* UV endonuclease with UV-irradiated DNA. We set as a minimum estimate for $k_1 \sim 10^7 \ \ell \ mol^{-1}sec^{-1}$, and for $k_2 \sim 10^1$ to 10^2 sec^{-1}. This order of magnitude for the dissociation rate constant suggests a rather unstable enzyme-substrate complex; i.e., that dissociation of the complex is more probable than completion of phosphodiester bond hydrolysis. This sort of property is not unreasonable for the specificity required of an enzyme which must seek out a few damaged sites in the presence of thousands of normal base pairs.

REFERENCES

Harm, W. (1968). *Mutat. Res.* **6**, 25.

Kushner, S. R., Kaplan, J. C., Ono, H. and Grossman, L. (1971). *Biochemistry* **10**, 3325.

Patrick, M. H. (1970). *Photochem. Photobiol.* **11**, 477.

Patrick, M. H. and Harm, W. (1971). *Biophys. Soc. Abstr.* **11**, 194.

Patrick, M. H. and Harm, H. (1973). *Photochem. Photobiol.* **18**, 371.

Varghese, A. J. and Day, R. S. (1970). *Photochem. Photobiol.* **11**, 511.

Unrau, P., Wheatcroft, R., Cox, B. and Olive, T. (1973). *Biochim. Biophys. Acta* **312**, 626.

27

Two Temperature-Sensitive *polA* Mutants: An Approach to the Role *in Vivo* of DNA Polymerase I

Dennis G. Uyemura, E. B. Konrad, and I. R. Lehman

Department of Biochemistry
Stanford University School of Medicine
Stanford, California 94305

ABSTRACT

DNA polymerase I (Pol I) purified from a strain carrying *polA12* demonstrates a defect primarily in the polymerase portion of the molecule. The mutant enzyme is altered in its structure and template specificity. Pol I isolated from strains carrying *polAex1* are primarily defective in $5' \rightarrow 3'$ exonuclease. The *polAex1* lesion renders strains conditionally lethal.

As an approach to defining the role of DNA polymerase I (Pol I) in the repair and replication of DNA *in vivo*, we have examined Pol I isolated from two *polA* mutants of *Escherichia coli* K12. One, *polA12*, has a major defect in the polymerase portion of the molecule (Lehman and Chein, 1973); and the other, *polAex1*, is defective in its $5' \rightarrow 3'$ exonuclease (Konrad and Lehman, 1974).

 The *E. coli* K12 *polA12* mutant was isolated by Monk and Kinross (1972) by screening for sensitivity to the alkylating reagent methyl methanesulfonate (MMS) at 42°C but not at 30°C. Although polymerase activity was reported to be undetectable in crude extracts at either restrictive or permissive temperature (Monk and Kinross, 1972), we find that maintenance of salt concentrations at or above 0.1 M allows detection of the activity and hence its purification. Indeed, we have purified the activity 1700-fold to about 90% homogeneity. The mutant enzyme comigrates with wild-type Pol I in polyacrylamide gels containing sodium dodecyl sulfate or urea. However, the mutant enzyme migrates with a 4% lower relative mobility than wild-type enzyme in native discontinuous gels. In addition, mutant Pol I has a lower sedimentation coefficient than wild type in 5–20% sucrose density gradients. Since *polA12* enzymatic activity comigrates with the protein in both the native gels and sucrose gradients, altered mobility could not have been due to a general denaturation of the mutant protein. The data suggest that one consequence of the *polA12* lesion is a structural alteration in the mutant protein.

Such a structural perturbation might be expected to at least partially destabilize the mutant activity. Indeed, *polA12* polymerase activity is abnormally thermosensitive when compared to wild-type Pol I. Furthermore, the *polA12* enzyme, in the absence of salt, is abnormally thermolabile to preincubation at $30°C$. In the presence of 0.2 M NH_4Cl, the mutant enzyme is as stable to preincubation at $30°C$ as wild type. Thus, the effect of NH_4Cl is to enhance the stability of a native conformation of the mutant protein; it has no effect on the lowered catalytic rate at $43°C$. Finally, on nicked calf thymus DNA, mutant and wild-type enzymes have similar profiles in terms of pH optima and optimal specific activities. However, when supplied with a $d(pA)_{800} \cdot d(pT)_{10}$ homopolymer pair containing large single-stranded regions, *polA12* enzyme shows a lower pH optimum and a drastically lowered maximal specific activity.

Lehman and Chien (1973) reported *polA12* $5' \rightarrow 3'$ exonuclease activity as having the same ratio of activity at 42 versus $30°C$ as wild type. With highly purified material we find a slightly lower ratio of activity at 43 versus $30°C$ when compared with wild-type enzyme. Although the polymerase activity is affected to a greater extent than the exonuclease activity, the increased thermosensitivity of both activities is consistent with a general structural perturbation of the mutant protein rather than a localized aberration within one of the two active sites.

To complement studies on a mutant Pol I with a major defect in its polymerase activity, we now report the isolation of a mutant containing a major defect in its $5' \rightarrow 3'$ exonuclease activity (Konrad and Lehman, 1974). This mutant, designated *E. coli* K12 *polAex1*, is phenotypically similar to other *polA* mutants in that it is sensitive to MMS or ultraviolet irradiation; it will support the growth of wild-type bacteriophage lambda, but not lambda phage deficient in general recombination; and it accumulates Okazaki fragments. The most striking characteristic of this strain that distinguishes it from other *polA* mutants, including the *polA107* strain which also has a $5' \rightarrow 3'$ exonuclease defect (Glickman et al., 1973), is that *polAex1* strains are conditionally lethal. Only 1% of logarithmically growing cells survive to form colonies on tryptone-yeast extract plates at $43°C$. On tryptone plates or glucose minimal medium plates, where growth rate is lower, the proportion of survivors is increased.

Sucrose density gradients of partially purified material indicate that the *polAex1* strain examined contains levels of polymerase I activity comparable to those of wild-type preparations. In contrast, it has only 3% of wild-type levels of $5' \to 3'$ exonuclease activity when assayed at 30°C. At 43°C this activity drops to about 1% of wild-type levels. Thus, the exonuclease activity is abnormally thermosensitive. Moreover, this exonuclease activity is sensitive to antiserum directed against purified Pol I and is resistant to the reagent *N*-ethylmaleimide. These data support the conclusion that the activity detected is that of the mutant Pol I specified by the *polAex1* lesion.

A revertant of *polAex1* selected for resistance to MMS also acquired the capacity for growth at 43°C, contained a nearly normal level of $5' \to 3'$ exo-nuclease activity, and joined Okazaki fragments normally. In addition, bacterio-phage P1 transduction of *polAex1* into five different wild-type backgrounds showed no segregation of the traits of MMS sensitivity and temperature-sensitive conditional lethality. By these criteria, the phenotype of *E. coli polAex1* is due to a single mutation.

We conclude that the *polAex1* mutation results in a functional lesion distinct from other *polA* mutants, which are deficient in polymerase but not $5' \to 3'$ exonuclease. Its phenotypic similarity to other *polA* mutants suggests that both polymerase I and $5' \to 3'$ exonuclease activities may be required *in vivo* for normal DNA replication and repair. The conditional lethality of *polAex1* strains is novel and establishes that Pol I is essential for viability in *E. coli*. As with the *polA12* mutant enzyme, we are undertaking the purification of the *polAex1* mutant enzyme. With these, as well as other purified mutant preparations, we hope to gain further insight into each of the activities associated with Pol I, how they are coordinated, and a definition of their function *in vivo*.

REFERENCES

Glickman, B. W., van Sluis, C. A., Heijneker, H. L. and Rörsch, A. (1973). *Mol. Gen. Genet.* **124**, 69–82.
Konrad, E. B. and Lehman, I. R. (1974). *Proc. Nat. Acad. Sci. U.S.A.* **71**, 2048–2051.
Lehman, I. R. and Chien, J. R. (1973). *J. Biol. Chem.* **248**, 7717–7723.
Monk, M. and Kinross, J. (1972). *J. Bacteriol.* **109**, 971–978.

28

The Role of DNA Polymerase I in Excision-Repair

Barry W. Glickman

Laboratory for Molecular Genetics
Leiden State University
Wassenaarseweg 64
Leiden, The Netherlands

ABSTRACT

The ability of three different DNA polymerase I mutants of *Escherichia coli* to carry out excision-repair was examined. Strains having the same genetic origin but carrying either the *polA1, polA107, resA1,* or *pol*⁺ alleles were compared. The rate of ultraviolet-induced dimer excision was slightly reduced, relative to that found in Pol⁺ strains, in the PolA1 strain; greatly reduced in the PolA107 strain; and found not to occur in the ResA1 strain. Ultraviolet-light-induced repair synthesis as determined by the ultraviolet-stimulated incorporation of ^{3}H-labeled 5-bromo-2′-deoxyuridine into DNA of the parental density showed that the *polA1* mutation resulted in an increase in repair replication, while the presence of the *polA107* allele caused a reduction in the amount of repair synthesis relative to that of the Pol⁺ strain. The ResA1 strain, however, showed no ultraviolet stimulation of the incorporation of the density label. These observations indicate that DNA polymerase I plays a key role in the excision-repair process in *E. coli.*

In order to clarify the relationship between DNA polymerase I and the repair of damage caused by ultraviolet (UV) irradiation, three DNA polymerase I mutations (*polA*) of *Escherichia coli* K12 were examined for their effects upon excision-repair. The mutations studied were *polA1*, isolated by de Lucia and

Cairns (1969); *polA107*, isolated in our laboratory (Glickman et al., 1973); and *resA1*, a *polA* mutation originally isolated in *E. coli* B by Kato and Kondo (1967).

Extracts from both the PolA1 and ResA1 strains are deficient in the polymerizing activity of polymerase I but retain the $5' \rightarrow 3'$ exonucleolytic activity normally associated with polymerase I (Lehman and Chien, 1973). Extracts of the PolA107 strain, however, lack the $5' \rightarrow 3'$ exonucleolytic activity normally associated with polymerase I, while retaining the polymerizing activity of polI (Heijneker et al., 1973). Some properties of these strains are summarized in Table 1 (Glickman et al., 1973; Glickman, 1974). The present study concentrates on the effect of the *polA* mutations on the removal of pyrimidine dimers and the resynthesis of the excised region (repair synthesis) following UV irradiation.

The relative rates of pyrimidine dimer removal in three PolA strains following a UV dose of 1000 ergs/mm^2 are presented in Fig. 1. In agreement with the results of Boyle et al. (1970), it was found that the *polA1* mutation only slightly reduced the rate of dimer excision. The *polA107* mutation, however, results in a much reduced rate of dimer excision, while the *resA1* mutation, as previously reported by Kato (1972), results in the loss of the cells' excision capability, and during the postirradiation incubation period the relative amount of acid-precipitable radioactivity present in the dimer fraction actually increases. These results suggest that polymerase I plays an important role in the removal of pyrimidine dimers following UV irradiation. The influence of the *polA* mutations upon the resynthesis step of excision-repair was determined by examining the UV-stimulated incorporation of ^{3}H-labeled 5-bromo-2'dexoxyuridine (BrdUrd) into DNA retaining the normal parental density, as described by Hanawalt and Cooper (1971). The fractions from the first CsCl equilibrium density gradient which

Table 1. Some Properties of the PolA Strains

	PolA1	PolA107	ResA1
PolI polymerizing activity	−	+	−
PolI $5' \rightarrow 3'$ exonucleolytic activity	+	−	+
UV sensitivity	High	Moderate	High
DNA degradation after UV	Extreme	Moderate	Extreme
Host cell reactivation (λuv)	Intermediate	Intermediate	Intermediate
Host cell reactivation (λMMS)	Deficient	Deficient	Deficient
Replication of λred	Defective	Defective	Normal
Viable with *recA, B,* or *C*	No	No	No
Viable with *uvrE*[a]	No	Yes	No
Joining of Okazaki fragments[b]	Slow	Slow	Slow

[a]Mattern (personal communication).
[b]Glickman and Heijneker (unpublished results).

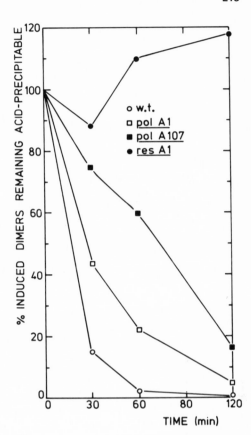

Fig. 1. The rate of pyrimidine dimer excision in three otherwise isogenic PolA strains of E. coli following a UV dose of 1000 ergs/mm². The results are the average of two experiments. The cells were grown and labeled as described by Glickman (1974), and dimer determination was carried out as described by Carrier and Setlow (1971). ○, KMBL 1788 (Pol⁺); □, KMBL 1787 (PolA1); ■, KMBL 1789 (PolA107); ●, KMBL 1791 (ResA1).

Table 2. Results of the Repair Replication Experiments, Including Data on DNA Degradiation[a]

Strain	Dose (ergs/mm²)	Ratio(^{3}H/ ^{32}P)	Relative increase	DNA breakdown(%)
KMBL 1788	0	0.101	–	–
(Pol⁺)	250	0.268	2.7	3
KMBL 1787	0	0.101	–	–
(PolA1)	250	0.245	2.5	14
KMBL 1789	0	0.100	–	–
(PolA107)	250	0.195	1.95	4
KMBL 1791	0	0.102	–	–
(ResA1)	250	0.106	1.03	13

[a]The relative increase was obtained by dividing the ^{3}H/ ^{32}P ratio after UV irradiation by the unirradiated control values.

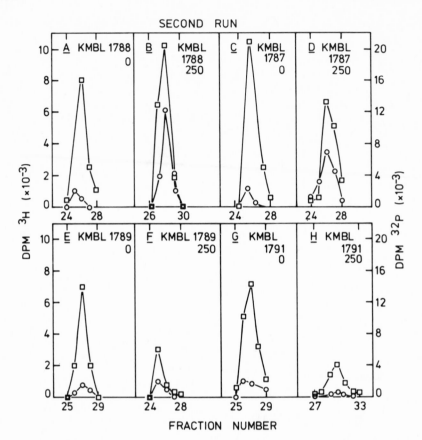

Fig. 2. Repair replication measured for 10 min in the presence of [³H] BrdUrd in unirradiated cultures (A, C, E, and G) and following a UV dose of 250 ergs/mm² (B, D, F, and H). Centrifugation was carried out at 37,000 rpm in a Beckman 50 Ti rotor for 44 hr at 25°C. Shown here are the results of the second run, in which the fractions containing DNA of the normal density were pooled and recentrifuged. Strains: KMBL 1787 (PolA1), KMBL 1788 (Pol⁺), KMBL 1789 (PolA107), and KMBL 1791 (ResA1). ○, [³H] BrdUrd label; □, ³²P-labeled DNA.

contained DNA of the parental density were pooled and rebanded in a second CsCl equilibrium density gradient (Fig. 2). Since DNA degradation would result in the dilution of the BrdUrd pool, the amount of DNA breakdown occurring in these strains was determined. These results, along with those of repair replication, are summarized in Table 2.

As shown by our results, as well as those of Cooper and Hanawalt (1972), when DNA breakdown is taken into consideration, the presence of the *polA* allele results in an increased amount of repair synthesis following UV irradiation. We believe the extensive repair synthesis to be the result of excessive 5'→3' exonucleolytic activity in the absence of the polymerizing activity of the enzyme, creating single-stranded regions which are then resynthesized by another polymerase. This would account for the rapid excision of dimers and the extensive degradation found to occur in the PolA1 strain following UV irradiation.

In the PolA107 strain, dimers are excised at a greatly reduced rate and little DNA degradation is observed. This is accompanied by reduced levels of repair replication. We believe that, in the PolA107 strain, dimers are removed by the displacement of the strand containing the dimer, followed by nicking by means of a nuclease acting on single-stranded DNA. Displacement synthesis by polymerase I has been described by Masume and Richardson (1971) and has been shown to occur extensively (Heijneker et al., 1973) with the enzyme isolated from the PolA107 strain.

These experiments show that DNA polymerase I is intimately involved in the excision-repair processes following UV irradiation. It also appears that this process is most efficient when the $5' \rightarrow 3'$ exonucleolytic and polymerizing activities of polymerase I are present in one molecule, suggesting that the cell normally relies on a "patch and cut" rather than a "cut and patch" mechanism (Setlow, 1966) for the removal of pyrimidine dimers. The lack of dimer excision in the ResA strain indicates that, in *E. coli,* DNA polymerase I is solely responsible for dimer removal by the excision-repair pathway. On the other hand, there is evidence that DNA polymerase III, the enzyme involved in DNA replication (Gefter et al., 1971), can play a role in dark repair, since this enzyme has been shown capable of gap filling, leading to strand rejoining in a PolA1 strain following UV irradiation (Youngs and Smith, 1973; Tait et al., 1974).

It is, however, not clear whether these results are dependent upon the particular properties associated with the *polA1* mutation present in the strain used in these studies. Similar studies using strains containing the *resA1* mutation are in progress.

REFERENCES

Boyle, J. M., Paterson, M. C. and Setlow, R. B. (1970). *Nature* **226**, 708–710.
Boyce, R. P. and Howard-Flanders, P. (1964). *Proc. Nat. Acad. Sci. U.S.A.* **51**, 293–300.
Carrier, W. L. and Setlow, R. B. (1971). *Methods Enzymol.* **210**, 230–237.
Cooper, P. K. and Hanawalt, P. C. (1972). *Proc. Nat. Acad. Sci. U.S.A.* **69**, 1156–1160.
de Lucia, P. and Cairns, J. (1969). *Nature* **224**, 1164–1166.
Gefter, M. L., Hirota, Y., Kornberg, T., Weschler, J. A. and Barnoux, C. (1971). *Proc. Nat. Acad. Sci. U.S.A.* **68**, 3150–3153.
Glickman, B. W. (1974). *Biochim. Biophys. Acta* **335**, 115–122.
Glickman, B. W., van Sluis, C. A., Heijneker, H. L. and Rörsch, A. (1973). *Mol. Gen. Genet.* **124**, 69–82.
Hanawalt, P. C. and Cooper, P. K. (1971). *Methods Enzymol.* **21D**, 221–230.

Heijneker, H. L., Tjeerde, R. H., Glickman, B. W., van Dorp, B., Ellens, D. J. and Pouwels, P. H. (1973). *Mol. Gen. Genet.* **124**, 83–96.
Kato, T. (1972). *J. Bacteriol.* **112**, 1234–1246.
Kato, T. and Kondo, S. (1967). *Mutat. Res.* **4**, 253–263.
Lehman, I. R. and Chien, J. (1973). *J. Biol. Chem.* **248**, 7717–7723.
Masume, Y. and Richardson, C. C. (1971). *J. Biol. Chem.* **246**, 2692–2701.
Setlow, R. B. (1966). *Radiat. Res.* **6** (Suppl.), 220–228.
Tait, R. C., Harris, A. L. and Smith, D. W. (1974). *Proc. Nat. Acad. Sci. U.S.A.* **71**, 675–679.
Youngs, D. A. and Smith, K. C. (1973). *Nature New Biol.* **244**, 240–241.

29

Involvement of *Escherichia coli* DNA Polymerase-I-Associated 5′→3′ Exonuclease in Excision-Repair of UV-Damaged DNA

Herbert L. Heyneker

Laboratory of Molecular Genetics
Leiden State University
Wassenaarseweg 64
Leiden, The Netherlands

and

Hans Klenow

Biokemiske Institut B
University of Copenhagen
Juliane Mariesvej 30
Copenhagen Ø, Denmark DK-2100

ABSTRACT

From comparative studies between *Escherichia coli* PolA107 cells (lacking 5′→3′ exonucleolytic activity associated with DNA polymerase I) and the isogenic wild-type strain, and between the purified DNA polymerase I preparations isolated from these strains, it can be concluded that the 5′→3′ exonuclease is involved in excision of pyrimidine dimers in *E. coli*. Evidence is presented that the *polA107* mutation is located on that part of the DNA polymerase I gene coding for the small fragment on which 5′→3′ exonucleolytic activity is found.

The discovery by de Lucia and Cairns (1969) of a mutant of *Escherichia coli* (*polA1*) lacking DNA polymerase I initiated many studies on the role of this enzyme in dark repair of UV-irradiated DNA both *in vivo* (Gross and Gross,

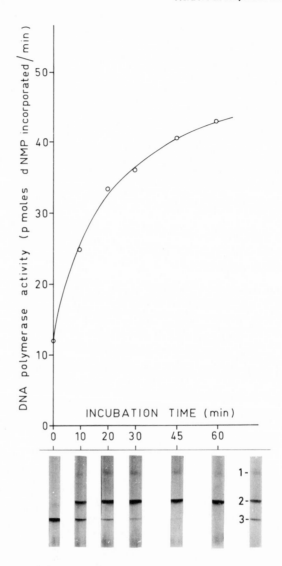

Fig. 1. The effect of subtilisin treatment on the rate of polymerization by DNA polymerase
I. Treatment with subtilisin was carried out at 37°C in 0.13 M potassium phosphate buffer,
pH 6.5, containing 40 μg/ml calf thymus DNA (Sigma, type V), approximately 7 μg/ml
DNA polymerase I (spec. act. > 20,000 U/mg), and 75 ng/ml subtilisin (type Carlsberg). At
times 0, 10, 20, 30, 45, and 60 min, aliquots of 2.5 ml were withdrawn from the reaction
mixture and the reaction was terminated by the addition of 0.3 ml 50% TCA. After 30 min
on ice, the precipitated protein was collected by centrifugation (5 min, 20,000 rpm, 0°C,
Spinco rotor 40) and dissolved in 0.1 ml 0.1 M tris-acetate buffer, pH 7.5, containing 1%
SDS, 0.02 M dithiothreitol, and 25% sucrose. The mixture was incubated for 1 hr at 37°C
and subjected for electrophoresis in polyacrylamide gels in the presence of SDS for 16 hr at

1969; Boyle et al., 1970; Cooper and Hanawalt, 1972; Tait et al., 1974; Glickman, 1974) and *in vitro* (Kelly et al., 1969; Heyneker et al., 1971). DNA polymerase I consists of one polypeptide chain of molecular weight 109,000 (Jovin et al., 1969), which by mild proteolytic treatment (Brutlag et al., 1969; Klenow and Overgaard-Hansen, 1970) can be split into two fragments: a large fragment of molecular weight 76,000 containing DNA polymerizing and $3' \rightarrow 5'$ exonucleolytic activities, and a small fragment of molecular weight 36,000 containing $5' \rightarrow 3'$ exonucleolytic activity.

The elegant experiments performed by Brutlag and Kornberg (1972) and Setlow and Kornberg (1972) demonstrated that the multiple functions of DNA polymerase I give the enzyme the necessary characteristics to catalyze repair reactions on damaged DNA. DNA polymerase I binds at nicks introduced by a UV-specific endonuclease in UV-irradiated DNA (Braun et al., this volume). Polymerization then proceeds in the $5' \rightarrow 3'$ direction, while the $5' \rightarrow 3'$ exonucleolytic activity associated with DNA polymerase I simultaneously releases nucleotides, including pyrimidine dimers, from the primer strand ahead of the nick. This reaction is called "nick translation."

In addition to PolA1, other PolA mutants have been isolated such as the PolA12 mutant with a thermosensitive DNA polymerase I (Monk and Kinross, 1972) and the ResA mutant (Kato and Kondo, 1970). All of these mutants retain the $5' \rightarrow 3'$ exonucleolytic activity unaffected by the mutation (Lehman and Chien, 1973; Lehman, personal communication). Glickman et al. (1973) described the phenotypic characteristics of a PolA mutant with normal levels of DNA polymerizing activity but deficient in the $5' \rightarrow 3'$ exonucleolytic activity (Heyneker et al., 1973). This mutant (PolA107) is moderately UV sensitive and excises pyrimidine dimers from UV-irradiated DNA more slowly than isogenic wild-type or PolA1 strains (Glickman, 1974). It has been shown that DNA polymerase I purified from *E. coli* cells carrying the *polA107* mutation is able to perform "strand-displacement synthesis" instead of "nick translation" (Heyneker et al., 1973). Since the PolA107 mutant is only moderately sensitive to UV irradiation, one possibility is that DNA polymerase I binds to nicks intro-

---

1.5 mA per gel at 22°C as described by Klenow et al. (1971). The gels are shown below their respective sampling times except for the reference gel (right frame), which represents the electrophoretic pattern of a mixture of small fragment (1), large fragment (2), and intact DNA polymerase I (3), purified from wild-type *E. coli* cells. At the indicated times, aliquots of 25 μl were removed from the reaction mixture and mixed with 25 μl 1% albumin (Armour Pharmaceutical Co.) containing 25 μg/ml phenylmethane sulfonylfluoride. DNA polymerase I activity was subsequently measured in 0.05 M potassium phosphate buffer, pH 7.4, supplemented with the four deoxyribonucleoside triphosphates: 17 μM each [TTP was tritium-labeled (NEN), 100 mCi/mmole]; calf thymus DNA activated with pancreatic DNase, 300 μg/ml; 3 mM $MgCl_2$; and 0.3 mM dithiothreitol. Samples were taken at 5, 10, 15, and 20 min and assayed for acid-precipitable radioactivity, and the results were used to calculate the reaction rate shown in the figure.

duced by a UV-specific endonuclease into irradiated DNA, then catalyzes polymerization in the $5' \rightarrow 3'$ direction by displacement of the strand ahead of the nick without the excision of dimers. The displaced strands may be subsequently digested by exonuclease VII, which degrades single-stranded DNA in the $5' \rightarrow 3'$ direction (Chase and Richardson, this volume). Other possibilities are that some residual $5' \rightarrow 3'$ exonucleolytic activity is associated with the mutant DNA polymerase I, or that the displaced strand is nicked by a single-strand-specific endonuclease and further degraded by exonuclease I. Recently, Konrad and Lehman (1974; communicated by Uyemura et al., this volume) isolated another mutant (PolAex1) also lacking $5' \rightarrow 3'$ exonucleolytic activity. Unlike the *polA107* mutation, the *polAex1* mutation is reported to result in temperature-dependent lethality, suggesting that the $5' \rightarrow 3'$ exonuclease has multiple functions in the repair and synthesis of DNA.

To determine whether the *polA107* mutation is due either to a mutation in the large fragment and influencing the structure of the enzyme in such a way that the $5' \rightarrow 3'$ exonucleolytic activity is affected or to a mutation in the small fragment affecting the active site of the $5' \rightarrow 3'$ exonuclease or influencing the tertiary structure of the small fragment, the mutant enzyme was treated with subtilisin. Klenow et al. (1971) have found that the action of subtilisin on DNA polymerase I, resulting in the splitting of the enzyme into a small and a large fragment, can be monitored by the concomitant 3- to 4-fold increase in the reaction rate of DNA polymerizing activity. When PolA107 DNA polymerase I is split with subtilisin, an almost 4-fold stimulation of polymerizing activity is obtained under the same conditions which are optimal for stimulation of wild-type polymerizing activity (Fig. 1). Furthermore, the size of the fragment obtained by splitting the mutant enzyme is indistinguishable from the size of the fragments of wild-type DNA polymerase I (Fig. 1). The $5' \rightarrow 3'$ exonucleolytic activity of the PolA107 polymerase, measured by the formation of acid-soluble material formed from tritium-labeled poly[d(A-T)], however, was less than 5% of the wild-type $5' \rightarrow 3'$ exonuclease both before and after the subtilisin treatment. From these experiments we conclude that the *polA107* mutation is located in the region of the *polA* gene coding for the small fragment of DNA polymerase I and is probably a point mutation.

It might be of interest to investigate which difference in amino acid composition is responsible for the inactivation of the $5' \rightarrow 3'$ exonuclease associated with DNA polymerase I.

REFERENCES

Boyle, J. M., Paterson, M. C. and Setlow, R. B. (1970). *Nature* **226**, 708–710.
Brutlag, D. L. and Kornberg, A. (1972). *J. Biol. Chem.* **247**, 241–248.

Brutlag, D. L., Atkinson, M. R., Setlow, P. and Kornberg, A. (1969). *Biochem. Biophys. Res. Commun.* **37**, 982–989.

Cooper, P. K. and Hanawalt, P. C. (1972). *Proc. Nat. Acad. Sci. U.S.A.* **68**, 2954–2957.

de Lucia, P. and Cairns, J. (1969). *Nature* **224**, 1164–1166.

Glickman, B. W. (1974). *Biochim. Biophys. Acta* **335**, 115–122.

Glickman, B. W., van Sluis, C. A., Heyneker, H. L. and Rörsch, A. (1973). *Mol. Gen. Genet.* **124**, 69–82.

Gross, J. and Gross, M. (1969). *Nature* **224**, 1166–1168.

Heyneker, H. L., Pannekoek, H., Oosterbaan, R. A., Pouwels, P. H., Bron, S., Arwert, F. and Venema, G. (1971). *Proc. Nat. Acad. Sci. U.S.A.* **68**, 2967–2971.

Heyneker, H. L., Ellens, D., Tjeerde, R. H., Glickman, B. W., van Dorp, B. and Pouwels, P. H. (1973). *Mol. Gen. Genet.* **124**, 83–96.

Jovin, T. M., Englund, P. T. and Bertsch, L. L. (1969). *J. Biol. Chem.* **244**, 2996–3008.

Kato, T. and Kondo, S. (1970). *J. Bacteriol.* **104**, 871–881.

Kelly, R. B., Atkinson, M. R., Huberman, J. A. and Kornberg, A. (1969). *Nature* **224**, 495–501.

Klenow, H. and Overgaard-Hansen, K. (1970). *FEBS Lett* **6**, 25–29.

Klenow, H., Overgaard-Hansen, K. and Patkar, S. A. (1971). *Eur. J. Biochem.* **22**, 371–381.

Konrad, E. B. and Lehman, I. R. (1974). *Proc. Nat. Acad. Sci. U.S.A.* **71**, 2048–2051.

Lehman, I. R. and Chien, J. (1973). *J. Biol. Chem.* **248**, 7717–7723.

Monk, M. and Kinross, J. (1972). *J. Bacteriol.* **109**, 971–978.

Setlow, P. and Kornberg, A. (1972). *J. Biol. Chem.* **247**, 232–240.

Tait, R. C., Harris, A. L. and Smith, D. W. (1974). *Proc. Nat. Acad. Sci. U.S.A.* **71**, 675–679.

30

Exonuclease VII of *Escherichia coli*

John W. Chase and Charles C. Richardson

Department of Biological Chemistry
Harvard Medical School
Boston, Massachusetts 02115

ABSTRACT

A new exonuclease of *Escherichia coli* K12, exonuclease VII, has been purified
1700-fold and characterized. The enzyme is specific for single-stranded DNA and
can initiate hydrolysis at both 5' and 3' termini. It is also capable of thymine-
dimer excision *in vitro*. The limit products of the reaction are oligonucleotides,
predominantly in the range of tetramers to dodecamers. DNA is hydrolyzed by
the enzyme in a processive fashion. Mutants of *E. coli* have been isolated having
reduced levels of exonuclease VII activity in crude extracts. Mapping studies
place the exonuclease VII locus between 45 and 56 minutes on the *E. coli* K12
linkage map.

During examination of DNA polymerase I deficient extracts of *Escherichia coli*
for $5' \rightarrow 3'$ exonuclease activities, exonuclease VII was identified and subsequently purified 1700-fold from *E. coli* K12 (Chase and Richardson, 1974*a*).
The enzyme is an exonuclease specific for single-stranded DNA. It degrades
denatured DNA, single-stranded regions extending from the ends of duplex
DNA, or displaced single-stranded regions (Chase and Richardson, 1974*b*).
Hydrolysis is initiated at both 3' and 5' termini. Exonuclease VII can also excise
thymine dimers from duplex DNA following treatment of the DNA with the
Micrococcus luteus endonuclease specific for thymine dimers. The enzyme has
no detectable activity on RNA or DNA-RNA hybrid molecules. The products of

the reaction are oligonucleotides which vary in size with the extent of the reaction; no mononucleotide products have been observed. Although the products of the reaction are large oligonucleotides, the fact that a DNA terminus is required for activity indicates that the mechanism of hydrolysis is exonucleolytic. The enzyme acts in a processive fashion, initially releasing large, acid-insoluble oligonucleotides which can be degraded further to produce a limit product of acid-soluble oligonucleotides.

PURIFICATION AND PROPERTIES

The purification procedure for exonuclease VII has been described (Chase and Richardson, 1974a). The most highly purified fraction of the enzyme is free of contamination by all known *E. coli* exo- and endonucleases (Chase and Richardson, 1974a).

The purified enzyme does not require divalent cations for activity, as shown by the high level of activity in the presence of 8.3 mM EDTA in the standard assay. Only a 25% stimulation of the activity occurs in the presence of 6.7 mM Mg^{2+}. The ability of exonuclease VII to function in the absence of divalent cations is an important property of this nuclease, since it distinguishes it from all of the previously characterized *E. coli* exonucleases and makes possible a specific assay of activity in cell extracts.

When 2-mercaptoethanol is omitted from the standard reaction mixture and enzyme diluent, 83% of the optimal activity remains. In the absence of 2-mercaptoethanol in the enzyme diluent and reaction mixture, the addition of 10 mM *N*-ethylmaleimide results in a 47% inhibition. The enzyme has a sharp pH optimum at 7.8–7.9 in both tris and potassium phosphate buffers.

Exonuclease VII has been shown to be entirely specific for single-stranded DNA (Chase and Richardson, 1974b). The enzyme is more active on sonically irradiated, denatured DNA than on DNA which has only been denatured. In addition, the method used to denature high-molecular-weight DNA appears to influence its substrate activity. Thermally denatured DNA is a more active substrate than DNA denatured gently in alkali. Since the conditions used for thermal denaturation lead to partial breakage of the DNA, this suggests that exonuclease VII is sensitive to the presence of any secondary structures or is stimulated by the higher concentration of termini present in the preparation of broken DNA. This may also explain the higher activity of sonically irradiated DNA.

Several polydeoxyribonucleotides and polyribonucleotides were tested as substrates for exonuclease VII. Only polydeoxyribonucleotides were found to be active substrates. The synthetic polymer $[^3H] d(T)_{140}$, which exists as a random coil, was hydrolyzed at the same rate as sonically irradiated, denatured T7 $[^3H] DNA$. Ribosomal RNA, poly(U), and poly(A) were not active as substrates,

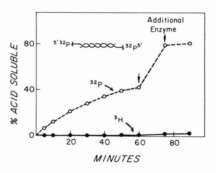

Fig. 1. Activity of exonuclease VII on exonuclease III-treated T7 [³H,5'−³²P] DNA. A standard reaction mixture (1.65 ml) containing 11 nmoles of exonuclease III-treated T7 DNA was prepared and incubated with 0.014 exonuclease III-treated DNA unit of exonuclease VII. At the times indicated, 0.15 ml portions were assayed for acid-soluble radioactivity. Where indicated, 0.085 unit of additional enzyme was added. See Chase and Richardson (1974*b*) for definition of enzyme units. (Reprinted from Chase and Richardson, 1974*b*.)

indicating that exonuclease VII is a specific deoxyribonuclease. Homopolymer pairs (including ribo-deoxyribo hybrids), prepared by mixing together equimolar amounts of homopolymers, did not serve as substrates for the enzyme.

5'-EXONUCLEASE ACTIVITY

When the exonuclease III-treated T7 [³H,5'- ³²P]DNA[1] substrate was treated with exonuclease VII (Fig. 1), 80% of the ³²P was made acid-soluble under conditions where only a few percent of the ³H became acid-soluble. The small amount of ³H released is approximately equal to the amount removed by exonuclease III in the preparation of the substrate, suggesting that only the short single-stranded regions at the 5' termini of the molecule can be hydrolyzed. This indicates that the enzyme can hydrolyze DNA in a 5'→3' direction.

3'-EXONUCLEASE ACTIVITY

To determine whether exonuclease VII can initiate hydrolysis at 3' termini, [³H]d(T)$_{140}$ − [³²P]d(C)$_4$ was prepared and annealed to poly(dA), producing a duplex substrate containing a short displaced 3' single-stranded region [i.e. d(C)$_4$]. The enzyme was able to render 52% of the ³²P acid-soluble under conditions where less than 0.5% of the ³H was made acid-soluble (Fig. 2), indicating that exonuclease VII is able to remove nucleotides from 3' single-

---

[1] Exonuclease III-treated [³H,5'-³²P]DNA was prepared by shearing T7 [³H]DNA by sonic irradiation and labeling the 5' termini of these molecules using [γ-³²P]ATP and polynucleotide kinase. Limited treatment of these molecules with *E. coli* exonuclease III produces a substrate for exonuclease VII containing short 5' single-stranded regions of DNA attached to a duplex region. The detailed preparation of this substrate has been described (Chase and Richardson, 1974*a*).

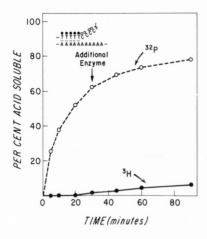

Fig. 2. 3'-Exonuclease activity of exonuclease VII. A standard reaction mixture (1.5 ml) containing 3.6 nmoles of $[^3H]d(T)_{140} - [^{32}P]$ $d(C)_4$ and 7.2 nmoles of poly(dA) was incubated with 0.29 exonuclease III-treated DNA unit of enzyme at 20°C. At the times indicated, 0.15-ml portions were assayed for acid-soluble radioactivity. Where indicated, an additional 0.14 unit of enzyme was added. (Reprinted from Chase and Richardson, 1974b.)

stranded termini. Since these termini contained an average of only four nucleotides, the enzyme can attack very short single-stranded regions. However, as more ^{32}P was made acid-soluble, an increasing amount of ^3H was also made acid-soluble, suggesting that additional ^{32}P was being removed due to "slipping" or "breathing" of the DNA, allowing a large enough piece of the DNA to become single-stranded long enough for the enzyme to act upon it. This conclusion is supported by the fact that, during incubations carried out at 37°C rather than at 20°C, a larger proportion of the ^3H was made acid-soluble. It therefore seems unlikely that exonuclease VII can completely remove a single-stranded region from a piece of DNA. It must be pointed out that the 5' and 3' substrates used in these studies (Figs. 1 and 2) are not identical in structure.

EXONUCLEOLYTIC MECHANISM OF HYDROLYSIS

Although exonuclease VII hydrolyzes single-stranded DNA to yield products which are exclusively oligonucleotides (Chase and Richardson, 1974b), the mechanism of hydrolysis is clearly exonucleolytic, since the enzyme can attack single-stranded linear but not circular DNA molecules. Two preparations of single-stranded ϕX174 DNA were prepared, each labeled with different radioactive isotopes. One of these preparations was nicked with pancreatic DNase to produce linear molecules, and then, after inactivation of the DNase, it was combined with the intact circular molecules and treated with exonuclease VII. The extent of the reaction was monitored by determining the amount of acid-soluble radioactivity at various times during the reaction. The integrity of the DNA molecules in the reaction mixture was monitored by sucrose gradient analysis. The circular DNA molecules remained intact, while the linear molecules were totally degraded, proving that a DNA terminus is required in order for the enzyme to initiate hydrolysis, and therefore that the mechanism of hydrolysis of the enzyme is exonucleolytic (see Fig. 5 in Chase and Richardson, 1974a).

ABSENCE OF ENDONUCLEASE ACTIVITY

In order to determine whether the purified enzyme contained endonuclease activity, studies were carried out with both double-stranded and single-stranded covalently closed DNAs (Chase and Richardson, 1974a). Using sucrose gradient sedimentation analysis in alkali, no conversion of covalently closed double-stranded DNA into the nicked form could be detected. It was calculated that the endonuclease activity on these molecules represented less than 0.02% of the activity on single-stranded DNA.

The level of endonuclease activity on single-stranded DNA was determined as described by Frenkel and Richardson (1971), using ϕX174 DNA as the substrate for the enzyme. It was estimated that the total endonuclease activity on covalently closed single-stranded DNA is 0.005% of the activity on linear single-stranded DNA (Chase and Richardson, 1974a).

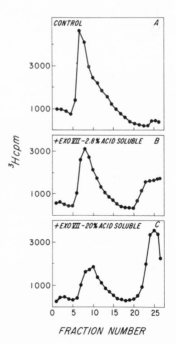

Fig. 3. Processive action of exonuclease VII. Three reaction mixtures (0.15 ml) containing 5.5 nmoles of alkali denatured T7 [³H]DNA were treated with exonuclease VII according to standard conditions.

The first contained no enzyme (*A*); the second contained 0.6 [³H]DNA unit of enzyme (*B*); and the third contained 1.9 units of enzyme (*C*). After 30 min incubation at 37°C, 0.005 ml of 1 N NaOH was added to stop the reactions. The radioactivity made acid-soluble was then determined on a portion of each, and 0.1 ml was applied to an alkaline sucrose gradient and centrifuged 2 3/4 hr at 5 × 10⁴ rpm at 2°C in the Beckman SW50.1 rotor. Ten-drop fractions were collected directly into scintillation vials, and the radioactivity was determined in a Triton-toluene scintillation fluid. (Reprinted from Chase and Richardson, 1974b.)

PROCESSIVE MECHANISM OF HYDROLYSIS

The question of whether exonuclease VII was acting in a processive or a random fashion was investigated in the following manner. Denatured T7 DNA was treated with exonuclease VII until either 2.8% or 20% of the DNA was made acid-soluble, and then it was analyzed by sucrose gradient sedimentation in alkali (Fig. 3). At the point where 2.8% of the DNA was acid-soluble (*B*), 64% of the T7 DNA molecules remained nearly intact; and when 20% was acid-soluble (*C*), 36% of the DNA was still of high molecular weight. The amount of material remaining near the top of the gradients for the enzyme-treated samples (*B, C*) does not agree with the amount made acid-soluble, suggesting that the initial products may be of sufficient size to remain acid-insoluble. This point will be considered in more detail later.

An exonuclease acting randomly should have degraded 36% of all the DNA in *B* and 64% of all the DNA in *C*. A processive exonuclease should have totally degraded 36% of the DNA in *B*, leaving 64% intact, and in *C* 64% of the DNA should have been totally degraded with 36% remaining intact. The data indicate that there has not been extreme degradation of the molecules of high molecular weight in *B* and *C*. The results, therefore, are consistent with processive exonuclease action.

EXCISION OF THYMINE DIMERS

Exonuclease VII, acting in conjunction with the UV endonuclease from *M. luteus,* was able to render ^{3}H from UV-irradiated ϕX174 RF DNA acid-soluble but not that from unirradiated ϕX174 RF DNA (Table 1). Neither of these enzymes, acting alone, was able to release into acid-soluble form significant amounts of ^{3}H from either DNA. We have shown that the production of acid-soluble radioactivity represents excision of thymine dimers by chromatographic analysis of the dimer content of the acid-insoluble DNA. Since the *M. luteus* endonuclease is known to incise adjacent to a thymine dimer (Kushner et al., 1971), this result shows that exonuclease VII can excise thymine dimers, thus acting in a manner similar to that of the *M. luteus* UV exonuclease (Kaplan et al., 1971). (We are grateful to Dr. L. Grossman for collaborating with us in these experiments.)

PRODUCTS OF REACTION

The limit products of exonuclease VII action on single-stranded substrates and substrates with long single-stranded regions extending from duplex DNA are oligonucleotides varying in length from 2 to greater than 25 nucleotides. Most of

Table 1. Excision of Thymine Dimers from ^{3}H-Labeled ϕX174 RF DNA
by Exonuclease VII[a]

Treatment	Unirradiated ϕX174 RF DNA	UV-irradiated ϕX174 RF DNA
	pmoles acid-soluble ^{3}H	
M. luteus endonuclease	0.89	0.93
M. luteus exonuclease	3.2	3.8
M. luteus endonuclease + *M. luteus* exonuclease	4.6	38
Exonuclease VII	1.2	1.8
M. luteus endonuclease + exonuclease VII	1.9	12.5

[a]Reaction mixtures (0.15 ml) contained 33 mM potassium phosphate buffer (pH 7.9), 6.7 mM MgCl$_2$, 10 mM 2-mercaptoethanol, and 3.3 nmoles of ϕX174 RF DNA, or ϕX174 DNA containing about 200 thymine dimers per molecule, produced by ultraviolet irradiation. Incubations were at 37°C for 30 min. We are grateful to Dr. L. Grossman for providing the ϕX174 DNAs and the *M. luteus* UV exo- and endonucleases.

the products are in the range of tetramers to dodecamers. No significant amounts of mononucleotides have been detected from hydrolysis of any substrates tested so far. It has been shown that there is a shift in product size distribution as the reaction progresses (Chase and Richardson, 1974*b*); and, taken together with the alkaline sucrose gradient sedimentation studies (Fig. 3), this implies that the initial products of exonuclease VII hydrolysis are acid-insoluble oligonucleotides which can be further degraded to acid-soluble oligonucleotides. More extensive product characterization, including end-group analysis, has been published (Chase and Richardson, 1974*b*).

MODEL FOR EXONUCLEASE VII HYDROLYSIS

The data that have been presented suggest a model for the action of exonuclease VII (Fig. 4). Processive hydrolysis begins from either terminus of a single-stranded piece of DNA and progresses at approximately equal rates in both directions (Fig. 4.1). The initial products of the reaction are mainly acid-insoluble oligonucleotides (Fig. 4.2). When enzyme molecules become free, many of the initial products can be further degraded (Fig. 4.3). This process may be repeated many times until finally a limit digest contains mostly acid-soluble products (Fig. 4.4).

The size and shape of exonuclease VII are consistent with this model. The molecular weight of the enzyme is 88,000. The Stokes' radius of the enzyme is 89 Å, and the frictional coefficient is 3.07. These physical properties suggest that exonuclease VII is extremely asymmetric, being more accurately represented as a rigid rod or an ellipsoid with a large major semi-axis than as a

1. Hydrolysis initiated at 5' and 3' termini.

2. Initial products are acid-soluble oligonucleotides (a) and acid insoluble oligonucleotides (b).

3. Acid-insoluble oligonucleotides further degraded.

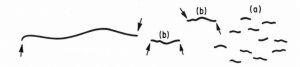

4. *Limit digest contains > 95 % acid-soluble products, $\bar{n}=6$.*

Fig. 4. Model for the hydrolysis of DNA by exonuclease VII.
(Reprinted from Chase and Richardson, 1974b.)

globular protein. If a prolate ellipsoid model is assumed, exonuclease VII could cover a piece of single-stranded DNA more than 100 bases in length (Chase and Richardson, 1974a). If exonuclease VII attaches to the DNA and progresses along its length releasing oligonucleotides, the size of the molecule is consistent with the production of large oligonucleotide products.

The function of exonuclease VII *in vivo* is not yet known, although the properties of the enzyme *in vitro* suggest possible roles in both recombination and the repair of ultraviolet damage in DNA. Models of recombination can be envisioned as requiring the removal of a 5' or 3' single-stranded region from the itermediate recombinant molecules (Signer, 1971). The fact that exonuclease VII can excise thymine dimers *in vitro* makes it a likely candidate for the excision-repair system. Until now the only activity known in *E. coli* capable of excising thymine dimers was the 5'→3' hydrolytic activity associated with DNA polymerase I. These possibilities are even more intriguing when one considers that both the recombination and excision-repair systems are believed to function in the repair of DNA damaged by ultraviolet irradiation (Rupp and Howard-Flanders, 1968; Cooper and Hanawalt, 1972), raising the possibility of an involvement of exonuclease VII in the process of dark repair and recombination.

MUTANT OF EXONUCLEASE VII

Numerous existing E. coli mutants have been assayed for exonuclease VII activity, including most of the available recombination-deficient mutants and mutants deficient in the repair of DNA damage due to ultraviolet (UV) irradiation, as well as other UV-sensitive strains. Exonuclease VII was found to be present in nearly normal levels in all of these strains. It has been necessary, therefore, to search for a mutant lacking exonuclease VII activity by a more direct approach.

Because exonuclease VII can specifically remove 5' single-stranded regions from duplex DNA and function in the absence of divalent cations, the enzyme can be assayed directly in crude extracts. This has made possible the rapid screening of thousands of extracts from a stock mutagenized by treatment with N-methyl-N'-nitro-N-nitrosoguanidine using the techniques outlined by Milcarek and Weiss (1972). Several mutants of independent origin have now been produced. Preliminary mapping studies place the exonuclease VII locus between 45 and 56 minutes on the E. coli K12 linkage map of Taylor and Trotter (1972). More detailed mapping studies and characterization of the exonuclease VII mutation are now in progress. The mutants deficient in exonuclease VII activity that have now been produced should greatly aid in the elucidation of the function(s) of exonuclease VII in vivo.

ACKNOWLEDGMENTS

This work was supported by grants from the United States Public Health Service (AI06045, GM-13,634, GM50564) and from the American Cancer Society, Incorporated (NP-1C and PF-825).

REFERENCES

Chase, J. W. and Richardson, C. C. (1974a). J. Biol. Chem. **249**, 4545.
Chase, J. W. and Richardson, C. C. (1974b). J. Biol. Chem. **249**, 4553.
Cooper, P. K. and Hanawalt, P. C. (1972). J. Mol. Biol. **67**, 1.
Frenkel, G. D. and Richardson, C. C. (1971). J. Biol. Chem. **246**, 4839.
Kaplan, J. C., Kushner, S. R. and Grossman, L. (1971). Biochemistry **10**, 3315.
Kushner, S. R., Kaplan, J. C., Ono, H. and Grossman, L. (1971). Biochemistry **10**, 3325.

Milcarek, C. and Weiss, B. (1972). *J. Mol. Biol.* **68**, 303.

Rupp, W. D. and Howard-Flanders, P. (1968). *J. Mol. Biol.* **31**, 291.

Signer, E. (1971). In *The Bacteriophage Lambda* (Hershey, A. D., ed.), p. 139. Cold Spring Harbor Laboratory, Cold Spring Harbor, New York.

Taylor, A. L. and Trotter, C. D. (1972). *Bacteriol. Rev.* **36**, 504.

31

Enzymatic Repair of UV-Irradiated DNA *in Vitro*

Lester D. Hamilton, Inga Mahler, and Lawrence Grossman

Graduate Department of Biochemistry
Brandeis University
Waltham, Massachusetts 02154

ABSTRACT

Excision repair of UV-damaged *Bacillus subtilis* transforming DNA has been carried out by a sequential enzyme system *in vitro*. Incision adjacent to the pyrimidine dimer in the DNA strand by correndonuclease II-initiated excision of the damage by the $5' \rightarrow 3'$-directed exonuclease of the *Micrococcus luteus* DNA polymerase. Reinsertion of nucleotides into the gap in the strand by the DNA polymerase at $10°C$ terminated in a single-strand break which was sealed by a polynucleotide ligase, thereby repairing the DNA strand. This restored biological activity to damaged DNA up to doses resulting in 60% inactivation of transforming activity. At higher doses, less repair was achieved, due to the development of double-strand breaks during the *in vitro* incision and excision steps.

Excision-repair of cellular DNA after exposure of cells to UV radiation requires an enzyme system which removes photoproducts, reinserts the excised nucleotides, and restores continuity of the DNA strand (Setlow, 1970). Isolation of pure enzymes from bacterial cells with the required specificities for individual steps allows construction of repair systems *in vitro*. Such repair systems make it possible to study the efficiency and specificity of individual enzymes and their functions in DNA repair (Grossman, 1973).

A DNA polymerase isolated from *Micrococcus luteus,* in common with the *Escherichia coli* DNA polymerase I, has an associated $3' \rightarrow 5'$-directed exonuclease active against single-stranded DNA and a $5' \rightarrow 3'$ exonuclease specific for double-stranded DNA. Like the *E. coli* enzyme, the $5' \to 3'$ exonuclease excises pyrimidine dimers from UV-irradiated DNA incised with pancreatic DNase (Kelly et al., 1969) or correndonuclease II (Hamilton et al., 1974). The enzyme could, therefore, function in both the excision and reinsertion steps of DNA repair. This was tested in an *in vitro* system in which the individual steps were characterized as follows:

1. Incision: The formation of a single-strand break with a $3'$-OH terminus $5'$ to the pyrimidine dimer by correndonuclease II (correndo II) isolated from *M. luteus* specific for UV-irradiated DNA.
2. Excision: Removal of the pyrimidine dimers from the "incised" DNA by the $5' \to 3'$ exonuclease of the DNA polymerase at $30°C$, which in the absence of synthesis (no deoxyribonucleoside triphosphates) allowed control of the size of the gap in the DNA strand.
3. Reinsertion: Stoichiometric incorporation of radioactive-labeled nucleotides by the DNA polymerase at $10°C$ in the presence of 0.5 mM Mg^{2+}.
4. Ligation: Closure of the resultant break in the "repaired" strand by polynucleotide ligase from *E. coli* (Modrich and Lehman, 1970), which terminated the reinsertion process and restored the single-stranded molecular weight approaching that of the unirradiated DNA.
1. Biological repair: Restoration of biological activity to UV-irradiated *Bacillus subtilis* transforming DNA, selecting trytophan$^-$-to-tryptophan$^+$ transformants from a UV-sensitive, excision-defective host, B.D. 172, thr_5^-, trp_2^-.

RESULTS

Biochemical repair of UV-irradiated DNA was demonstrated in the following experiments. The DNA was first exposed to different UV fluences, incised with correndonuclease II, and excised at $30°C$ with the DNA polymerase for various times, which resulted in gaps of different lengths in the DNA strand. The excised templates were incubated with the DNA polymerase under synthesis conditions at $10°C$ such that reinsertion could be measured. Incubation was then continued in the presence and absence of polynucleotide ligase and its cofactor NAD^+ for 60 min. The temperature was raised to $30°C$ to estimate those $3'$-OH priming sites not utilized in ligation. Figure 1 shows there was a *stoichiometric* reinsertion of nucleotides into the gaps and that the presence of polynucleotide ligase prevented further incorporation when the temperature was raised to $30°C$, thus confirming that the extent of reinsertion was quantitatively completed.

The second series of experiments was designed to show that the nucleotides reinserted by the DNA polymerase were associated with the species of DNA undergoing molecular-weight restoration. The single-strand molecular weight of $[^3H]$ DNA was measured by alkaline sucrose gradient centrifugation (McGrath

UV DNA ⟶ Incision ⟶ Excision ⟶ Reinsertion ± Pol Ligase

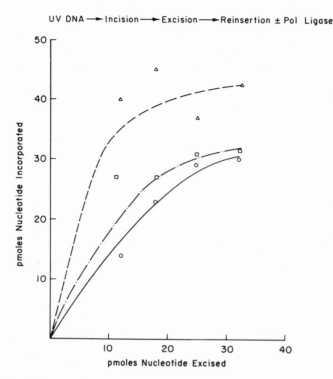

Fig. 1. Stoichiometry of reinsertion of nucleotides into excised DNA and its termination by polynucleotide ligase. Excised, UV-irradiated (6.6 × 10² and 1 × 10³ ergs/mm²) DNAs with various lengths of gaps were prepared as described. Reinsertion of nucleotides was carried out at 10°C in the presence and absence of polynucleotide ligase. Continued incorporation, which occurs on raising the temperature to 10°C and adding fresh DNA polymerase, reflects the presence of 3'-OH priming sites. Control reinsertion at 10°C (○); incorporation on raising the temperature to 30°C after incubation at 10°C with NAD⁺(△) or polynucleotide ligase + NAD⁺(□).

and Williams, 1966). [¹⁴C] TTP included in the DNA polymerase reaction mixture was used to identify the reinserted region. Figure 2 shows that incubation of the excised DNA with only a single ¹⁴C-labeled deoxyribonucleoside triphosphate, TTP, did not influence the molecular weight. Furthermore, there was no label associated with the high-molecular-weight components. Figure 3 confirms that, in the presence of the appropriate deoxyribonucleoside triphosphates, the labeled substrate was incorporated into repaired polymer in which there was a significant restoration of single-stranded molecular weight from 1.7 × 10⁶ to 3.3 × 10⁶, thus approaching the value of 4.6 × 10⁶ which was calculated for native *B. subtilis* DNA by the method of Burgi and Hershey (1963). The coincidence of the incorporated [¹⁴C] TTP with the "repaired" DNA indicated specific gap repair. Table 1 shows that this resulted in complete

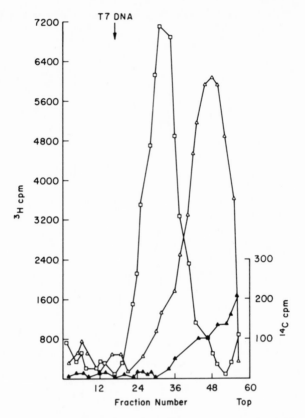

Fig. 2. Alkaline sucrose gradient sedimentation of B. subtilis transforming $[^3H]DNA$ $(0.6 \times 10^5$ cpm/nmole). The DNA was irradiated with 6.6×10^2 ergs/mm^2 and incised with UV endonuclease, 60 min at 37°C in 66 mM tris-HCl, pH 8.0, 50 mM NaCl, and 0.036 unit of correndo II per µg DNA. UV-irradiated DNA + polynucleotide ligase (□); UV-irradiated DNA + correndo II + DNA polymerase + [^{14}C]TTP only, (△); incorporated [^{14}C]TTP (40 mCi/mmole) (▲). T7 [^{14}C]DNA of single-strand molecular weight 1.26×10^7 daltons was used as a standard.

restoration of biological activity of *B. subtilis* transforming DNA exposed to a UV dose of 0.44×10^3 ergs/mm^2.

EFFECT OF HIGH UV DOSES

The dose response in Fig. 4 shows that the enzyme capabilities for complete repair declined dramatically above UV doses which resulted in 50–60% inactivation of transforming activity. A possible explanation for this reduced repair is the effective introduction of double-strand breaks arising from repair at the incision and excision steps. This was substantiated by following the changes in double-stranded molecular weight by neutral sucrose gradient centrifugation. After incision, a dose-dependent decline in the duplex molecular weights of *E. coli* DNA from 2.36×10^7 to 2.04×10^7 at 0.7×10^3 ergs/mm^2 and to 1.44×10^7 at 1.92×10^3 ergs/mm^2 was found. A mechanism by which this degradation could be initiated by incision is shown in Fig. 5. Correndonuclease II-cata-

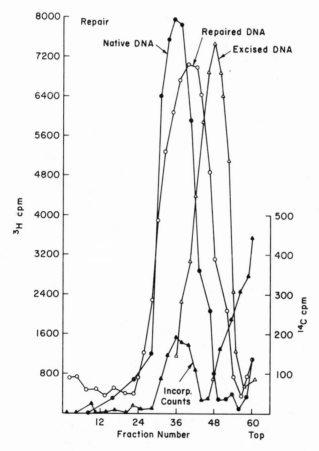

Fig. 3. *Alkaline sucrose gradient sedimentation of* B. subtilis *transforming [^{3}H]DNA irradiated with 6.6 $\times$ 10^2 ergs/mm^2.* Restoration of molecular weight after reinsertion of nucleotides and ligase action. Native DNA (o); incised UV-irradiated DNA + polynucleotide ligase (△); incised UV-irradiated DNA with complete repair conditions (o); incorporated [^{14}C]TTP (40 mCi/mmole) (▲).

Table 1. Transforming Activities of Native and
UV-Irradiated *B. Subtilis* DNA at the Individual Steps
of DNA Repair

Conditions	Native	UV irradiated 4.4 × 10² ergs/mm²
No treatment	100	43
+ ligase		60
+ Correndonuclease II	96	39
+ DNA polymerase	93	43
+ Correndo II + ligase	100	60
+ Correndo II + DNA polymerase		30
+ DNA polymerase + ligase		60
+ Correndo II + DNA polymerase + ligase[a]	96	96

[a]Two-step repair process described in the text.

lyzed single-strand breaks would generate double-strand breaks at low ionic strength, if they resulted in "nicks" in opposite strands less than 15 nucleotide pairs apart (Freifelder and Trumbo, 1969). This could arise *in vitro* either by concentration of the pyrimidine dimers in (A+T)-rich regions of the UV-irradiated DNA (Shafranovskaya et al., 1973; Brunk, 1973) or by the introduction of breaks into the complementary strands close to the dimers due to some other form of damage (Gaudin and Yielding, 1972). During excision, strand antipolarity would result in the $5' \rightarrow 3'$ exonuclease acting to decrease the distances between chain interruptions and eventually lead to double-strand breaks.

The formation of a double-strand break during incision by this mechanism would inevitably yield a DNA molecule with the pyrimidine dimer on a single-strand end, which is refractory to the $5' \rightarrow 3'$ exonuclease of the DNA polymerase (Hamilton et al., 1974). The excision of pyrimidine dimers from incised UV-irradiated DNA *in vitro* was, in fact, found to be incomplete at high UV doses. Figure 6 shows that 40–50% of the pyrimidine dimers were not excised from DNA irradiated with doses above 10^4 ergs/mm². There is, however, the alternative explanation that these dimers represent a fraction in the intact DNA which require for excision special UV exonucleases (correxonucleases), as described by Kaplan and Grossman (1971) and Chase and Richardson (1974).

The decreased repair of UV-irradiated DNA at high doses raises the question of stabilization of DNA intermediates during *in vitro* manipulations and *in vivo* repair (Burrel et al., 1971).

One possibility being explored is the use of polyamines, which are present in both prokaryotic and eukaryotic cells at significant concentration levels. These di-, tri-, and tetraamines can interact specifically with nucleic acids and have

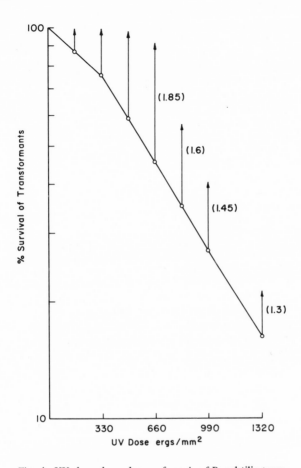

Fig. 4. UV dose dependence of repair of B. subtilis *transforming DNA* in vitro. Transforming activity remaining after UV irradiation (○); extent of repair (△). Enhancement of UV-irradiated activity is shown in parentheses.

been shown to stabilize DNA to heat and shearing damage *in vitro* (Mahler and Mehrotra, 1963; Steven, 1967; Kaiser et al., 1963).

ACKNOWLEDGMENTS

We acknowledge with pleasure the purification of the UV endonuclease used in these studies by Dr. Howard Ono and Dr. Sheikh Riazuddin, also the expert assistance of Mrs. Marguerite Cahoon.

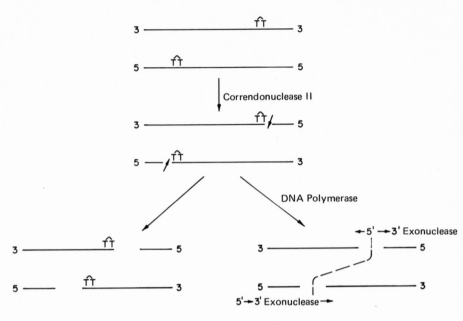

Fig. 5. *A possible mechanism for the formation of double-strand breaks during incision and excision of UV-irradiated DNA* in vitro.

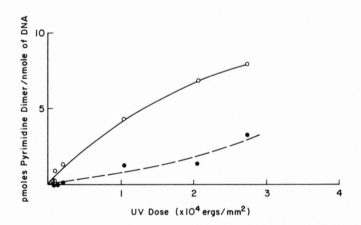

Fig. 6. *The excision of pyrimidine dimers from UV-irradiated DNA as a function of dose.* Total pyrimidine dimers introduced into DNA by UV irradiation (○); pyrimidine dimers remaining after incubation with DNA polymerase 15 min at 30°C under standard assay conditions (●).

REFERENCES

Brunk, C. F. (1973). *Nature New Biol.* **241**, 74.

Burgi, E. and Hershey, A. D. (1963). *Biophys. J.* **3**, 309.

Burrel, A. D., Feldschreiber, P. and Dean, C. J. (1971). *Biochim. Biophys. Acta* **247**, 38.

Chase, J. W. and Richardson, C. C. (1974). *J. Biol. Chem.* **249**, 4553.

Freifelder, D. and Trumbo, B. (1969). *Biopolymers* **7**, 681.

Gaudin, D. and Yielding, K. L. (1972). *Biochem. Biophys. Res. Commun.* **47**, 1396.

Grossman, L. (1973). *Advan. Radiat. Biol.* **4**, 77.

Hamilton, L., Mahler, I. and Grossman, L. (1974). *Biochemistry* **13**, 1886.

Kaiser, D., Tabor, H. and Tabor, C. W. (1963). *J. Mol. Biol.* **6**, 141.

Kaplan, J. C. and Grossman, L. (1971). *Methods Enzymol.* **21**, 249.

Kelly, R. B., Atkinson, M. R., Huberman, J. A. and Kornberg, A. (1969). *Nature* **224**, 495.

Mahler, H. R. and Mehrotra, B. D. (1963). *Biochim. Biophys. Acta* **68**, 211.

McGrath, R. A. and Williams, R. W. (1966). *Nature* **212**, 534.

Modrich, P. and Lehman, I. R. (1970). *J. Biol. Chem.* **245**, 3626.

Setlow, R. B. (1970). *Prog. Nucl. Acid Res. Mol. Biol.* **8**, 257.

Shafranovskaya, N. N., Trifonov, E. N., Lazurkin, Yu. S. and M. D. Frank-Kamenetskii (1973). *Nature New Biol.* **241**, 58.

Steven, L. (1967). *Biochem. J.* **103**, 811.

32

Repair Replication in Permeabilized
Escherichia Coli

Warren E. Masker,[1] Thomas J. Simon, and Philip C. Hanawalt

Department of Biological Sciences
Stanford University
Stanford, California 94305

ABSTRACT

We have examined the modes of DNA synthesis in *Escherichia coli* strains made permeable to nucleoside triphosphates by treatment with toluene. In this quasi *in vitro* system, polymerase-I-deficient mutants exhibit a nonconservative mode of synthesis with properties expected for the resynthesis step of excision-repair. This UV-stimulated DNA synthesis can be performed by either DNA polymerase II or III and it also requires the *uvrA* gene product. It requires the four deoxynucleoside triphosphates; but, in contrast to the semiconservative mode, the ATP requirement can be partially satisfied by other nucleoside triphosphates. The ATP-dependent *recBC* nuclease is not involved. The observed UV-stimulated mode of DNA synthesis may be part of an alternate excision-repair mechanism which supplements or complements DNA-polymerase-I-dependent repair *in vivo*.

According to present models of excision-repair, the removal of a pyrimidine dimer or other structural distortion from a damaged DNA molecule is accompanied by a step in which a DNA polymerase resynthesizes the excised

[1] Present address: Biology Division, Oak Ridge National Laboratory, Oak Ridge, Tennessee 37830.

region using the complementary strand as template. In *Escherichia coli*, DNA polymerase I, with its associated $5' \to 3'$ exonuclease, has ideal properties for both the excision and resynthesis steps of the repair process (Kelly et al., 1969). However, the observation that *polA* mutants, deficient in DNA polymerase I, are less sensitive to ultraviolet radiation than are the excision-deficient *uvrA* mutants suggests that the *polA* strains are able to complete some dark repair (Kanner and Hanawalt, 1970; Monk et al., 1971). Also, it has been shown that repair replication does occur in UV-irradiated *polA* strains and that, in fact, this synthesis is more extensive in *polA* mutants, presumably because in these strains the average length of the repair patches is greater (Cooper and Hanawalt, 1972*a,b*). These observations suggest either that residual DNA polymerase I still present in *polA* mutants (Lehman and Chien, 1973) is sufficient for the repair process or that one (or both) of the other two known *E. coli* DNA polymerases operate in an alternate repair pathway.

To examine DNA synthesis in *polA* mutants which have suffered UV radiation damage we have used a quasi *in vitro* system in which bacteria are made permeable to nucleoside triphosphates by exposure to toluene. In this system intracellular pools of low-molecular-weight precursors and cofactors can be controlled. A number of other workers have shown that DNA synthesis in toluene-treated cells closely mimics the *in vivo* replication process (Moses and Richardson, 1970; Mordoh et al., 1970; Burger, 1971; Matsushita et al., 1971). Our experiments with toluenized *E. coli* have characterized a mode of DNA synthesis which resembles *in vivo* repair replication in that it is nonconservative, is stimulated by ultraviolet irradiation, requires the *uvrA* gene product, and persists under conditions which inhibit semiconservative replication. This mode of synthesis requires the four deoxyribonucleoside triphosphates and millimolar concentrations of ATP or another nucleoside triphosphate (Masker and Hanawalt, 1973, 1974*b*). This UV-stimulated synthesis is distinct from the ATP-*independent* repair-like synthesis observed in toluenized cells with normal levels of DNA polymerase I (Moses and Richardson, 1970).

In a typical experiment, the mutant under investigation is grown for several generations in minimal medium containing [^{14}C] thymine to prelabel the DNA. The exponentially growing bacteria are harvested, and a portion of the resuspended culture is irradiated with UV. Then, the cells are concentrated and treated with toluene for a sufficient period of time to maximize the amount of ATP-dependent DNA synthesis in the unirradiated control. The toluene-treated cells are diluted in an assay mixture which includes [^{3}H] dATP and BrdUTP (in place of dTTP). After incubation for an hour, the reaction is stopped, the cells are lysed, and their DNA is examined in an isopycnic gradient as described by Hanawalt and Cooper (1971). Since the repair patches are likely to be relatively short, the small amount of BrdUMP incorporated during repair replication should not appreciably alter the buoyant density of the DNA fragments carrying the repaired regions. Thus, the amount of repair replication can be experimentally determined from the amount of newly synthesized (^{3}H-labeled) DNA sedimenting to the same buoyant density as the ^{14}C-labeled parental DNA in a CsCl gradient.

The amount of repair synthesis completed by toluene-treated preparations of UV-irradiated cells is small (<5%) compared to the amount of semiconservative DNA replication performed by unirradiated control cells under identical conditions (Masker and Hanawalt, 1973). This means that a convincing and quantitative determination of the amount of UV-stimulated DNA synthesis is difficult if the normal density region of the isopycnic gradient containing DNA from the unirradiated control preparation is contaminated by even a small amount of UV-independent DNA synthesis. Interpretation of repair experiments is facilitated by examining repair resynthesis under conditions where normal semiconservative replication can not continue. Previous *in vivo* studies have used a "quick stop" DNA replication mutant (*dnaB*) to study repair in the absence of chromosome replication (Couch and Hanawalt, 1967). We have extended this approach to a study of DNA synthesis in toluene-treated preparations of another DNA replication mutant, TG169 (*polA dnaB*).

Figure 1 shows the profiles from isopycnic gradient analysis of DNA from a culture of TG169 subjected to the experimental procedure described above. Figure 1*A* shows that DNA synthesis by toluene-treated preparations of unirradiated cells incubated at the permissive temperature (33°C) is primarily semiconservative. In agreement with the findings of Moses and Richardson (1970), this synthesis is strongly ATP-dependent (Fig. 1*B*), and it is almost entirely eliminated by incubating the permeable cells at the restrictive (44°C) temperature (Fig. 1*C*). Cells subjected to an incident dose of 200 ergs/mm^2 UV and incubated at 44°C show a pronounced stimulation of nonconservative DNA synthesis, presumably due to repair resynthesis (Fig. 1*D*). This UV-stimulated synthesis is further enhanced (by approximately 5/2) in cells exposed to 500 ergs/mm^2 UV (Fig. 1*E*). Figure 1*F* shows that, like semiconservative replication, this UV-stimulated synthesis cannot proceed if ATP is missing from the assay mixture.

Certain chemical agents such as nalidixic acid and 1-β-D-arabinofuranosylcytosine triphosphate (araCTP) also inhibit semiconservative replication while leaving repair synthesis essentially unaffected (Masker and Hanawalt, 1974*a*; Simon et al., 1974). AraCTP has a pronounced inhibitory effect on ATP-dependent DNA synthesis in toluenized cells (Rama Reddy et al., 1971), but our studies, which include experiments performed on toluenized preparations of a mutant deficient in DNA polymerase I and III, show little or no inhibition of repair synthesis by this agent (Masker and Hanawalt, 1974*a*). This selective inhibition by araCTP has proved useful in measuring repair replication in strains which are not temperature sensitive for DNA replication.

As shown in Fig. 1*F*, UV-stimulated DNA synthesis in toluene-treated cells requires ATP. This requirement is distinct from the ATP requirement for

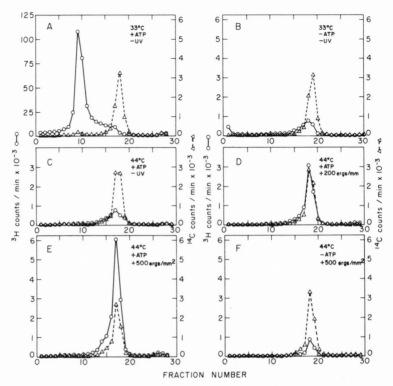

Fig. 1. Isopycnic analysis of DNA from strain TG169. A culture of TG169 (*polA1 dnaB266*) was uniformly labeled with [^{14}C]thymine, divided, irradiated with UV, and toluene-treated for 2 min at 37°C. The cells were diluted into an assay mixture containing BrdUTP and [^{3}H]dATP (14 μCi/20nmoles/ml) as previously described (Masker and Hanawalt, 1973). Incubation was for 1 hr at the permissive (33°C) or restrictive (44°C) temperature in the presence or absence of 1.2 mM ATP as indicated in the individual frames. The profiles show trichloroacetic-acid-precipitable material recovered from CsCl gradients. ○, [^{3}H]dATP; △, [^{14}C]thymine. (Reprinted from Masker and Hanawalt, 1973.)

semiconservative replication, since several other nucleoside triphosphates can partially substitute for ATP as a requirement for repair synthesis (Masker and Hanawalt, 1974*b*), whereas semiconservative replication shows a stringent requirement for ATP (Pisetsky et al., 1972). For example, we have found that if GTP is substituted for ATP in the assay mixture, semiconservative synthesis is reduced by about 95%, whereas repair synthesis is reduced by only 30%. A comparison of two nearly isogenic mutants, JG138 (*polA*) and JG136 (*polA uvrA*), has been made possible by replacing ATP with GTP in the assay mixture to reduce the background of semiconservative replication. This comparison has shown that UV-stimulated synthesis requires the *uvrA* gene product (Masker and Hanawalt, 1974*b*).

RecB and *recC* mutants are sensitive to UV irradiation (Smith, 1971), and it is known that the *recB* and *recC* genes code for a nuclease that requires ATP (Wright et al., 1971; Goldmark and Linn, 1972). It was of interest to learn whether the ATP dependence of UV-stimulated DNA synthesis could be correlated with the *recB* mutation. Since the *recB* and *polA1* mutations are incompatible (Gross et al., 1971), we used a *polA12* mutant, temperature sensitive for DNA polymerase I (Monk and Kinross, 1972), to examine the effect of *recB* in a DNA-polymerase-I-deficient background. When toluene-treated preparations of a *recB21 polA12* mutant were treated with araCTP to selectively block semiconservative replication and incubated at a temperature which inactivated DNA polymerase I, UV-stimulated DNA snythesis still occurred. Thus, there is no evident connection between the *recB* genotype and the nucleoside triphosphate dependence of repair replication in toluene-treated cells (Masker and Hanawalt, 1974*b*).

In an effort to identify the DNA polymerase responsible for repair resynthesis, we examined DNA synthesis in toluene-treated preparations of two nearly isogenic DNA replication mutants, HS434 (*end polA polC$_{ts}$*) and HS432 (*end polA polB polC$_{ts}$*). These strains were constructed by H. Shizuya, who transduced the *polB* (deficient in DNA polymerase II) mutation reported by Campbell et al. (1972) into BT1026 (*end polA polC*). When these strains were treated by the experimental procedure outlined above and then incubated under permissive conditions (Fig. 2*A, E*), both mutants exhibited semiconservative DNA synthesis, while at the restrictive temperature (Fig. 2*B, F*) little synthesis occurred because of thermal inactivation of DNA polymerase III (Gefter et al., 1971). Although the double mutant HS434 performed UV-stimulated DNA synthesis (Fig. 2*C*), this synthesis was not seen in strain HS432, which is additionally deficient in DNA polymerase II (Fig. 2*G*). This result demonstrates that the repair resynthesis performed by this *polA polC* strain is due to DNA polymerase II and not caused by residual DNA polymerase I or polymerase III (Masker et al., 1973). A similar result was found using another triple mutant, YH5 (*end polA polB polC*), isolated by Hirota et al., (1972) (Masker and Hanawalt, unpublished).

To determine whether the UV-stimulated synthesis observed in *polA* mutants is due exclusively to DNA polymerase II, we examined a mutant, HMS83 (Campbell et al., 1972), which is *polA polB* but has normal DNA polymerase III activity. Figure 3 shows the result of isopycnic gradient analysis of DNA from toluenized cells of HMS83 subjected to the usual experimental procedure. In this experiment GTP was used in place of ATP in order to reduce semiconservative synthesis in the unirradiated control (Fig. 3*A, B*). As seen in Fig. 3*C*, UV-stimulated synthesis persists in this strain in spite of reduced levels of both DNA polymerases I and II.

Our studies with toluenized cells suggest that in *polA* mutants both DNA polymerases II and III are capable of repair resynthesis. Studies using alkaline sucrose gradient sedimentation analysis to measure *in vivo* restoration of high-

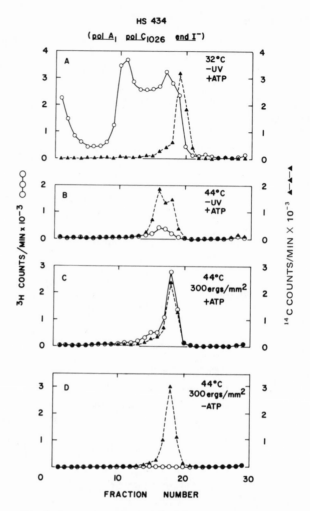

*Fig. 2. Isopycnic analysis of DNA from strains HS434
and HS432.* Cultures labeled with [^{14} C] thymine were
subjected to the experimental procedure outlined in
the text and incubated in assay mixture containing
BrdUTP and [^{3} H] dATP (22μCi/20 nmoles/ml). Experimental details are given by Masker et al. (1973).

molecular-weight DNA after UV irradiation have implicated DNA polymerase III
in excision-repair and have shown that some residual repair, perhaps due to DNA
polymerase II, occurs at the restrictive temperature in *polA polC* strains (Youngs
and Smith, 1973; Tait et al., 1974). The *polC* mutants suffer extensive postirradiation breakdown of their DNA at both the permissive and restrictive temperatures (Youngs and Smith, 1973; Masker et al., 1973; Tait et al., 1974). This

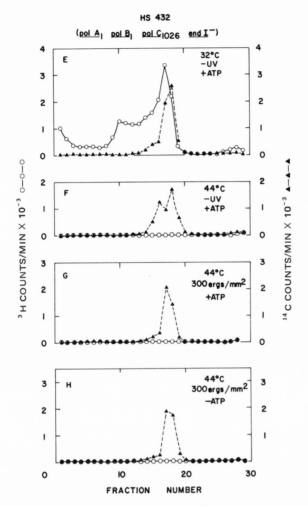

Results with HS434 (*polA1 polC1026 endI⁻*) are shown in frames *A–D*. Results with HS432 (*polA1 polB1 polC1026 endI⁻*) are shown in frames *E* and *F*. Incubation of irradiated cells and controls was as indicated in the individual frames. ○, [³H] dATP; ▲, [¹⁴C] thymine. (Reprinted from Masker et al., 1973.)

UV-induced degradation is not a problem in experiments using toluene-treated preparations, but it does make interpretation of *in vivo* experiments difficult. *In vivo* determinations of repair replication using isopycnic gradient analysis indicate that at the permissive temperature some repair synthesis occurs in the triple mutant HS432 (*polA polB polC*); and this synthesis is presumably due to DNA polymerase III. However, because of serious postirradiation DNA degradation in

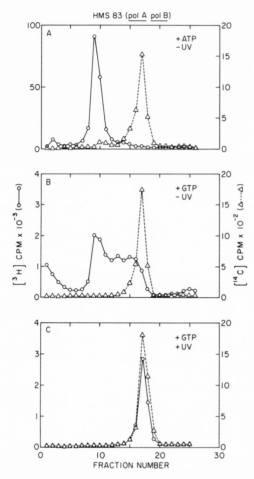

Fig. 3. Isopycnic analysis of DNA from strain HMS83. A culture of HMS83 (polA1 polB1) was uniformly labeled with [¹⁴C] thymine and subjected to the usual experimental procedure. Toluene treatment was for 3 min at 37°C. The cells were incubated in an assay mixture (Masker and Hanawalt, 1973) which included BrdUTP, [³H] dATP (39 μCi/20 nmoles/ml), and 1.2 mM ATP (A) or 1.2 mM GTP (B, C). Incubation was for 60 min at 37°C. The incident UV dose to the irradiated cells was 500 ergs/mm². ○, [³H] dATP; △, [¹⁴C]-thymine.

this mutant, a quantitative measurement of the amount of repair replication is impossible (Masker and Hanawalt, unpublished). A similar problem is inherent in the application of alkaline sucrose gradient sedimentation analysis to *polC* strains. Since this technique measures completion of the repair process (i.e. the joining of the newly synthesized repair patch to the contiguous parental strand), it is a more demanding test for repair. In the *polC* strains it is possible that the extensive DNA breakdown may in some way interfere with completion of repair by DNA polymerase II at temperatures restrictive for DNA polymerase III, or that the newly synthesized repair regions may suffer degradation which precludes ligation. Because of the intrinsic difficulties involved in working with the *polC* mutants, it is presently impossible to make an accurate assessment of the relative importance of DNA polymerases II and III in the excision-repair process. The available experimental evidence suggests that both DNA polymerases II and III can function in repair, a result which is not surprising in view of the known biochemical properties of these enzymes. Also, if the various polymerases can indeed substitute for one another in repairing UV damage, this might explain why *polA polB* strains are no more UV sensitive than are *polA* mutants (Campbell et al., 1972; Hirota et al., 1972).

ACKNOWLEDGMENTS

This work was supported by a postdoctoral fellowship from the Helen Hay Whitney Foundation to W. E. M., a Research Grant (GM 09901) from the National Institutes of Health, and a contract AT(04-3)326-7 with the U.S. Atomic Energy Commission.

REFERENCES

Burger, R. M. (1971). *Proc Nat. Acad. Sci. U.S.A.* **68**, 2124–2126.
Campbell, J. L., Soll, L. and Richardson, C. C. (1972). *Proc. Nat. Acad. Sci. U.S.A.* **69**, 2090–2094.
Cooper, P. K. and Hanawalt, P. C. (1972a). *J. Mol. Biol.* **67**, 1–10.
Cooper, P. K. and Hanawalt, P. C. (1972b). *Proc. Nat. Acad. Sci. U.S.A.* **69**, 1156–1160.
Couch, J. and Hanawalt, P. (1967). *Biochem. Biophys. Res. Commun.* **29**, 779–784.
Gefter, M., Hirota, Y., Kornberg, T., Barnoux, C. and Wechsler, J. (1971). *Proc. Nat. Acad. Sci. U.S.A.* **68**, 3150–3153.
Goldmark, P. J. and Linn, S. (1972). *J. Biol. Chem.* **247**, 1849–1860.
Gross, J. D., Grunstein, J. and Witkin, E. M. (1971). *J. Mol. Biol.* **58**, 631–634.
Hanawalt, P. C. and Cooper, P. K. (1971). *Methods Enzymol.* **21**, 221–230.
Hirota, Y., Gefter, M. and Mindich, L. (1972). *Proc. Nat. Acad. Sci. U.S.A.* **69**, 3238–3242.
Kanner, L. and Hanawalt, P. (1970). *Biochem. Biophys. Res. Commun.* **39**, 149–155.
Kelly, R. B., Atkinson, M., Huberman, J. and Kornberg, A. (1969). *Nature* **244**, 495–501.
Lehman, I. R. and Chien, J. R. (1973). *J. Biol. Chem.* **248**, 7717–7723.
Masker, W. E. and Hanawalt, P. C. (1973). *Proc. Nat. Acad. Sci. U.S.A.* **70**, 129–133.
Masker, W. E. and Hanawalt, P. C. (1974a). *Biochim. Biophys. Acta* **340**, 229–236.
Masker, W. E. and Hanawalt, P. C. (1974b). *J. Mol. Biol.* **88**, 13–23.
Masker, W. E., Hanawalt, P. C. and Shizuya, H. (1973). *Nature New Biol.* **244**, 242–243.
Matsushita T., White, K. P. and Sueoka, N. (1971). *Nature New Biol.* **232**, 111–114.
Monk, M. and Kinross, J. (1972). *J. Bacteriol.* **109**, 971–978.
Monk, M., Peacey, M. and Gross, J. D. (1971). *J. Mol. Biol.* **58**, 623–630.
Mordoh, J., Hirota, Y. and Jacob, F. (1970). *Proc. Nat. Acad. Sci. U.S.A.* **67**, 773–778.
Moses, R. E. and Richardson, C. C. (1970). *Proc. Nat. Acad. Sci. U.S.A.* **67**, 674–681.
Pisetsky, D., Berkower, I., Wickner, R. and Hurwitz, J. (1972). *J. Mol. Biol.* **71**, 557–571.
Rama Reddy, G. V., Goulian, M. and Hendler, S. S. (1971). *Nature New Biol.* **234**, 286–288.

Simon, T. J., Masker, W. E. and Hanawalt, P. C. (1974). *Biochim. Biophys. Acta* **349**, 271–274.

Smith, K. C. (1971). In *Photophysiology* (Giese, A. C., ed.), vol. 6, pp. 210–278. Academic Press, New York.

Tait, R. C., Harris, A. L. and Smith, D. W. (1974). *Proc. Nat. Acad. Sci. U.S.A.* **71**, 675–679.

Wright, M., Buttin, G. and Hurwitz, J. (1971). *J. Biol. Chem.* **246**, 6543–6555.

Youngs, D. A. and Smith, K. C. (1973). *Nature New Biol.* **244**, 240–241.

33

Requirement for *uvrAB* Function for Postirradiation DNA Synthesis *in Vitro*

D. Bowersock and R. E. Moses

Marrs McLean Department of Biochemistry
Baylor College of Medicine
Houston, Texas 77025

ABSTRACT

Toluene-treated *Escherichia coli* cells from *uvrA⁻* and *uvrB⁻* strains do not show an ultraviolet-induced, DNA polymerase-III-dependent DNA synthesis.

Replicative DNA synthesis is inhibited in toluene-treated cells of *Escherichia coli* which have been exposed to ultraviolet radiation (Masker and Hanawalt, 1973; Moses, 1973). There is a residual level of DNA synthesis which, we have concluded, is due to DNA polymerase III action (Bowersock and Moses, 1973).

DNA polymerases I and II do not appear to be required for this synthesis. In all studies reported here, DNA polymerase I-deficient mutants were used because the presence of polymerase I activity in some instances markedly decreases the postirradiation plateau synthesis.

Since the plateau of DNA synthesis is dependent on irradiation, it seems reasonable that it represents a facet of cellular repair. The known requirements of DNA polymerase III suggest that a gap would be required for it to function. These observations suggest that a second activity would be required for such synthesis.

We have investigated several repair-deficient mutants of *E. coli* for postirradiation residual DNA synthesis. If a repair function is needed to form an incision and/or gap prior to DNA polymerase activity, then a mutant deficient in the

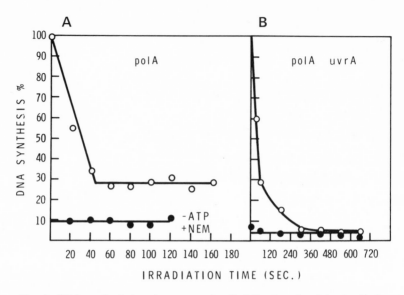

Fig. 1. DNA synthesis in toluene-treated cells following ultraviolet irradiation. Cells were grown and toluene-treated at 2×10^{10} cells/ml. Irradiation of toluene-treated cells was done at $4°C$, and measurement of DNA synthesis in the presence of ATP was done as described earlier (Bowersock and Moses, 1973). (A) D110 (*polA1 EndI⁻*), (B) M148 (*polA12 uvrA⁻*). With strain M148 the same result was obtained at 32 and $42°C$ incubation. Reactions with *N*-ethylmaleimide (NEM) contained 5 mM NEM. Each assay contained 8×10^8 cells/0.3 ml. 100% synthesis represents 72 pmoles for (*a*) and 127 pmoles for (*b*).

required function should show no residual synthesis. *uvrA⁻* mutants do not show a residual DNA synthesis after irradiation, in contrast to *uvr⁺* strains (Fig. 1*a,b*).

polA₁₂ uvrB⁻ mutants (Shizuya and Dykhuizen, 1972) also fail to show a plateau of DNA synthesis after irradiation. However, *uvrD⁻* mutants do show a normal plateau. *recA⁻* and *recB⁻* mutants also have a wild-type response. From these results we conclude that an early *uvr* function (*A* and *B*) is required for the ultraviolet irradiation stimulation of DNA synthesis *in vitro*. Our results suggest that DNA polymerase III may play a role in synthesis involved in gap closure following pyrimidine dimer excision and that an action by early *uvr* functions is required for such synthesis to occur. The postirradiation synthesis we observe is non-semiconservative, as originally reported by Masker and Hanawalt (1973).

ACKNOWLEDGMENTS

This study was supported by National Institutes of Health Grant No. GM–19122, American Cancer Society Grant No. VC–97, and Robert A. Welch

Foundation Grant No. Q–543. R. E. M. is the recipient of Career Development Award No. GM–70314.

REFERENCES

Bowersock, D. and Moses, R. E. (1973). *J. Biol. Chem.* **248**, 7449.
Masker, W. E. and Hanawalt, P. C. (1973). *Proc. Nat. Acad. Sci. U.S.A.* **70**, 129.
Moses, R. E. (1973). *Fed. Proc.* **32**, 498 (abstract).
Shizuya, H. and Dykhuizen, D. (1972). *J. Bacteriol.* **112**, 676.

34

DNA Polymerase II-Dependent DNA Synthesis in Toluenized *Bacillus subtilis* Cells

Tatsuo Matsushita

Division of Biological and Medical Research
Argonne National Laboratory
Argonne, Illinois 60439

ABSTRACT

An ATP-dependent DNA-polymerase-II-mediated repair can be demonstrated in *Bacillus subtilis* toluenized cells using 6-(*p*-hydroxyphenylazo)uracil (HPUra) and exogenous DNase I.

Toluenized cells have been used to study semiconservative DNA replication in both *Escherichia coli* (Moses and Richardson, 1970) and *Bacillus subtilis* (Matsushita et al., 1971). In *B. subtilis*, this ATP-dependent process represents only elongation of preexisting replication forks. There is no chromosome initiation (Sueoka et al., 1973) or background repair under these conditions. However, DNA repair can be induced in toluenized *B. subtilis* by introducing exogenous DNase I to nick the chromosome while simultaneously inhibiting DNA replication with sulfyhydryl-group reagents such as *p*-chloromercuribenzoic acid (PCMB) (Matsushita and Sueoka, 1974). Under this "repair condition," the *polA*$^+$ strain 168TT (Farmer and Rothman, 1965) shows DNA repair in both the presence and absence of ATP, but *polA*$^-$ mutant strains NB841 (Neville and Brown, 1972) and HA101(59)F (Gass et al., 1971) show no DNA synthesis. This suggests a *polA* repair function in *B. subtilis* analogous to that in *E. coli*.

Attempts to duplicate this repair synthesis assay by substituting 6-(p-hy-droxyphenylazo)uracil (HPUra), a known replication inhibitor (Brown, 1971), for PCMB unexpectedly revealed residual ATP-stimulated DNA synthesis in the NB841 polA⁻ mutant (Fig. 1). Since 770 μM HPUra completely inhibits ATP-stimulated DNA synthesis (replication) in the absence of DNase I, this residual DNA synthesis could not be due to incomplete inhibition of DNA replication. We conclude that this HPUra-resistant, DNase-I-stimulated activity in toluenized polA⁻ cells is exclusively due to DNA polymerase II activity, since HPUra is known to inhibit DNA replication in toluenized *B. subtilis* cells (Brown et al., 1972) and inhibits only DNA polymerase III *in vitro* (Neville and Brown, 1972; Bazill and Gross, 1972; Gass et al., 1973). Thus, in the absence of polymerase I and by inhibiting polymerase III with HPUra, polymerase-II-mediated DNA

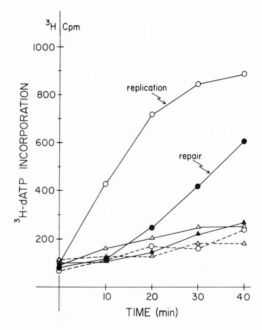

Fig. 1. *DNA replication and repair in s,polA⁻ toluene-treated* B. subtilis. Strain NB841 was grown, collected, and toluenized as described previously (Matsushita et al., 1971). The reaction mixture (0.5 ml) contained 70 mM KPO₄ buffer (pH 7.4), 13 mM MgSO₄, 1.3 mM ATP, 33 μM dGTP, dCTP, dATP, and TTP, 5 μCi [³H] dATP (spec. act. 8.48 Ci/mmole) and 4 × 10⁸ toluene-treated cells. Reactions were carried out at 37°C. At timed intervals, 50-μl samples were withdrawn and the TCA-insoluble counts were determined as described previously (Matsushita and Sueoka, 1974). o———o, △———△–control (+ 1.9 mM NaOH); o- - -o, △- - -△—+ 770 μM HPUra (+ 1.9 mM NaOII + 10 mM dithiothreitol); •———•, ▲———▲—+ 770 μM HPUra + 1.5 μg/ml DNase I (+ 1.9 mM NaOH + 10 mM dithiothreitol); o, •—+ATP; △, ▲— –ATP.

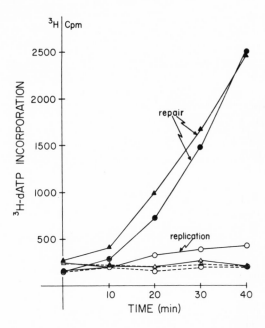

Fig. 2. DNA replication and repair in recA⁻ *toluene-treated* B. subtilis. Same conditions as Fig. 1, except strain GSY 1006 (*his*-25, *trp*-16, *recA*) was used. ○———○, △———△—control (+ 1.9 mM NaOH + 10 mM dithiothreitol); ○--○, △---△-+ 770 μM HPUra (+ 1.9 mM NaOH + 10 mM dithiothreitol); ●———●, ▲———▲-+ 770 μM HPUra + 1.5 μg/ml DNase I (+ 1.9 mM NaOH + 10 mM dithiothreitol); ○,●—+ATP; △,▲—-ATP.

synthesis can be demonstrated in toluenized *B. subtilis polA⁻* cells. This polymerase II synthesis occurs when exogenous DNase and ATP are added. It appears that the polymerase-II-mediated repair in toluenized *B. subtilis* is similar to the UV-stimulated polymerase-II-mediated repair demonstrated by Masker and Hanawalt (1973) in toluenized *E. coli,* which is also ATP-dependent and which they suggest is responsible for "long patch" repair.

Preliminary studies in a *B. subtilis* mutant GSY 1006 (*his*-25, *trp*-16, *recA*) (Hoch et al., 1967; Hoch and Anagnostopoulos, 1970) show a much higher toluenized cell DNA repair in the presence of HPUra and exogenous DNase I, correlating with the "reckless" phenotype of this mutant (Fig. 2). The DNA repair in this case is probably mediated by both polymerase I and II since this strain is *polA⁺*. Under these conditions there appears to be little difference in the amount of DNA repair in the presence or absence of ATP in this mutant. Studies are now in progress to determine the mechanism behind this unusual behavior of DNA repair in toluenized *recA* cells.

REFERENCES

Bazill, G. W. and Gross, J. D. (1972). *Nature New Biol.* **240**, 82.

Brown, N. C. (1971). *J. Mol. Biol.* **59**, 1.

Brown, N. C., Wisseman, C. L., III and Matsushita, T. (1972). *Nature New Biol.* **237**, 72.

Farmer, J. L. and Rothman, F. (1965). *J. Bacteriol.* **89**, 262.

Gass, K. B., Hill, T. C., Goulian, M., Strauss, B. S. and Cozzarelli, N. R. (1971). *J. Bacteriol.* **108**, 364.

Gass, K. B., Low, R. L. and Cozzarelli, N. R. (1973). *Proc. Nat. Acad. Sci. U.S.A.* **70**, 103.

Hoch, J. A. and Anagnostopoulos, C. (1970). *J. Bacteriol.* **103**, 295.

Hoch, J. A., Barat, M. and Anagnostopoulos, C. (1967). *J. Bacteriol.* **93**, 1925.

Masker, W. E. and Hanawalt, P. C. (1973). *Proc. Nat. Acad. Sci. U.S.A.* **70**, 129.

Matsushita, T. and Sueoka, N. (1974). *J. Bacteriol.* **118**, 974.

Matsushita, T., White, K. P. and Sueoka, N. (1971). *Nature New Biol.* **232**, 111.

Moses, R. E. and Richardson, C. C. (1970). *Proc. Nat. Acad. Sci. U.S.A.* **67**, 674.

Neville, M. M. and Brown, N. C. (1972). *Nature New Biol.* **240**, 80.

Sueoka, N., Matsushita, T., Ohi, S., O'Sullivan, M. A. and White, K. P. (1973). In *DNA Synthesis In Vitro* (Wells, R. D. and Inman, R. B., eds.), p. 385. University Park Press, Baltimore, Maryland.

V

Repair by Genetic Recombination in Bacteria

35

Repair by Genetic Recombination in Bacteria: Overview

Paul Howard-Flanders

Department of Therapeutic Radiology
Yale University School of Medicine
New Haven, Connecticut 06520

ABSTRACT

DNA molecules that have been damaged in both strands at the same level are not subject to repair by excision but instead can be repaired through recombination with homologous molecules. Examples of two-strand damage include postreplication gaps opposite pyrimidine dimers, two-strand breaks produced by X-rays, and chemically induced interstrand cross-links. In ultraviolet-irradiated bacteria, the newly synthesized DNA is of length equal to the interdimer spacing. With continued incubation, this low-molecular-weight DNA is joined into high-molecular-weight chains (postreplication repair), a process associated with sister exchanges in bacteria. Recombination is initiated by pyrimidine dimers opposite postreplication gaps and by interstrand cross-links that have been cut by excision enzymes. The free ends at the resulting gaps presumably initiate the exchanges. Postreplication repair in *Escherichia coli* occurs in $recB^-$ and $recC^-$ but is greatly slowed in $recF^-$ mutants. *RecB* and *recC* are the structural genes for exonuclease V, which digests two-stranded DNA by releasing oligonucleotides first from one strand and then from the other. The postreplication sister exchanges in ultraviolet-irradiated bacteria result in the distribution of pyrimidine dimers between parental and daughter strands, indicating that long exchanges involving both strands of each duplex occur.

The R1 restriction endonuclease from *E. coli* has been used to cut the DNA of a bacterial drug-resistance transfer factor with one nuclease-sensitive site, and also DNA from the frog *Xenopus* enriched for ribosomal 18S and 28S genes. The fragments were annealed with the cut plasmid DNA and ligated, producing a new larger plasmid carrying the eukaryotic rDNA and able to infect and replicate in *E. coli.*

DETERMINATION OF THE CHOICE OF REPAIR MECHANISM BY THE NATURE OF THE DAMAGE IN DNA

DNA molecules that have been damaged in both strands at the same level are not subject to repair by excision, there being no intact complementary strand to serve as template for excision and rejoining (Howard-Flanders et al., 1968; Kelly et al, 1969; George and Devoret, 1971; Cole, 1971). DNA molecules that have suffered two-strand damage can, however, be repaired through recombination with homologous molecules. The base sequences needed accurately to reconstruct the molecule at sites of two-strand damage are available in the homolog and apparently can be brought into use through genetic exchanges.

Several radiation-induced or chemically induced damages are known to affect both DNA molecules at the same level. These include postreplication gaps opposite pyrimidine dimers, two-strand breaks produced by X-rays, and interstrand cross-links produced by two-armed alkylating agents, by mytomycin C, or by the photosensitizing agent psoralen. DNA molecules carrying two-strand damage are presumably blocked in transcription in the affected region and cannot be fully replicated. Two-strand damage is more injurious to cells than single-strand damage, as is shown by the greater effectiveness of two-armed as compared to one-armed alkylating agents in killing cells (Brookes and Lawley, 1961).

A ROLE FOR RECOMBINATION IN REPAIR PROCESSES

A hint of a possible involvement of recombination in the processes of repair after ultraviolet irradiation was first obtained when multiplicity reactivation was discovered. Several lethally irradiated T2 phages infecting a single host cell cooperate to produce viable progeny phages (Luria, 1947). Similarly, ultraviolet-irradiated λν phages exhibit a higher level of survival when plated on λ lysogens than on nonlysogenic host cells. The higher level of survival on lysogens is associated with genetic exchanges between the infecting phage and prophage (Jacob and Wollman, 1953; Devoret et al., this volume). These experiments point to a relationship between recombination and survival after ultraviolet irradiation but give no detailed insight at a molecular level. Results significant to understanding repair processes were then obtained in experiments on the abnormally small phage T4 particles that carry only two-thirds of the normal length of phage DNA. Two incomplete phages infecting the single cell produce viable progeny if they complement each other and carry the whole phage genome into the host cell (Mosig, 1963). One interpretation is that the free ends of the partial phages initiate genetic recombination, leading to the reconstruction of the entire phage genome. The interaction between the two partial phages is analogous to the interaction between DNA molecules containing two-strand breaks.

The relationship between recombination and survival after irradiation is further strengthened by the high ultraviolet and X-ray sensitivity of recombination-deficient mutants of *E. coli* (Clark and Margulies, 1965; Howard-Flanders and Theriot, 1966).

DNA METABOLISM IN ULTRAVIOLET-IRRADIATED CELLS

Why does the survival of bacteria after ultraviolet irradiation depend upon some or all of the machinery of genetic recombination? Studies on the DNA metabolism in ultraviolet-irradiated cells have provided revealing information. The DNA synthesized at early times after ultraviolet irradiation is of molecular weight approximately equal to that of the DNA chains between the ultraviolet-induced pyrimidine dimers (Rupp and Howard-Flanders, 1968). This result suggests that the newly synthesized DNA chains terminate at pyrimidine dimers in the template DNA, and that new chains are started at initiation sites further along the template strands. The resulting gaps opposite pyrimidine dimers, generally referred to as postreplication caps, can be 500 to 1000 nucleotides in width (Iyer and Rupp, 1971). These gaps are not immediately repaired but remain open for 15 min or more in bacteria. Presumably the pyrimidine dimers in the single-stranded regions prevent rapid repair by DNA polymerase and polynucleotide ligase. The termination of newly synthesized DNA strands at photoproducts can be observed more directly by centrifugation methods and by electron microscopy in single-stranded bacteriophage systems (Francke and Ray, 1971; Benbow et al., 1974).

In cultured animal cells also, the newly synthesized DNA strands are shorter after ultraviolet irradiation than in untreated cells (Cleaver and Thomas, 1969; Meyn and Humphrey, 1971; Rupp et al., 1970). The lengths of the newly synthesized strands are about equal to the interdimer spacing in mouse lymphoma cells and in human fibroblasts (Lehmann, 1972; Buhl et al., 1972; Zipser, 1973). In certain other cell lines, the newly synthesized DNA is of higher molecular weight, even at the end of the shortest labeling period employed (Chiu and Rauth, 1972; Rauth et al., 1974), as if some joining occurred before the end of the labeling period and so escaped detection (Lehmann, this volume). In spite of complications arising from the origins of replication being spaced only about 10^8 daltons apart in animal cells, it seems likely that the newly synthesized DNA strands are terminated at damaged bases such as pyrimidine dimers, and that the resulting gaps remain open long enough to be detected in many, though not all, animal cell lines in culture.

POSTREPLICATION REPAIR

If ultraviolet-irradiated bacteria are further incubated, the newly synthesized DNA increases in molecular weight, becoming comparable to that of DNA from unirradiated cells (Rupp and Howard-Flanders, 1968). This increase in molecular weight of the newly synthesized DNA, referred to as postreplication repair, suggests that the daughter strands synthesized on damaged templates might be used to construct high-molecular-weight chromosomal DNA. If relatively free from damage, the daughter chromosomes so constructed might contribute to cell survival and the production of viable descendents.

In situations where, at a single replication fork, several newly synthesized strands lie along one template strand, separated by postreplication gaps, an increase in molecular weight can be taken as evidence for gap filling. There has been considerable recent interest in the mechanism of postreplication gap filling and effort to elucidate the mechanism in bacteria and animal cells.

The postreplication increase in molecular weight and presumptive gap filling occur at about the normal rate in most strains of bacteria, including those carrying *recB*, *recC*, *recA*, and *polA*, but not those carrying *recA* (Smith and Meun, 1970; Howard-Flanders et al., 1971). Similarly, in *Haemophilus influenzae* it occurs in recombination-deficient mutants such as strain DB117 but not in others such as strain DB112 (LeClerc and Setlow, 1972). A number of genes are required for the efficient acceptance of DNA and integration to form recombinants. Some but not all of these genes are also required for postreplication repair.

Many recombination-deficient mutants in *E. coli* K12 have been investigated. *RecA*⁺ is needed for virtually all recombination. *RecB*⁺ and *recC*⁺ are needed for only certain types of recombination and between them determine the ATP-dependent exonuclease V. This enzyme has been highly purified, and its properties have been investigated in considerable detail (Goldmark and Linn, 1972; Linn and MacKay, this volume). Moreover, *recB* and *recC* mutants with a temperature-sensitive phenotype have been used to prove that these are the structural genes for exonuclease V (Kushner, this volume). It has also been found that *recB*⁺ and *recC*⁺ appear to provide an alternative function to *recF*⁺ and hence to be involved in alternative pathways for recombinant formation (Horii and Clark, 1973). It is therefore of interest to investigate the roles of *recB*, *recC*, and *recF* in postreplication repair, and this has now been explored (Rothman et al., this volume). Mutants carrying *recB*⁻ or *recC*⁻ are about as ultraviolet sensitive as mutants carrying *recF*⁻. *RecB*⁻ *recC*⁻ *recF*⁻ triple mutants are several times more ultraviolet sensitive. These results are again consistent with an alternative pathway concept. As *recB*⁻ and *recC*⁻ mutations do not inhibit postreplication repair, it might be predicted that *recF*⁻ would do so. This has now been tested, and *recF*⁻ strains show a reduced although still detectable rate of increase in molecular weight in postreplication repair.

Exonuclease V contributes to repair after X irradiation, since the lack of this enzyme in recB⁻ and recC⁻ mutants causes increased X-ray sensitivity. In *E. coli* λ lysogens lacking exonuclease V, X-ray resistance can be restored if at any time up to an hour after X irradiation the *red*⁺ genes which control λ exonuclease and β protein are transitorily expressed. The X-ray resistance of λ lysogens can be increased above the wild-type level and DNA breakdown can be reduced if the phage λ gene *gam*⁺ is expressed for a short time immediately following irradiation (Rupp et al., this volume).

Further investigations into the ways in which particular mutations either singly or in combination can affect cell survival following irradiation point to areas of ignorance. The mutant alleles *uvrD⁻*, *recB⁻* and *exrA⁻*, when present in excision-defective strains carrying *uvrB⁻*, all cause increased ultraviolet sensitivity, and hence must act through mechanisms other than excision (Youngs et al., this volume). (*ExrA* is a gene in *E. coli* B which is similar to the gene *lexA* in *E. coli* K12, the strain in which genetic analysis can most readily be carried out. All the other genes discussed here are from strain K12.)

If we are to reach a sophisticated understanding of cellular repair processes, we will need to understand the roles of the several genes and their gene products in DNA repair. It is in most cases the lack of an assay for the gene products that impedes progress. Hopefully we will be able to devise new substrates for these unknown gene products and so be able to investigate the roles of genes such as *recA, recF, lexA, lexB,* and *uvrD*.

As already noted, the lengths of the newly synthesized strands in ultraviolet-irradiated animal cells are initially shorter than in controls, being often comparable to the interdimer spacing. If the incubation is continued, the molecular weight of the newly synthesized DNA increases to equal that in control cells in virtually all the animal cell lines so far tested (Cleaver and Thomas, 1969; Rupp et al., 1970; Meyn and Humphrey, 1971; Lehmann, 1972; Chiu and Rauth, 1972; Buhl et al., 1972, 1973). Thus postreplication repair occurs in animal cell lines.

THE MECHANISM OF POSTREPLICATION REPAIR

If postreplication repair plays an active role in promoting cell survival, it is likely that the gap filling and strand joining are generally accurate. The nucleotide sequences inserted into the gaps opposite pyrimidine dimers must generally be the correct ones. If the two duplexes formed by the replication of DNA containing pyrimidine dimers is considered, it is readily seen that at each point where there is a dimer and postreplication gap in one duplex, the homologous sister duplex will usually be intact and carry the correct base sequences necessary to fill in the postreplication gap. Thus gap filling through sister exchanges can lead to the insertion of the correct nucleotide sequences and the duplex carrying two-strand damage can be accurately repaired (Rupp and Howard-Flanders, 1968).

There is no direct proof that gap filling is accurate, but the closely related process of gene conversion has been found capable of inserting accurate nucleotide sequences. For example, in yeast, accurate conversion between *amber* and *ochre* codons can occur (Fogel and Mortimer, 1970), while in the fruit fly *Drosophila*, accurate conversion occurs between electrophoretic enzyme variants (McCarron et al., 1974).

There is much interest in the mechanism of gap filling, and several possibilities can be considered: (1) by repair synthesis without exchange, (2) by sister exchange involving insertion of a newly synthesized strand formed on a template on a sister duplex, (3) by a sister exchange involving preexisting DNA with repair synthesis in the join, or (4) by an exchange of preexisting DNA without repair synthesis.

Alternative 1 has been proposed as a mechanism for ultraviolet mutagenesis (Witkin, 1969, and this volume) but does not provide for accurate repair. Alternative 4 can be excluded, since some repair synthesis is required to ensure that all strands are rejoined. This leaves alternatives 2 and 3 as being most likely in postreplication repair, assuming it to be accurate. Another possibility is that gap filling is usually accurate, but gap filling by *de novo* synthesis may occur ocassionally.

An exchange of material between sister duplexes occurs in bacteria in response to ultraviolet irradiation (Rupp et al., 1971). Using a density-labeling method, approximately one recombinant molecule can be detected for every pyrimidine dimer in the DNA undergoing replication. If one exchange occurs for each pyrimidine dimer and postreplication gap, these structures must be very efficient at initiating sister exchanges.

Sister exchanges require the removal of segments from the parental chains and insertion into the gap of the opposite duplex. Repair of the resulting gaps in the parental strands can also be detected if cells are incubated after the ultraviolet irradiation in medium containing bromodeoxyuridine. Late after irradiation, segments of bromodeoxyuridine-containing DNA of the order of 10^4 nucleotides in length can be detected in excision-defective mutants by photolysis with 313-nm light (Ley, 1973, and this volume; Cooper and Hanawalt, 1972).

SISTER OR DOUBLE EXCHANGES BETWEEN SISTER DUPLEXES

We can ask whether exchanges occurring between sister duplexes in response to pyrimidine dimers and postreplication gaps are generally single or double exchanges. A double exchange would provide for the transfer from the sister duplex of a single-strand segment more than sufficient to fill the gap, without requiring that all four strands be cut and rejoined. On the other hand, a single exchange could also fill the gap opposite a pyrimidine dimer but would involve the exchange of a substantial length of duplex. In particular, single exchanges would result in the distribution of photoproducts between all four strands of the sister duplexes, while double exchanges might leave the photoproducts mainly in parental strands.

Experiments on the rate of recovery of the capacity to synthesize high-molecular-weight DNA in bacteria after ultraviolet irradiation revealed the synthesis of high-molecular-weight material in increasing amounts, starting at about 1 hr after ultraviolet irradiation. Since these experiments were performed in excision-defective bacteria, they were taken as evidence that high-molecular-weight templates were becoming available 1–2 hr after irradiation for the direct synthesis of high-molecular-weight chains (Rupp et al., 1971; Ganesan and Smith, 1971).

The development of more powerful methods for the detection of pyrimidine dimers in DNA as endonuclease-sensitive sites (Ganesan, 1973, and this volume) has provided further insight into the nature of these exchanges and cast doubts on the interpretation just given. Using the ν-gene endonuclease from phage T4, it has been shown that the endonuclease-sensitive sites do not remain in the parental strands but become distributed approximately equally between all the daughter strands at each DNA replication. Evidently, postreplication gaps can cause single exchanges of considerable length (Ganesan, 1974, and this volume). These experiments show with unusual clarity that recombination with exchange of strands carrying dimers can be caused by dimers and postreplication gaps. It is remarkable that symmetrical exchanges involving the cutting and joining of both strands of each duplex and the exchange of long segments of DNA can be initiated in this unsymmetrical way. Several aspects of the recombination de-tected in these endonuclease experiments (Ganesan, 1974) bear a remarkable similarity to spontaneous recombination occurring during meiosis in fungi, for which unsymmetrical models recently been developed (Fogel and Mortimer, 1969; Stadler and Towe, 1972; Sigal and Alberts, 1972; Meselson and Radding, personal communication).

While the exact way in which spontaneous recombination is initiated may be difficult to investigate, we know of several structures that initiate recombination efficiently. The first is a dimer and postreplication gap. The second structure is produced through the action of excision enzymes on cross-links in DNA. The resulting half-excised cross-link has properties of great interest.

GENETIC EXCHANGES CAUSED BY CROSS-LINKS AND THE REPAIR OF CROSS-LINKED DNA

Two-armed alkylating agents (but not one-armed agents) have been found to cause genetic exchanges in fungi and bacteria phage systems (Morpurgo, 1963; Yamamoto, 1967). Similarly, the cross-linking agent mytomycin C is recombinagenic in fungi and animal cells (Holliday, 1964; Shaw and Cohen, 1965; German and LaRock, 19690. In *E. coli*, the damage-specific endonuclease controlled by the *uvrA,B* genes (Braun and Grossman, 1974; Braun et al., this volume) is required for the excision of mytomycin C and nitrogen-mustard cross-links (Boyce and Howard-Flanders, 1964; Kohn et al., 1965; Lawley and Brookes, 1965). Excision enzymes appear to make two cuts in one strand, releasing one arm of each cross-link (Kohn et al., 1965; Cole, 1973; Cole and Sinden, Part B of this book). Moreover, the resulting structure comprising a single-strand gap with the half-excised cross-link still attached to the other strand appears to initiate recombination with high efficiency (Cole, 1973, and this volume; Howard-Flanders and Lin, 1973; Howard-Flanders et al., this volume).

By following the changes in molecular weight of the DNA in *E. coli* treated with the photo cross-linking agent psoralen, some insight has been gained into the repair of cross-linked DNA. After cross-linking, a fall in the molecular weight of the DNA is observed, indicative of one arm of each cross-link being excised. This is followed by a $recA^+$-dependent increase in molecular weight toward the normal value for untreated controls. It is postulated that the gap opposite the half-excised cross-link is filled by a genetic exchange with a homolog in the same cell and that the other arm of the cross-link is then subject to removal by the excision enzymes. Thus the accurate repair of interstrand cross-links in the DNA of cells depends upon the presence of a homologous molecule and may occur in three steps, involving excision, recombination, and finally a second excision event (Cole, 1973; Cole and Sinden, Part B of this book; Howard-Flanders and Lin, 1973).

The efficiency with which psoralen cross-links cause recombination is high in λ phage-prophage recombination. Under conditions in which the cross-linked phage DNA does not undergo replication, phage-prophage recombination occurs in wild-type but not excision-defective strains lacking the *uvrA B* endonuclease. Presumably the structure causing recombination is produced through the action of the *uvrA uvrB* endonuclease on cross-links.

Recombination is carried out with high efficiency and appears to occur at distances of up to 5% of the page genome away from cross-links. There are strong marker affects, as if recombination involves heteroduplex formation and subsequent marker correction, much as is visualized in the genetics of fungi (Howard-Flanders et al., this volume). Thus, in addition to half-excised cross-links with gaps in the opposite strands, dimers with postreplication gaps and two-strand breaks also appear to cause exchanges with an efficiency on the order of magnitude unity.

CONSTRUCTION OF BACTERIAL PLASMIDS CARRYING DNA SEGMENTS FROM OTHER SPECIES

Bacterial restriction enzymes cut DNA at specific sequences and can be used for producing specific fragments of virus DNA molecules (Danna and Nathans,

1971). The *E. coli* R1 endonuclease makes cuts, leaving single-stranded ends four nucleotides long carrying a specific sequence of bases (Hedgepeth et al., 1972). DNA molecules cut by this enzyme can be annealed together at ice temperature and rejoined by polynucleotide ligase. This procedure has been used to cut a bacterial drug-resistance transfer factor which has one nuclease-sensitive site and the repetitive rDNA genes coding for the 18S and 28S ribosomal RNA from the frog *Xenopus,* which can be isolated (Dawid et al., 1970). The plasmid and rDNA fragment have been annealed together, and a chimeric plasmid carrying rDNA has been isolated and, after transformation, grown in *E. coli* (Morrow et al., 1974; Cohen, this volume). This demonstration of the covalent joining of selected eukaryotic genes to a bacterial plasmid presents a remarkable feat and opens new prospects for the analysis of selected eukaryotic genes and gene products.

REFERENCES

Benbow, R., Zuccarelli, A., Davis, G. and Sinsheimer, R. L. (1974). *J. Virol.* 13, 898–907.

Boyce, R. P. and Howard-Flanders, P. (1964). *Z. Vererbungsl.* 95, 345–350.

Braun, A. and Grossman, L. (1974). *Proc. Nat. Acad. Sci. U.S.A.* 71, 1838–1842.

Brookes, P. and Lawley, P. D. (1961). *Biochem. J.* 80, 496–503.

Buhl, S. N., Stillman, R. M., Setlow, R. B. and Regan, J. D. (1972). *Biophy. J.* 12, 1183–1191.

Buhl, S. N., Setlow, R. B. and Regan, J. D. (1973). *Biophys. J.* 13, 1265–1275.

Chiu, S. F. H. and Rauth, A. M. (1972). *Biochim. Biophys. Acta* 259, 164–174.

Clark, A. J. and Margulies, A. D. (1965). *Proc. Nat. Acad. Sci. U.S.A.* 53, 451–459.

Cleaver, J. E. and Thomas, G. H. (1969). *Biochem. Biophys. Res. Commun.* 36, 203–208.

Cole, R. S. (1971). *J. Bacteriol.* 106, 143–149.

Cole, R. S. (1973). *Proc. Nat. Acad. Sci. U.S.A.* 70, 1064–1068.

Cooper, P. K. and Hanawalt, P. C. (1972). *Proc. Nat. Acad. Sci. U.S.A.* 69, 1156–1160.

Danna, K. and Nathans, D. (1971). *Proc. Nat. Acad. Sci. U.S.A.* 12, 2913–2917.

Dawid, I. B., Brown, D. D. and Reeder, R. H. (1970). *J. Mol. Biol.* 51, 341–360.

Fogel, S. and Mortimer, R. K. (1969). *Proc. Nat. Acad. Sci. U.S.A.* 62, 96–103.

Fogel, S. and Mortimer, R. K. (1970). *Mol. Gen. Genet.* 109, 177–189.

Francke, B. and Ray, D. S. (1971). *J. Mol. Biol.* 61, 565–586.

Ganesan, A. K. (1973). *Proc. Nat. Acad. Sci. U.S.A.* 70, 2753–2756.

Ganesan, A. K. (1974). *J. Mol. Biol.* 87, 103–119.

Ganesan, A. K. and Smith, K. C. (1971). *Mol. Gen. Genet.* 113, 285–296.

George, J. and Devoret, R. (1971). *Mol. Gen. Genet.* 111, 103–119.

German, J. and LaRock, J. (1969). *Tex. Rep. Biol. Med.* 27, 409–418.

Goldmark, P. J. and Linn, S. (1972). *J. Biol. Chem.* 247, 1849–1860.

Hedgepeth, J., Goodman, H. M. and Boyer, H. W. (1972). *Proc. Nat. Acad. Sci. U.S.A.* 69, 3448–3452.

Holliday, R. (1964). *Genet. Res.* 5, 282–304.

Horii, Z. I. and Clark, A. J. (1973). *J. Mol. Biol.* 80, 327–344.

Howard-Flanders, P. and Lin, P.-F. (1973). *Genetics* 73 (suppl.), 85–90.

Howard-Flanders, P. and Theriot, L. (1966). *Genetics* 53, 1137–1150.

Howard-Flanders, P., Rupp, W. D., Wilkins, B. M. and Cole, R. S. (1968). *Cold Spring Harbor Symp. Quant. Biol.* 33, 195–208.

Howard-Flanders, P., Rupp, W. D., Wilde, C. E. and Reno, D. L. (1971). In *Recent Advances in Microbiology*. Tenth International Congress of Microbiology, pp. 271–282, Mexico D.F.

Iyer, V. N. and Rupp, W. D. (1971). *Biochim. Biophys. Acta* **228**, 117–126.

Jacob, F. and Wollman, E. L. (1953). *Cold Spring Harbor Symp. Quant. Biol.* **18**, 101–121.

Kelly, R. B., Atkinson, M. R., Huberman, J. A. and Kornberg, A. (1969). *Nature* **224**, 495–501.

Kohn, K. W., Steigbigel, N. H. and Spears, C. L. (1965). *Proc. Nat. Acad. Sci. U.S.A.* **53**, 1154–1161.

Lawley, P. D. and Brookes, P. (1965). *Nature* **206**, 480–481.

LeClerc, J. and Setlow, J. (1972). *J. Bacteriol.* **110**, 930–934.

Lehmann, A. R. (1972). *J. Mol. Biol.* **66**, 319–337.

Ley, R. D. (1973). *Photochem. Photobiol.* **18**, 87–95.

Luria, S. E. (1947). *Proc. Nat. Acad. Sci. U.S.A.* **9**, 253–264.

McCarron, M., Gelbart, W. and Chovnick, A. (1974). *Genetics* **76**, 289–299.

Meyn, R. E. and Humphrey, R. M. (1971). *Biophys. J.* **11**, 295–301.

Morpurgo, G. (1963). *Genetics* **48**, 1259–1263.

Morrow, J. F., Cohen, S. N., Chang, A. C. Y., Boyer, H. W., Goodman, H. M. and Helling, R. B. (1974). *Proc. Nat. Acad. Sci. U.S.A.* **71**, 1743–1747.

Mosig, G. (1963). *Cold Spring Harbor Symp. Quant. Biol.* **28**, 35–41.

Rauth, A. M., Tammemagi, M. and Hunter, G. (1974). *Biophys. J.* **14**, 209–220.

Rupp, W. D. and Howard-Flanders, P. (1968). *J. Mol. Biol.* **31**, 291–304.

Rupp, W. D., Zipser, E., von Essen, C., Reno, D. L., Prosnitz, L. and Howard-Flanders, P. (1970). In *Time and Dose Relationship in Radiation Biology as Applied to Radiotherapy*, pp. 1–13. Brookhaven National Laboratory, Long Island, New York.

Rupp, W. D., Wilde, C. E., Reno, D. L. and Howard-Flanders, P. (1971). *J. Mol. Biol.* **61**, 25–44.

Shaw, M. W. and Cohen, M. M. (1965). *Genetics* **51**, 181–191.

Sigal, N. and Alberts, B. (1972). *J. Mol. Biol.* **71**, 789–793.

Smith, K. C. and Meun, D. H. C. (1970). *J. Mol. Biol.* **51**, 459–472.

Stadler, D. R. and Towe, A. M. (1972). *Genetics* **68**, 401–413.

Witkin, E. (1969). *Ann. Rev. Genet.* **3**, 525–552.

Yamamoto, N. (1967). *Biochem. Biophys. Res. Commun.* **27**, 263–269.

Zipser, E. (1973). Ph.D. Thesis, Yale University.

36

Genetic Exchanges Induced by Structural Damage in Nonreplicating Phage λ DNA

P. Howard-Flanders, P.-F. Lin, and E. Bardwell

*Department of Molecular Biophysics and Biochemistry
and Department of Therapeutic Radiology
Yale University
New Haven, Connecticut 06520*

ABSTRACT

Genetic recombination between irradiated λ phage and the unirradiated λ prophage in homoimmune lysogens has been studied under conditions in which phage DNA replication and repair were controlled. The λ phage were exposed to one of three treatments before infecting the lysogens: (a) 254-nm light, which produces pyrimidine dimers and other photoproducts; (b) 313-nm light with acetopheneone D, which produces thymine dimers and a different spectrum of other photoproducts; (c) 360-nm light with trimethylpsoralen, which produces monoadducts and cross-links. With both replication and excision-repair of the damaged phage DNA blocked, treatment b (acetophenone D) caused no significant increase in recombination, indicating that thymine dimers do not cause recombination if the DNA in which they are contained is not replicated. Treatment a (254 nm), producing the same total number of pyrimidine dimers, caused a marked increase in recombination. This indicates that photoproducts other than pyrimidine dimers produced by 254-nm light can cause recombination in the absence of replication. Treatment c (psoralen) caused a marked increase in recombination in wild type but not in *uvrA* and *uvrB* mutants. The frequency of recombination in two-factor crosses varied with marker separation in such a way as to suggest that cross-links can act over distances of at least 5% of the λ genome to cause exchanges between pairs of relatively closely spaced markers. The psoralen photo cross-links and monoadducts initiate recombination only following the action of excision enzymes, which appear to release one arm of each cross-link, producing a gap with free strand ends. It may be these strand ends which induce recombination. The action at a distance of 5% of the λ genome may reflect heteroduplex formation and the subsequent reduction to homozygosity of mismatched base pairs at genetic markers. Recombination between closely spaced markers in the *P* gene is reduced in strains carrying *polA*.

The frequency of recombination in genetic crosses in various microorganisms can be increased by treating them with radiation or chemical mutagens (Jacob and Wollman, 1955; Manney and Mortimer, 1964; Zimmerman, 1971). In most published experiments of this type, the crosses were carried out under conditions in which the damaged DNA molecules continued to undergo replication. Under these conditions, DNA repair, replication, and recombination take place simultaneously, and it is difficult to sort out these processes or to interpret the results.

Genetic systems can be set up in phage λ and its host *Escherichia coli* K12 in which recombination, repair, and replication can be controlled independently. Meselson (1964) discovered that recombinant λ phages can be formed from preexisting phage DNA molecules, even if the DNA has not undergone replication. Recombination can take place while bacterial and phage DNA replication is blocked, although local DNA synthesis may take place in association with recombination (Stahl et al., 1972*a*,*b*).

An indication of the value of recombination in promoting survival after ultraviolet irradiation is seen in the levels of survival of ultraviolet-irradiated phage λ, which are higher when the treated phage are plated on heteroimmune lysogenic bacteria than on nonlysogenic cells (Jacob and Wollman, 1963; Hart and Ellison, 1970; Yamamoto, 1967). The increased level of survival in the lysogenic hosts requires *recA*[+] and appears to depend upon recombination between homologous DNA of the damaged phage and the intact resident prophage (Blanco and Devoret, 1973).

Studies such as these suggest that genetic recombination can be induced by structural damage in DNA, but they would have been easier to interpret if carried out under conditions in which the damaged DNA molecules were not replicating. We have used a system in which phage λ was irradiated or experimentally treated in various ways and allowed to infect a homoimmune lysogen. Under these conditions the DNA of the infecting phage particle is repressed by the λ immunity in the host cell and replicates little if at all. The prophage, however, replicates as part of the bacterial chromosome. The phage-prophage recombination system has three useful features. First, if homoimmune phages are used, recombination is induced by structural damage in the DNA of the infecting phage and can be studied under conditions in which the damaged DNA is never called upon to replicate. Recombination can be studied in the prophage, the viability of which is essentially 100%, regardless of damage to the phage. Second, spontaneous phage-prophage recombination is very low, and damage in the phage DNA can result in a 100-fold increase in the frequency of prophage recombination. Third, parallel experiments can be performed in which replication of the infecting phage DNA is permitted, using a heteroimmune lysogen or thermoinducible lysogens carrying *cI857.* These systems have been used in a study of the extent to which different types of structural damage in the phage DNA initiate recombination.

Suspensions of λ*cI857 P3* phages were exposed to one of three treatments: (1) irradiation with 254-nm light, producing pyrimidines dimers, thymine adducts, pyrimidine hydrates, DNA protein cross-links (Setlow and Carrier, 1966; Vargahese and Wang, 1968; Grossman and Rogers, 1968; Smith, 1967), and possibly other unknown photoproducts; (2) irradiation with 313-nm light in the presence of acetophenone D, producing thymine dimers and other unknown photoproducts (Meistrich and Lamola, 1972); and (3) treatment with trimethylpsoralen and 360-nm light, which produces interstrand cross-links and monoadducts (Cole, 1970, 1971). The treated phages were then permitted to infect the homoimmune lysogen *E. coli su⁻*(λ *cI857 P80*) and were incubated at 32°C until the lysogens had replicated 10 to 15 times. During the growth, the DNA of the infecting phage was diluted out, and the lysogens, by then almost free of the infecting phages, were induced by heating to 42°C. The frequency of P^+ recombinant prophages was assayed by plating on *su⁻* indicator bacteria, while the total number of lysogens was determined from the numbers forming colonies on agar. The results of experiments of this type are shown in Fig. 1.

The pyrimidine dimers produced by 313-nm light and acetophenone D cause very little increase in the frequency of recombination in wild-type and no significant increase in excision-defective lysogens (Fig. 1*B*). Assuming that the phage DNA was injected normally in these crosses, as is indicated by control experiments in heteroimmune hosts (not described here), it can be concluded that pyrimidine dimers do not appreciably increase the frequency of recombination, while the DNA in which they are contained does not undergo replication.

Treatment with germicidal light of 254-nm wavelength caused recombination to a much greater extent (Fig. 1*A*) than was produced by acetopheneone D and 360-nm light (Fig. 1*B*). Since the number of pyrimidine dimers induced by these two treatments is comparable, it appears that the 254-nm light produces photoproducts other than pyrimidine dimers capable of causing recombination in the absence of replication. However, the nature of the 254-nm photoproducts causing recombination in the absence of DNA replication has yet to be determined.

A different spectrum of photoproducts and cross-links is produced in the phage DNA by exposure to 360-nm light and trimethylpsoralen. As seen in Figure 1*C*, this treatment results in a substantial increase in the frequency of

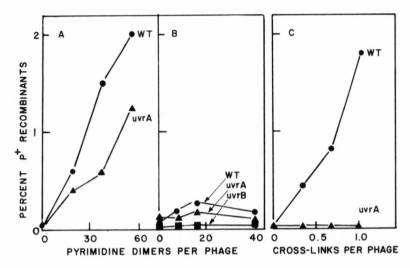

Fig. 1. *The frequency of wild-type recombinant prophage as a function of the structural damage in the DNA of infecting phages.* Lysogens of *E. coli su⁻* (λ *cI857 P80*) that were wild type or carried *uvrA* or *uvrB* were infected with phage λ*cI P3* that had been irradiated in either of three ways: *(A)* 254-nm light—a dose of 1600 ergs/mm² produced 60 pyrimidine dimers per phage; *(B)* irradiation with 313-nm light in the presence of acetopheneone D-7 X 10⁴ ergs/mm² produced 40 pyrimidine dimers per phage; *(C)* 360-nm light in the presence of trimethylpsoralen—22,000 ergs/mm² produced one cross-link and about three monoadducts per phage. Lysogens were infected at 32°C with the treated phages, diluted 1/1000, and grown in broth to more than ten generations. The lysogens were plated or agar to determine the number of colony-forming units and also on soft agar seeded with *su⁻* indicator bacteria at 42°C to determine the number of infective centers produced by *P⁺* recombinant prophages.

recombination in wild-type cells but none in excision-defective mutants. Thus, excision appears to be a necessary step in generating the structures that cause recombination between phage and prophage. In earlier experiments it was noted that cross-links are acted on by excision enzymes that release one arm of each cross-link. This produces single-strand gaps with the cross-links still attached by one arm to the opposite single strand (Kohn et al., 1965; Cole, 1973). One interpretation is that the presence of the half-excised cross-links in single-stranded regions may prevent the gaps from being filled by repair synthesis. These structures may cause recombination by virtue of the free ends, which remain unjoined for a sufficiently long time to initiate recombination. They resemble dimers and postreplication gaps which are produced when DNA containing pyrimidine dimers undergoes replication. Here again, the free ends at the postreplication gap are not immediately rejoined and appear to initiate the exchanges with high efficiency (Rupp et al., 1971). Recombination produced in this way may be called free-end-induced recombination.

We next investigated the relationship between the numbers of cross-links in phage DNA after treatment with 360-nm light and psoralen and the frequency of recombination in two-factor crosses. The frequency of wild-type recombinants increased with increasing numbers of cross-links in each cross (Figs. 1*C* and 2). However, when a series of phage-prophage crosses were carried out with genetic markers separated by different distances in the genetic map, it was found that the frequency of wild-type recombinants increased in an irregular way with increasing marker spacing (Fig. 3*A*). The lowest frequency of recombination was obtained with the closest-spaced markers in the *P* gene, but disproportionally higher values were obtained for markers in the *O* gene. In reciprocal crosses, the greatest departures from reciprocity were seen in crosses involving *O8* and *O29*, possibly due to specific marker effects. Complementation is probably not important, as high values were obtained in *O8* × *O29*.

A comparison can be made between the observed frequency of recombination between any two markers and the frequency with which randomly spaced cross-links are likely to fall between them. For closely spaced markers such as *P3* and *P80*, the observed frequency of recombination exceeds by 10-fold the frequency at which cross-links can be expected to fall between them. This suggests that the psoralen photoproducts can cause recombination between markers at distances of up to 5% of the phage λ genome. This estimate is based upon the assumption that one recombination event is initiated by each cross-link in the excision-proficient strain. If this efficiency is less than unity, the cross-links must be capable of initiating exchanges between pairs of genetic markers situated at distances even greater than 5% of the genome away. Heteroduplex formation and the subsequent reduction to homozygosity of mismatched base pairs has been observed in fungal and phage systems (Holliday, 1964; Fincham

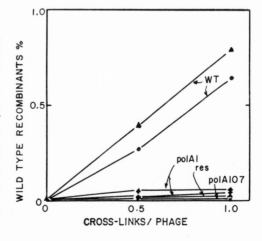

Fig. 2. The fraction of wild-type recombinant prophages plotted as a function of the dose of 360-nm light given in the presence of trimethylpsoralen to λ phages infecting lysogens. Phage λ *cI857 P3* was treated with 11,000 or 22,000 ergs/mm² of 360-nm light in the presence of trimethylpsoralen and allowed to infect *E. coli su⁻* (λ *cI857 P80*) that carried either wild type, *polA1, res,* or *polA107.* The frequency of wild-type prophages was determined as in Fig. 1, the multiplicity being corrected to unity.

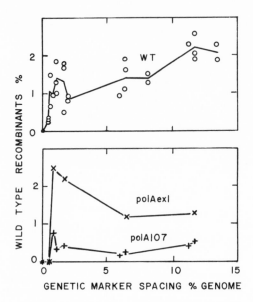

Fig. 3. The fraction of P^+ wild-type recombinant prophages induced by psoralen photoproducts as a function of the distance between the genetic markers in two-factor crosses. Phage λ *cI857 amber 1* was exposed to 11,000 ergs/mm² of 360-nm light in the presence of trimethylpsoralen and allowed to infect the lysogens *E. coli su⁻* (λ *cI857 amber 2*). The multiplicity was corrected to 1.0. *Amber 1* and *amber 2* were two different mutations in the series *O29, O8, P3, P80, Q73,* or *A11.* (*A*) Wild-type lysogens, (*B*) lysogens carrying *polA107* or *polAex1.*

and Holliday, 1970; White and Fox, 1974) and may be involved in the recombination between the closely spaced *P* mutations caused by distant cross-links.

The availability of mutants lacking exonuclease II activity (Glickman et al., 1973; Glickman, this volume; Konrad and Lehman, 1974; Uyemura et al., this volume) makes possible a test of the possible roles of exonuclease II in psoralen-plus-light-induced recombination. Exonuclease II could play roles either by modifying the photo cross-links in the DNA prior to initiating recombination or by affecting the metabolism of the DNA strands participating in the exchange. When phage-prophage crosses of the type described were carried out with phages treated with psoralen and 360-nm light in mutant lysogens carrying *polA1, res, polA107,* or *polAex1,* it was found that the frequency of induced recombination between *P3* and *P80* markers was only a few percent of normal (Fig. 2). With genetic markers at greater spacing, however, the frequencies were more comparable to the values in wild-type lysogens (Fig. 3B compared to 3A). It can be concluded from these data that the lack of exonuclease II activity does not block psoralen-plus-light-induced recombination, which is thought to be due to the psoralen cross-links. However, the recombination between the two closely spaced mutations in the *P* gene is severely depressed by each of the several mutations affecting polymerase I or exonuclease II activity. These mutations may perhaps affect the metabolism of strands carrying the closely spaced markers in heteroduplex configuaration. It is of interest that these mutants exhibit relatively normal yields of recombinants when crossed with suitable Hfr donors.

In summary, these experiments show that certain types of structural damage in phage DNA are capable of initiating genetic exchanges with homologous prophage molecules and can increase recombination frequencies 100-fold above the spontaneous level. Structures such as psoralen photo cross-links that have been acted on by excision enzymes and that have free ends at damage-stabilized single-strand gaps are especially efficient, but other unknown 254-nm photoproducts also cause recombination. Recombination between closely linked markers is depressed by *polA* mutations.

ACKNOWLEDGMENTS

This work was supported by United States Public Health Service grants CA 06519, AMK69397, and GM 11014.

REFERENCES

Blanco, M. and Devoret, R. (1973). *Mutat. Res.* 17, 293–305.
Cole, R. S. (1970). *Biochim. Biophys. Acta* 217, 30–39.
Cole, R. S. (1971). *Biochim. Biophys. Acta* 254, 30–39.
Cole, R. S. (1973). *Proc. Nat. Acad. Sci. U.S.A.* 70, 1064–1068.
Fincham, J. R. S. and Holliday, R. (1970). *Mol. Gen. Genet.* 109, 309–322.
Glickman, B. W., van Sluis, C. A., Heijneker, H. L. and Rorsch, A. (1973). *Mol. Gen. Genet.* 124, 69–82.
Grossman, L. and Rogers, E. (1968). *Biochem. Biophys. Res. Commun.* 33, 975–983.
Hart, M. G. R. and Ellison, J. (1970). *J. Gen. Virol.* 8, 197–208.
Holliday, R. (1964). *Genet. Res.* 5, 282–304.
Howard-Flanders, P. and Lin, P.-F. (1973). *Genetics* 73 (suppl.), 85–90.
Jacob, F. and Wollman, E. L. (1955). *Ann. Inst. Pasteur,* 88, 724–749.
Jacob, F. and Wollman, E. L. (1963). *Cold Spring Harbor Symp. Quant. Biol.* 18, 101–121.
Kohn, K. W., Steigbigel, N. H. and Spears, C. L. (1965). *Proc. Nat. Acad. Sci. U.S.A.* 53, 1154–1161.
Konrad, E. B. and Lehman, I. R. (1974). *Proc. Nat. Acad. Sci. U.S.A.* 71, 2048–2051.
Manney, T. R. and Mortimer, R. K. (1964) *Science* 143, 581–583.
Meistrich, M. L. and Lamola, A. A. (1972). *J. Mol. Biol.* 66, 83–95.
Meselson, M. (1964). *J. Mol. Biol.* 9, 734–745.
Rupp, W. D., Wilde, C. E. and Reno, D. L. (1971). *J. Mol. Biol.* 61, 25–44.
Setlow, R. B. and Carrier, W. L. (1966). *J. Mol. Biol.* 17, 237–254.
Smith, K. C. (1967). In *Radiation Research* (Silini, G., ed.), p. 756. Wiley, New York.
Stahl, F. W., McMilin, K. D., Stahl, M. M., Malone, R. E., Nozu, Y. and Russo, V.E.A. (1972a). *J. Mol. Biol.* 68, 49–55.
Stahl, F. W., McMilin, K. D., Stahl, M. M. and Nozu, Y. (1972b). *Proc. Nat. Acad. Sci. U.S.A.* 69, 3598–3601.
Vargahese, A. J. and Wang, S. Y. (1968). *Science,* 160, 186–187.
White, R. L. and Fox, M. S. (1974). *Proc. Nat. Acad. Sci. U.S.A.* 71, 1544–1548.
Yamamoto, N. (1967). *Biochem. Biophys. Res. Commun.* 27, 263–269.
Zimmerman, F. K. (1971). *Mutat. Res.* 11, 327–337.

37

The Beginning of an Investigation of the Role of *recF* in the Pathways of Metabolism of Ultraviolet-Irradiated DNA in *Escherichia coli*

Robert H. Rothman and Takesi Kato[1]

Department of Genetics
University of California
Berkeley, California 94720

and

Alvin J. Clark

Department of Molecular Biology
University of California
Berkeley, California 94720

The *recF* gene in *Escherichia coli* was discovered as a locus of mutations which block recombination in a $recB^- recC^- sbcB^-$ strain but not in a $recB^+ recC^+ sbcB^+$ strain (Horii and Clark, 1974). To understand the significance which was placed on this result it is necessary to know the functions of the *recB*, *recC*, and *sbcB* genes. The *recB* and *recC* genes determine exonuclease V (ExoV), an enzyme whose characteristics have been worked out in several laboratories (see e.g. Goldmark and Linn, 1972). Mutations in these genes inactivate ExoV and produce recombination deficiency (see e.g. Clark, 1973) which can be alleviated by *sbcB* mutations (Kushner et al., 1971). *sbcB* is thus a gene whose mutations indirectly suppress the recombination deficiency provoked by the absence of

---

[1] Present address: Department of Fundamental Radiology, Osaka University Medical School, Osaka, 530 Japan.

ExoV. The product of the *sbcB* gene is exonuclease I (ExoI) (Kushner et al., 1971; Yajko et al., 1974), an enzyme which degrades single-stranded DNA from 3'-OH termini (Lehman and Nussbaum, 1964). In *recB⁻ recC⁻* strains it is presumably the inactivation of ExoI by *sbcB* mutations which leads to recombination ability , while it is the presence of ExoI which produces recombination deficiency. To explain this, Horii and Clark (1974) hypothesized a pathway of recombination normally inhibited by ExoI which was released from inhibition by the *sbcB* mutations. This pathway, called the RecF pathway after the *recF* gene, was considered to be independent of the pathway of recombination involving the *recB* and *recC* genes, the RecBC pathway. The independence of the two pathways was adduced from the fact that *recF* mutations did not block recombination by the RecBC pathway, which occurred in *recB⁺recC⁺sbcB⁺*(or *sbcB⁻*) strains, while it did block recombination by the RecF pathway in *recB⁻ recC⁻ sbcB⁻* strains. Some representative results showing the effects of these genes on recombination are shown in Table 1. In addition, the table shows results indicating that *recA* mutations block both the RecBC and the RecF pathways of recombination.

Recombination in wild-type *E. coli* cells is therefore thought to be carried out mainly through the RecBC pathway, while the RecF pathway, which is inhibited by the presence of ExoI, is available but apparently not very effective. The question naturally raises as to what advantage, if any, the RecF pathway serves under these circumstances. One possibility is that there are certain types of recombination which the inhibited RecF pathway can carry out but the RecBC pathway cannot. At present there is no firm evidence to substantiate this possibility, but it would not be surprising to find that this is the case. Another possibility, compatible with the first, is that the RecF pathway is ineffective as far as detectable recombination is concerned but is quite effective in another

Table 1. Thr⁺ Leu⁺ [Sm^R] Recombinant Frequencies (in %)
Obtained by Crossing Various *rec* Mutants with
the Hfr JC158[a]

| | | ExoV: | + | − | − |
| | | ExoI: | + | + | − |
recA	*recF*		RecBC (mostly)	None (almost)	RecF (mostly)
+	+		6	3×10^{-2}	4
−	+		8×10^{-4}	3×10^{-5}	1×10^{-5}
+	−		6	3×10^{-4}	6×10^{-4}

[a]This Hfr transfers clockwise from 87 on the standard map of *E. coli* (Taylor and Trotter, 1972).

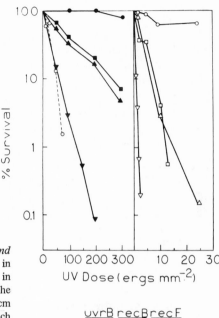

Fig. 1. UV sensitivity of various uvrB⁺ and uvrB⁻ strains. Cells were grown overnight in complete medium, washed and resuspended in buffer, and incubated at 37°C for 1 hr. The cells were irradiated at a distance of 103 cm from two G. E. G15T8 germicidal lamps, which produces a dose rate of 5.33 ergs mm⁻² sec⁻¹. Following irradiation, the cells were plated on minimum medium and allowed to incubate in the dark for two days.

function. In this paper we would like to summarize the evidence to date that the RecF pathway engages in the repair of UV-damaged DNA even in the presence of exonuclease I and supplements the RecBC pathways of repair previously hypothesized (Cooper and Hanawalt, 1972; Youngs et al., 1974; Youngs and Smith, personal communication).

The first piece of evidence is the survival curves of various mutants of *E. coli* irradiated with different fluences (i.e. doses) of UV light. The survival curves of *recB* and *recF* single-mutant strains as well as those of the double wild type and double-mutant strains are shown in Fig. 1a. It is clear that the single mutants are about equally UV sensitive, while the double mutant is much more sensitive than either single mutant. This is the type of effect that one would expect to observe if there were independent pathways of repair involving the *recB* and the *recF* genes. Note that the effect on UV sensitivity by the *recF* mutation is quite pronounced even though the *sbcB* gene is wild type and ExoI is functional. This was remarked upon by Horii and Clark (1974) because of its difference from the undetectable effect of the *recF* mutation on recombination in the same strain. There are two explanations of this phenomenon based on the idea that *in vivo*

ExoI can degrade intermediates in repair or recombination which have single-stranded portions terminated by a 3'OH group: (1) the RecF pathway of repair does not involve an intermediate sensitive to ExoI degradation while the RecF pathway of recombination does; or (2) the RecF pathway of repair does involve an intermediate sensitive to ExoI degradation but this intermediate is protected from ExoI, perhaps by being inaccessible to the enzyme. At present no evidence bearing on these explanations is available.

When an analysis of the interaction of the *recF* and *recB* mutations is carried out as outlined by Brendel and Haynes (1973), it is clear that the two mutations have a synergistic effect on the response of *E. coli* to UV irradiation. As pointed out by Brendel and Haynes, synergism is characteristic of mutations blocking pathways which act as competitors for the same substrate. We are led, thus, to expect that there are RecBC and RecF pathways of repair which act as alternative routes to the repair of at least one type of damage to DNA resulting from UV irradiation.

To eliminate the effects of excision-repair we selected a *uvrB⁻* deletion mutant and then constructed *recB⁻recF⁺*, *recB⁺recF⁻*, and *recB⁻recF⁻* derivatives of it. The survival curves of these strains are shown in Fig. 1*b*. Once again, the single-mutant derivatives have about the same sensitivity and the double-mutant derivative is very much more sensitive. The double mutant is so sensitive, in fact, that a lethal dose consists of about three dimers per genome. The sensitivity curves of the four strains also show synergistic action of the two mutations when analyzed by the method of Brendel and Haynes (1973). Since the strains were *uvrB⁻*, the RecBC and RecF pathways of repair may act as alternative routes of a non-excision-type repair of at least one type of damage. It seems possible that this type of damage consists of single-strand gaps, which are apparently produced in daughter-strand DNA when pyrimidine-dimer-containing DNA is replicated. Repair of these lesions is commonly called "postreplication repair."

Postreplication repair can be detected by examining, in alkaline sucrose gradients, the molecular weight distribution of DNA synthesized for 10 min in the presence of [³H]thymidine following exposure of the cells to UV. DNA from cells which have not been allowed to incubate for an additional period of time in nonradioactive medium is of low molecular weight and sediments slowly relative to that from an unirradiated control. Following increasing period of incubation in nonradioactive medium before centrifugation, however, the pulse-labeled DNA begins to increase in molecular weight until it cosediments with DNA from an unirradiated control (Rupp and Howard-Flanders, 1968). Using this technique, Smith and Meun (1970) were unable to detect successful postreplication repair in *recA* mutants. They did find, however, that *recB* and *recC* mutants could complete repair at doses below 180 ergs mm⁻², although at higher doses there may be some deficiency (Smith and Youngs, personal communication). This is the expected result if there is an alternative route of molecular-weight increase of the newly synthesized DNA through the RecF pathway of repair, which is blocked by *recA* mutations. We have begun, therefore, the test of our hypothesis by examining the molecular-weight increase of newly synthesized DNA in a UV-irradiated $recB^+$ $recC^+$ $sbcB^+$ $recF^-$ single mutant and a $recB^-$ $recC^-$ $sbcB^-$ $recF^-$ quadruple mutant. The data are shown in Fig. 2. In Figure 2a are shown the data for a wild-type strain, indicating that the molecular weight increase of the newly synthesized DNA in UV-irradiated cells has been completed within 30 min after the labeling period. The data for the *recF* single mutant show an incomplete molecular weight increase after 30 min and a complete molecular weight increase after 90 min. We interpret this to mean that the RecBC pathway of postreplication repair is operating slowly in the *recF* single mutant. The data for the quadruple mutant are shown in Fig. 2c. They reveal an incomplete molecular weight increase after 30 min following labeling and little further molecular weight increase at 90 min following labeling. (The data for the $sbcB^+$ $recB^-$ $recC^-$ $recF^-$ strain have been obtained and are not shown because they are the same as those for the $sbcB^-$ strain.) We are not certain at present how to interpret this. We expected to see no molecular weight increase at all, since we had presumably blocked both pathways of postreplication repair. Thus, we expected that the quadruple mutant would mimic a *recA* mutant in its inability to carry out the molecular weight increase. To explain the difference from expectation, we are considering three possibilities: (1) the mutational blocks to the RecBC or RecF pathways are "leaky" in our quadruple mutant; (2) there is a third alternative pathway of so-called postreplication repair, perhaps the SOS pathway hypothesized by Radman (1974) and Witkin and George (1973), and (3) the molecular weight increase we see is characteristic of an intermediate in repair which accumulates under the mutant conditions.

To summarize our hypothesis, we include a diagram (Fig. 3) showing the relationships among the several recognized pathways of repair of UV damage. We have chosen to illustrate that at the time of UV irradiation any part of the chromosomal DNA in an *E. coli* cell may be in the state either of having its homolog present or not. If we limit ourselves to the consideration of repair events within a single chromosome, then the two states correspond respectively to replicated DNA lying on one side between the two replicating forks and unreplicated DNA lying on the other side between the same two forks. Pyrimidine dimers form in the DNA during UV irradiation, producing the structures we have called "di-DNA" for short. di-DNA may be repaired by enzymatic photoreactivation of the pyrimidine dimers *in situ*. This is shown appropriately on the diagram.

Excision-repair is dependent on the *uvrA uvrB* endonuclease (Braun et al., this volume) to nick di-DNA, producing a chemical structure we have called "ni-di-DNA" for short. Recent evidence (Seeberg and Rupp, Part B of this book) indicates that ni-di-DNA is converted back to di-DNA by polynucleotide ligase, thus frustrating repair. Occurrence of ligase closure of the nicks is minimized by *uvrC,* which converts ni-di-DNA to a repair intermediate of unkonwn structure

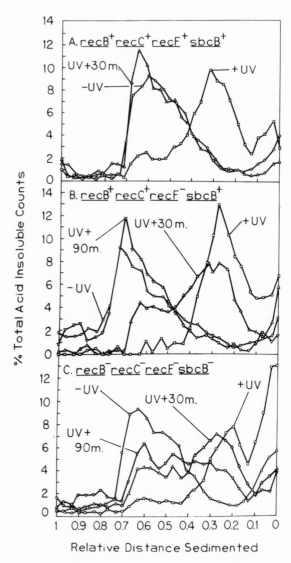

Fig. 2. Postreplication repair. Cells were grown to log phase (40 klett units) in K medium (Weigle et al., 1959), supplemented with 50 mg/ml thymidine and 0.25 mg/ml adenosine to facilitate incorporation of exogenous thymidine. The cells were then washed and resuspended in M-9 buffer, and an aliquot was illuminated with 74 ergs mm⁻² of UV radiation. K medium supplements and 100 mCi/ml [³H-*methyl*] thymidine (6.7 Ci/mmole, New England Nuclear Corp.) were then added to the illuminated and unilluminated cells, and the cells were allowed to incubate for 10 min at 37°C. Label was removed by filtration and the cells were resuspended in three times their original volume of nonradioactive K medium containing 50 μg/ml cold thymidine. They were then allowed to incubate for 30 min (△) and 90 min (▽) before sampling for sucrose gradient analysis. Irradiated but not

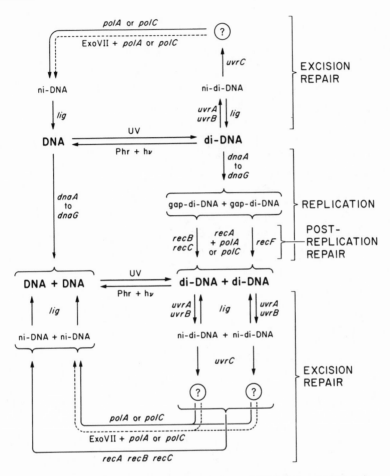

Fig. 3. Diagrammatic summary of the metabolism of UV-irradiated DNA. Information leading to this summary has been discussed and cited in the text with one exception. The involvement of *polA* and *polC* in postreplication repair has been adduced by Bridges and Sedgewick (1974) and Tait et al. (1974).

---

reincubated (□) and unirradiated but labeled (○) samples were used as controls. Spheroplast formation, sedimentation in 5–20% alkaline sucrose gradients, and collection and washing of fractions were performed as described by Smith and Meun (1970). Three different strains were tested: (*A*) AB1157, (*B*) JC9239, and (*C*) JC3904. The relevant genotypes of these strains are listed above. Samples were counted for 5 min each. The total amount of label distributed in the gradients is indicated as follows in counts per 5 min:

A: ○ = 35,516   □ = 9857   △ = 22,939
B: ○ = 19,689   □ = 3079   △ = 4199   ▽ = 5856
C: ○ = 2160   □ = 10,473   △ = 5771   ▽ = 3873.

(Seeberg and Rupp, this volume). This intermediate is probably the substrate for actual excision and replacement steps. Previous evidence had indicated that DNA polymerase I (i.e. PolI) is involved in these steps (Kelley et al., 1969; Setlow and Kornberg, 1972; Youngs and Smith, 1973a). It appears now that DNA polymerase III (i.e., PolIII) may substitute for PolI in performing both excision and replacement (Youngs and Smith, 1973b). During this meeting, Chase and Richardson reported that exonuclease VII (i.e. ExoVII) may also perform dimer excision and then require one of the polymerases to replace the polynucleotide excised. In addition to these excision-repair pathways, two others have been named "long patch repair" (Cooper and Hanawalt, 1972) and "growth medium dependent repair" (Youngs et al., 1974). Both of these depend on recA, recB, and RecC, and for diagrammatic purposes it is simplest to assume that they are the same. Because they depend on uvrA and uvrB, it seems warranted to include these pathways under excision-repair, as do their discoverers. In our diagram we have included the recA-, recB-, and recC-dependent excision-repair pathways only at the level of the di-DNA which is present along with its homolog. This seems reasonable to us because we think that rec genes operate in pathways in which interaction between homologous DNA elements is involved. We acknowledge that this may be found to be an incorrect assumption.

The third major repair pathway, commonly called "postreplication repair," differs from photoreactivation and excision in that it does not remove dimers from di-DNA. Rather, postreplication repairs secondary damage to the DNA caused when replication proceeds across a nonphotoreactivated or unexcised dimer. Presumably, replication stops at the dimer and resumes at the initiation point of the next Okazaki fragment, resulting in a gap in the newly synthesized daughter strand (Rupp and Howard-Flanders, 1968). We tend to view postreplication repair as being part of replication rather than subsequent to it, since completion of replication depends upon covalent closure of fragments of low-molecular-weight newly synthesized DNA. When dimers are present, the final covalent closures appear to occur by recombination between the damaged double helix containing a gap opposite a pyrimidine dimer (gap-di-DNA) and an undamaged region of its sister homolog (Rupp et al., 1971; Ley, 1973; Ganesan, 1974).

Our hypothesis is that there are at least two alternative pathways of this so-called postreplication repair, one involving recB and recC (a RecBC pathway of repair) and the other involving recF (the RecF pathway of repair).

ACKNOWLEDGMENTS

One of us (R. H. R.) is grateful to Dr. Kendric Smith for allowing him to visit his laboratory and learn how to run sucrose gradients, and to Dr. Klaus Martignoni, who bore the brunt of the tutelage. We are also indebted to Ann

Templin and Candace Newby for their help in strain construction and determination of UV-sensitivity curves.

R. H. R. was suppored by U. S. Public Health Service Training Grant No. GM 367-14. The research was supported by Public Health Service Research Grant No. AI 05371 from the National Institute of Allergy and Infectious Diseases.

REFERENCES

Brendel, M. and Haynes, R. H. (1973). *Mol. Gen. Genet.* **125**, 197–216.

Clark, A. J. (1973). *Ann. Rev. Genet.* **7**, 67–86.

Cooper, P. K. and Hanawalt, P. C. (1972). *Proc. Nat. Acad. Sci. U.S.A.* **69**, 1156–1160.

Goldmark, P. J. and Linn, S. (1972). *J. Biol. Chem.* **247**, 1849–1860.

Horii, Z. I. and Clark, A. J. (1974). *J. Mol. Biol.* **80**, 327–344.

Ganesan, A. K. (1974). *J. Mol. Biol.* **87**, 103–119.

Kelley, R. R., Atkinson, M. R., Huberman, J. and Kornberg, A. (1969). *Nature* **224**, 495–501.

Kushner, S. R., Nagaishi, H., Templin, A. and Clark, A. J. (1971). *Proc. Nat. Acad. Sci. U.S.A.* **68**, 824–827.

Lehman, I. R. and Nussbaum, A. L. (1964). *J. Biol. Chem.* **239**, 2628–2636.

Ley, R. P. (1973). *Photochem. Photobiol.* **18**, 87–95.

Radman, M. (1974). *Molecular and Environmental Aspects of Mutagenesis.* Thomas, Springfield, Ill.

Rupp, W. D. and Howard-Flanders, P. (1968). *J. Mol. Biol.* **31**, 291–304.

Rupp, W. D., Wilde, C. E., III, Reno, D. L. and Howard-Flanders, P. (1971). *J. Mol. Biol.* **61**, 25–44.

Setlow, P. and Kornberg, A. (1972). *J. Biol. Chem.* **247**, 232-240.

Smith, K. C. and Meun, D. H. C. (1970). *J. Mol. Biol.* **51**, 459–472.

Tait, R. C., Harris, A. C. and Smith, D. W. (1974). *Proc. Nat. Acad. Sci. U.S.A.* **71**, 675–679.

Taylor, A. L. and Trotter, C. D. (1972). *Bacteriol. Rev.* **36**, 504–524.

Weigle, J., Meselson, M. and Paigen, K. (1959). *J. Mol. Biol.* **1**, 379–386.

Witkin, E. M. and George, D. L. (1973). *Genetics* **73**, (suppl.) 91–108.

Yajko, D. M., Valentine, M. C. and Weiss, B. (1974). *J. Mol. Biol.* **85**, 323–343.

Youngs, D. A. and Smith, K. C. (1973*a*). *J. Bacteriol.* **116**, 175–182.

Youngs, D. A. and Smith, K. C. (1973*b*). *Nature New Biol.* **244**, 240–241.

Youngs, D. A., van der Schueren, E. and Smith, K. C. (1974). *J. Bacteriol.* **117**, 717–725.

38

The Degradation of Duplex DNA by the *recBC* DNase of *Escherichia coli*

Stuart Linn and Vivian MacKay[1]

Department of Biochemistry
University of California
Berkeley, California 94720

ABSTRACT

The catalytic reactions of the *recBC* enzyme of *Escherichia coli* with various substrates are reviewed, and a model for the sequence of events in the degradation of duplex DNA is presented. The potential of the enzyme to take part in genetic recombination and repair is discussed.

The *recBC* enzyme of *Escherichia coli* is the product of the *recB* and *recC* genes (Tomizawa and Ogawa, 1972). Similar DNases are widespread among bacterial species; the involvement of this class of enzymes in genetic recombination and DNA repair has been implied by a loss of the enzyme activity and a concomitant decrease in the frequency of genetic transformation in certain mutants of *Diplococcus pneumoniae* (Vovis and Buttin, 1970), *Haemophilus influenzae* (Greth and Chevallier, 1973), and *Bacillus subtilis* (Chestukhin et al., 1972) and by the recombination deficiency and radiation sensitivity of *recB* and *recC* mutants of *E. coli* (Clark, 1971). To date, this class of enzymes has not been reported in eukaryotic organisms. In this report we shall give a short synopsis of our previous observations in the *in vitro* activity of the enzyme, then present some of our ideas on the mechanism of degradation of duplex DNA by the enzyme. In addition, it should become apparent that at this point studies of the

[1] Present address: Waksman Institute of Microbiology, Rutgers University, New Brunswick, New Jersey 08903.

Table 1. Summary of the Catalytic Reactions
of the *recBC* DNase with Various Substrates

Linear DNA (duplex and single-stranded)
  ATP is absolutely required[a]
  Degradation is exonucleolytic[a]
  Degradation is processive, acting both $5' \rightarrow 3'$ and $3' \rightarrow 5'$
  Limit products are oligonucleotides (3–8)

Circular DNA (single-stranded)
  Random endonucleolytic breaks[b]
  ATP stimulates 7- to 9-fold[b]

Circular DNA (duplex)
  Resistant to digestion, unless gaps are present

ATP
  Hydrolyzed to ADP and P_i
  ATPase requires polydeoxyribonucleotide cofactor
  ATPase need not be coupled to DNA hydrolysis

---

[a]In the absence of ATP, single-stranded DNA is subject to random endonucleolytic scissions by the endonuclease function while duplex DNA is unaffected.
[b]When ATP is present the enzyme opens the circle by an endonucleolytic scission, then hydrolyzes the resulting linear DNA with the ATP-dependent exonuclease.

enzyme must begin to utilize other cellular components if we are to understand the *in vivo* action of the enzyme.

The enzyme has been purified roughly 20,000-fold to obtain a preparation which appears electrophoretically pure (Goldmark and Linn, 1972). Although the low enzyme yield (several μg enzyme per liter of bacterial culture) precludes extensive structural analysis, the enzyme was shown to consist of one each of two subunits of molecular weights approximately 140,000 and 130,000. These subunits are evidently the products of the *recB* and *recC* genes (Tomizawa and Ogawa, 1972).

As summarized in Table 1, the purified enzyme has been shown to catalyze several distinct enzymatic reactions (Goldmark and Linn, 1970, 1972; Karu et al., 1973). An ATP-dependent exonuclease hydrolyzes duplex or single-stranded DNA to a limit product of oligonucleotides with an average chain length of 4.5. The ATP requirement is complex, in that maximal DNase activity is observed with low ATP concentrations (roughly 30 μM for duplex DNA and 160 μM for single-stranded DNA) and higher ATP concentrations inhibit the reaction (Fig. 1). The reaction is processive but acts with both a $3' \rightarrow 5'$ and a $5' \rightarrow 3'$ polarity. It is unaffected by the presence or absence of phosphomonoester groups at the DNA termini. In addition, the enzyme has an endonuclease activity that is specific for single-stranded DNA. ATP stimulates this reaction sevenfold but is not absolutely required.

Duplex circular DNA, whether nicked or covalently closed, is resistant to hydrolysis. Removal of the phosphomonoester group at a nick does not affect the resistance, but extension of the nick to a gap of rougly five nucleotides renders the DNA susceptible to hydrolysis. A question which we are currently pursuing is whether this hydrolysis commences through endonucleolytic scission of the single-stranded portion of the gap or through extension of the gap by exonucleolytic hydrolysis from the exposed termini at the gap. Irradiation of circular duplex DNA with either 6 krads of X-rays or 2×10^4 ergs/mm^2 of ultraviolet light does not render it susceptible to any action by the *recBC* enzyme. (Linear DNA exposed to these levels of irradiation is hydrolyzed normally by the enzyme.)

During the exonuclease reaction, 2–40 molecules of ATP are hydrolyzed to ADP and inorganic phosphate per DNA phosphodiester bond cleaved, the degree of hydrolysis being dependent primarily upon the initial ATP concentration (see Fig. 1). Although the ATPase requires the presence of a polydeoxyribonucleotide, it need not be coupled to DNA hydrolysis, since the former reaction proceeds with the nondegradable polymers, RNA-DNA hybrid molecules, and duplex DNA with interstrand cross-links (Karu and Linn, 1972; Karu et al., 1973).

In order to explain some of the peculiar properties of the enzyme, we have examined in more detail the mechanism of digestion of duplex DNA (Karu et al., 1973; MacKay and Linn, 1974). Under optimal reaction conditions, only limit

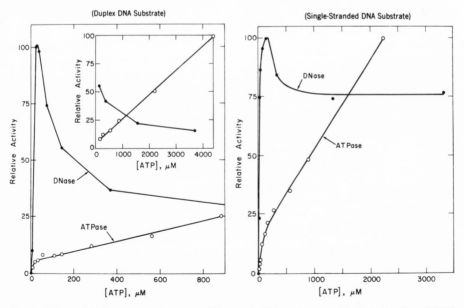

Fig. 1. *Effect of ATP concentration upon DNase and ATPase.* Assays utilized duplex *E. coli* DNA or single-stranded fd DNA which had been converted to a linear form with pancreatic DNase.

product material and undegraded DNA molecules can be distinguished during the hydrolysis (Fig. 2, *inset*). This effect is due to the processive nature of the enzyme and the fact that the initiation of the degradation of a molecule is slow relative to the continuation of degradation. However, under conditions of high ionic strength and ATP concentration, low temperature, brief time of incubation, and high enzyme-to-DNA ratios, the initiation of degradation can be partially synchronized and the rate of degradation can be slowed sufficiently to allow the observation and isolation of reaction intermediates by sedimentation through a neutral sucrose gradient (Fig. 2).

The material in the six regions of the gradient profile shown in Fig. 2 was characterized by sedimentation rates, isopycnic centrifugation (regions I–IV), and susceptibility to various deoxyribonucleases which are specific for single-stranded DNA. Regions I–IV (and the material sedimenting faster than untreated T7 DNA) consist of duplex molecules of varying length, containing single-stranded tails of up to several thousand nucleotides at both 3′ and 5′ termini. (Such molecules were also visualized by electron microscopy.) Region V mainly consists of single-stranded fragments several hundred nucleotides in length, whereas region VI contains shorter single-stranded pieces including the limit product, acid-soluble oligonucleotides. No free single-stranded DNA longer than several hundred nucleotides is detected, no nicks within duplex regions are observed, and, in general, during the digestion roughly one-fifth of the total single-stranded material is acid soluble.

In order to account for the presence of these intermediates, the processive

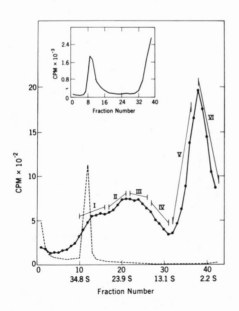

Fig.2. *Sedimentation profile of intermediates present during the hydrolysis of duplex T7 DNA.* ^{3}H-labeled T7 DNA was treated for 45 sec at 23°C with four molecules of *recBC* enzyme per DNA end in the presence of 20 mM $MgCl_2$, 100 mM NaCl, 5 mM ATP, 0.67 mM DTT, and 5 mM glycylglycine-NaOH, pH 7.0. Excess EDTA was then added and the digest was sedimented through a 5–20% linear sucrose gradient containing 10 mM tris-HCl, pH 8.2, 5 mM EDTA, and 0.25 M NaCl. ——, DNA incubated with DNase; - - -, DNA incubated without DNase. Inset: ^{3}H-labeled T7 DNA was treated for 30 min at 37°C with 0.4 molecules of *recBC* enzyme per DNA end in the presence of 10 mM $MgCl_2$, 0.33 mM ATP, 0.67 mM DTT, and 50 mM glycylglycine-NaOH, pH 7.0, then analyzed on a neutral sucrose gradient.

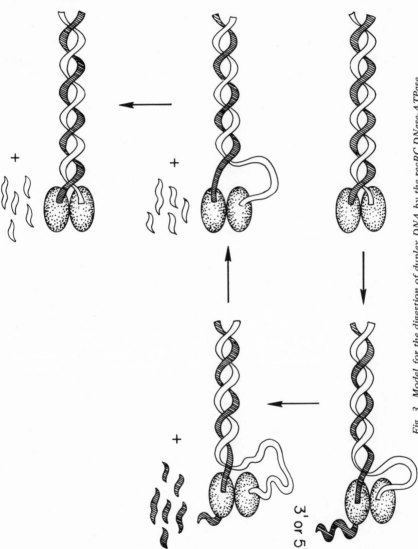

Fig. 3. Model for the digestion of duplex DNA by the recBC DNase-ATPase.

nature of the reaction, and the apparent ability of the enzyme to act both $3' \to 5'$ and $5' \to 3'$, the following sequence for the degradation of duplex DNA is proposed (Fig. 3). The enzyme binds the termini of both strands of a DNA duplex end and begins to unwind the DNA by tracking down one strand while remaining bound to the terminus of the other. Proceeding down the tracked strand it clips off fragments several hundred nucleotides in length until a tail of several thousand nucleotides is generated. The enzyme then begins to degrade the tail from its terminus until a shortened, wholly duplex molecule is formed. The cycle is then repeated until the molecule is entirely digested. As the small single-stranded fragments accumulate, they compete for the enzyme and get degraded to acid-soluble materials by the single-strand exonuclease activity. (Appropriate competition experiments are consistent with the last step.) This model suggests that ATP hydrolysis is required for DNA unwinding and perhaps for the tracking mechanisms, as suggested by Winder (1972). Interstrand cross-links would thus block degradation by preventing DNA unwinding. In essence this model states that the reaction unique to duplex DNA substrates is the unwinding of the DNA; DNA hydrolysis is carried out by a combination of the single-stranded exonucleolytic and endonucleolytic activities.

During repair, the enzyme might initiate its action at a preformed gap or large distortion. It could remove or unwind a large single-stranded piece of DNA, and the resulting large gap or unwound single-stranded DNA terminus could then initiate a recombination event. In order to study this hypothetical reaction, we must learn how to terminate DNA degradation by the *recBC* DNase once initiated. In this context a limiting factor is the undiscovered *recA* gene product, although naturally occurring cross-linking devices or RNA-DNA hybrid regions could conceivable control the enzyme action. In the immediate future we plan to return to the study of gapped, circular substrates in order to attempt to find conditions in which the enzyme might unwind but not extensively degrade the substrate. Ultimately, however, to reproduce the normal enzyme action *in vitro* we will have to study the enzyme in the presence of many cellular constituents [e.g., the *recA* gene product, DNA-binding or unwinding proteins, DNA poly-merase(s), or DNA ligase].

ACKNOWLEDGMENTS

The research discussed herein was supported by Contract AT(04-3)-34 from the U. S. Atomic Energy Commission and Grant GM19020 from the National Institutes of Health. V. M. was a postdoctoral fellow of the National Institute of General Medical Sciences (Grant GM55131) and of the American Cancer Society (Grant PF922).

REFERENCES

Chestukhin, A. V., Shemyakin, M. F., Kalinina, N. A. and Prozorov, A. A. (1972). *FEBS Letters*, **24**, 121–125.

Clark, A. J. (1971). *Ann. Rev. Microbiol.* **25**, 437–464.

Goldmark, P. J. and Linn, S. (1970). *Proc. Nat. Acad. Sci. U.S.A.* **67**, 434–441.

Goldmark, P. J. and Linn, S. (1972). *J. Biol. Chem.* **247**, 1849–1860.

Greth, M. L. and Chevallier, M. R. (1973). *Biochem. Biophys. Res. Commun.* **54**, 1–8.

Karu, A. E. and Linn, S. (1972). *Proc. Nat. Acad. Sci. U.S.A.* **69**, 2855–2859.

Karu, A. E., MacKay, V., Goldmark, P. J. and Linn, S. (1973). *J. Biol. Chem.* **248**, 4874–4884.

MacKay, V. and Linn, S. (1974). *J. Biol. Chem.* **249**, 4286–4294.

Tomizawa, J. and Ogawa, H. (1972). *Nature New Biol.* **239**, 14–16.

Vovis, G. F. and Buttin, G. (1970). *Biochem. Biophys. Acta* **224**, 29–41.

Winder, F. G. (1972). *Nature New Biol.* **236**, 75–76.

39

Analysis of Temperature-Sensitive *recB* and *recC* Mutations

Sidney R. Kushner[1]

Department of Biochemistry
Stanford University School of Medicine
Stanford, California 94305

ABSTRACT

The *in vivo* pleiotropic effects associated with the temperature-sensitive *recB270* and *recC271* mutations have been correlated with the *in vitro* behavior of the *recBC* nucleases coded for by these alleles. The ATP-dependent breakdown of double-stranded DNA is essential for cell viability, radiation repair, and genetic recombination. Temperature sensitivity can be suppressed *in vitro* and *in vivo*.

Mutations in the *recB* and/or *recC* genes result not only in reduced recombination proficiency but also in increased sensitivity to ultraviolet light and the cross-linking agent mitomycin C (Emmerson, 1968; Willetts and Mount, 1969). In addition, growing cultures of *recB⁻* and/or *recC⁻* deficient strains contain a large fraction of inviable cells (Haefner, 1968; Capaldo-Kimball and Barbour, 1971).

The *recB* and *recC* gene products yield an enzyme which catalyzes three classes of reactions: ATP-dependent exonucleolytic degradation of single- and double-stranded DNA; ATP-stimulated endonucleolytic digestion of single-stranded closed circles; and DNA-dependent hydrolysis of ATP (Goldmark and Linn, 1972). The recent isolation of temperature-sensitive mutations in *recB* and

---

[1] Present address: Department of Biochemistry, University of Georgia, Athens, Georgia 30602.

Table 1. Summary of Phenotypic Properties Associated with
recB270 and *recC271*

Property	Allele	Temperature	
		30°C	43°C
Ultraviolet	*recB270*	Resistant	Sensitive
sensitivity	*recC271*	Resistant	Sensitive
	recB270 recC271	Sensitive	Very sensitive
Mitomycin C	*recB270*	Resistant	Sensitive
sensitivity	*recC271*	Resistant	Sensitive
	recB270 recC271	Sensitive	Very sensitive
Recombination	*recB270*	Normal	Slightly reduced
	recC271	Normal	Slightly reduced
	recB270 recC271	Slightly reduced	Deficient
Effects of a	*recB270*	Viable	Lethal
polA mutation	*recC271*	Viable	Lethal
	recB270 recC271	Viable	Lethal
Viability	*recB270*	Normal	Reduced
	recC271	Normal	Reduced
	recB270 recC271	Normal	Sharply reduced

recC by Tomizawa and Ogawa (1972) provided a possible means of determining the *in vivo* role of this complex enzyme.

recB270 (formerly *recBts1*) and *recC271* (formerly *recCts1*) were transferred by P1 transduction into the genetically defined AB1157 (Bachmann, 1972) background. The two single mutants and the double mutant were analyzed for their sensitivity to ultraviolet light and to mitomycin C and for their ability to carry out genetic recombination. Furthermore, cell viability and their behavior in the presence of a *polA1* mutation were determined. The results of these experiments are summarized in Table 1. Sensitivity to ultraviolet light and to mitomycin C were partially suppressed by the addition of 1% NaCl to the plating medium.

Assuming that the multiple activities measured *in vitro* are functional *in vivo*, one explanation for the observed results is that in single *recB270* and *recC271* mutants one or more activities are thermosensitive but the remaining ones are not. In contrast, a second possibility involves residual levels of activity in the mutants. Partial inactivation at 43°C of one or more of the activities associated with the *recBC* nuclease could result in increased sensitivity to UV and mitomycin C but would be sufficient to allow almost normal levels of genetic recombination and cell viability.

In order to distinguish these two possibilitiés, exonuclease V from *recB270, recC271, recB270 recC271,* and wild-type strains was partially purified by the method of Goldmark and Linn (1972). The enzymes used in these experiments were absolutely dependent on ATP for hydrolysis of double- and single-stranded DNA and absolutely dependent on DNA for ATP hydrolysis. Deoxyribonuclease activity was obtained from all strains although in sharply reduced yields from the double mutant. The three mutant enzymes showed a marked decrease in their ability to degrade double-stranded DNA at 43°C with or without pre-incubation at that temperature (Fig. 1*A*, Fig. 2*A*). However, unlike *in vivo* where the two single mutants appear identical, *in vitro* the *recC271* enzyme is much less affected by the high temperature. A similar analysis of the ability to breakdown single-stranded DNA in either an exo- or endonucleolytic manner

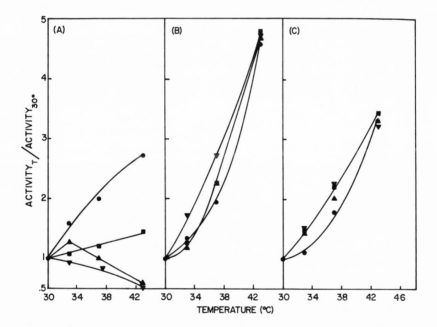

Fig. 1. Thermosensitivity of recBC deoxyribonuclease. Incubations were carried out at pH 7 for 10 min using assay procedures described by Goldmark and Linn (1972). The activity observed at a particular temperature was divided by the activity observed at 30°C. (*A*) Double-stranded DNA substrate; (*B*) single-stranded DNA substrate; (*C*) closed-circular single-stranded DNA substrate; ●———●, wild type; ▲———▲, *recB270;* ■———■, *recC271;* ▼———▼, *recB270 recC271.*

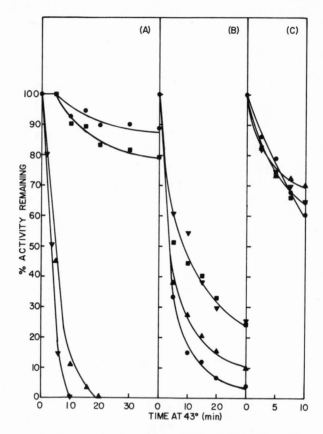

Fig. 2. Heat inactivation of the recBC deoxyribonuclease. The enzyme samples were preincubated in the presence of buffer (pH 8), dithiothreitol, and substrate for various time periods. The reactions were started by adding ATP and $MgCl_2$ and allowed to proceed for 20 min at 43°C. (*A*) Double-stranded DNA; (*B*) single-stranded DNA; (*C*) closed-circular single-stranded DNA; ●——●, wild type; ▲——▲, *recB270;* ■——■, *recC271;* ▼——▼, *recB270 recC271.*

showed that none of the mutant enzymes was either thermosensitive (Fig. 1*B,C*) or thermolabile (Fig. 2*B,C*).

It is concluded from these results that the rate-limiting step of exonuclease V *in vivo* is the ATP-dependent exonucleolytic degradation of double-stranded DNA. The enzyme's ability to digest single-stranded DNA in either an exo- or endonucleolytic manner could either be side reactions which are detected *in vitro* but are inoperative *in vivo* or activities which are necessary *in vivo* but occur after the action on double-stranded DNA.

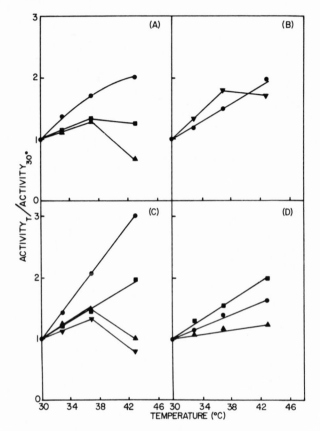

Fig. 3. Thermosensitivity of ATP-dependent double-stranded DNA hydrolysis at various ATP concentrations. Enzymes were assayed at pH 8 in the presence of double-stranded DNA for 10 min at various temperatures as described by Kushner and Lehman (1974), except for variation in ATP concentration. *(A)* 10 μM ATP; *(B)* 200 μM ATP; *(C)* 20 μM ATP; *(D)* 1mM ATP; ●————●, wild type, ▲————▲, *recB270;* ■————■, *recC271;* ▼————▼, *recB270 recC271.*

The *in vivo* suppression of temperature sensitivity by the addition of NaCl to the plating medium was examined *in vitro* with the partially purified enzymes. Various modifications of or additions to the reaction mixture proved ineffective in suppressing the temperature sensitivity. However, as shown in Fig. 3, specific alterations in the ATP concentration yielded *in vitro* suppression of temperature sensitivity. The relationship of this phenomenon to *in vivo* suppression is still not clear. A more detailed description of this work will be published elsewhere (Kushner, 1974*a,b*).

ACKNOWLEDGMENTS

This work was supported in part by a grant from the United States Public Health Service (GM-06196) and a Postdoctoral Fellowship from the National Institute of General Medical Sciences (GM-47038-03).

I wish to thank I. R. Lehman for his advice and encouragement and J. Tomizawa for graciously providing his strains.

REFERENCES

Bachmann, B. J. (1972). *Bacteriol. Rev.* **36**, 525.
Capaldo-Kimball, F. and Barbour, S. D. (1972). *J. Bacteriol.* **106**, 204.
Emmerson, P. T. (1968). *Genetics* **60**, 19.
Goldmark, P. J. and Linn, S. (1972). *J. Biol. Chem.* **247**, 1849.
Haefner, K. (1968). *J. Bacteriol.* **96**, 652.
Kushner, S. R. (1974*a*). *J. Bacteriol.* **120**, 1213.
Kushner, S. R. (1974*b*). *J. Bacteriol.* **120**, 1219.
Tomizawa, J. and Ogawa, H. (1972). *Nature New Biol.* **239**, 14.
Willetts, N. S. and Mount, D. W. (1969). *J. Bacteriol.* **100**, 923.

40

Recombination and Postreplication Repair

W. Dean Rupp, Alan D. Levine, and Zeljko Trgovcevic

Radiobiology Laboratories
Yale University School of Medicine
New Haven, Connecticut 06510

ABSTRACT

The available data concerning postreplication repair are summarized. In *Escherichia coli*, recombination is implicated in this repair because the *recA* [+] gene is necessary and because strand exchanges occur that extend over long regions. Other experiments involving phage-induced resistance also point to an interrelation between recombination and repair. In this phenomenon, gene products of λ bacteriophage are introduced into bacteria, resulting in an increased resistance of the cells when they are subsequently exposed to X rays.

POSTREPLICATION REPAIR IN *ESCHERICHIA COLI*

Studies in *Escherichia coli* have demonstrated the presence of a repair mechanism that allows the cells to survive even when damaged nucleotides are present in the template DNA strands at the time of replication. Although the precise molecular mechanism of postreplication repair is not yet fully elucidated, there is appreciable knowledge about DNA structures that are present during this process.

It is known that DNA synthesis proceeds past many pyrimidine dimers to form new strands with gaps, similar in size to Okazaki fragments, that are present at intervals similar to the spacing of dimers on the template strands (Rupp and Howard-Flanders, 1968; Iyer and Rupp, 1971). During subsequent incubation, repair of gaps occurs in *recA* [+] cells but not in *recA* mutants (Rupp and Howard-Flanders, 1968; Smith and Meun, 1970; Rupp et al., 1970). When cells are marked with appropriate density labels, UV-induced exchanges between old and new DNA strands are observed (Rupp et al., 1971). This is in accord with the interpretation that successful repair of dimer-gap structures in *E. coli* may require the exchange of homologous DNA strands, thus explaining the requirement for the *recA* [+] gene.

Several additional details are also known about this repair process in *E. coli*. Although it might be thought that photoproducts could be removed from the DNA during postreplication repair, measurements of pyrimidine dimers either by chromatography (Rupp, 1972) or with T4 UV endonuclease (Ganesan, 1974) show that the dimers remain in high-molecular-weight DNA after replication. The data also indicate that the exchanged regions are much longer than would be expected if only the gap opposite the dimer were filled in with an old strand (Rupp et al., 1971), and that the exchanges between old and new strands are limited to those old strands that have replicated after UV irradiation (Rupp, 1972). Ley (1973) found that substantial repair synthesis is involved in post-replication repair, and Ganesan (1974) has recently found that dimers are distributed into all strands during postreplication repair. These results can be explained by an intermediate consisting of two DNA duplexes with a single-strand crossover. The exchange could be driven by polymerase as proposed by Radding and Meselson (personal communication), resulting in synthesis of the type found by Ley (1973). Subsequent isomerization of this structure as suggested by Sigal and Alberts (1972) could then distribute the dimers into all the strands in accord with findings of Ganesan (1974).

The role of the various recombination genes in this repair is not yet understood. Whereas the *recA* [+] gene is required for successful postreplication repair, this repair takes place quite normally in *recB* and *recC* mutant strains, as measured by centrifugation in alkaline sucrose (Smith and Meun, 1970) and by density transfer experiments in our lab. Thus this mode of repair, including strand exchanges, can take place with near normal efficiency in some Rec⁻ strains. This situation also obtains in *Haemophilus influenzae*, where Leclerc and Setlow (1972) found Rec⁻ strains that were either normal or defective in postreplication repair.

IS POSTREPLICATION REPAIR INDUCIBLE?

Radman (1974) has suggested that a mutagenic postreplication repair can be induced by UV irradiation. We have measured postreplication repair in *uvrA* and in *uvrA recB* cells under conditions where either rifampicin or chloramphenicol was added just prior to irradiation. In these cases, the pulse-labeled DNA synthesized immediately after irradiation was converted into high-molecular-weight DNA during subsequent incubation, indicating that postreplication repair was not blocked by these inhibitors. There was an indication that the pulse-labeled DNA might be more susceptible to breakdown in the presence of these drugs, but the data were not adequate to be certain about this marginal effect. These results thus extend earlier findings by Ganesan and Smith (1972) and provide no support for the presence of inducible enzymes that are required for postreplication repair.

POSTREPLICATION REPAIR IN MAMMALIAN CELLS

This mode of repair is expected to be very important for mammalian cells, because much of their DNA is replicated at times when many photoproducts remain in the DNA. Using alkaline sucrose centrifugation, a number of investi-

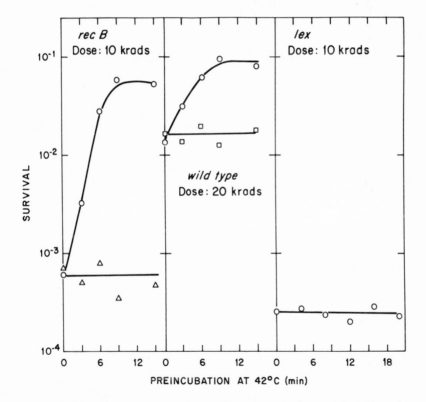

Fig. 1. *Effect of pulse-heating on the X-ray survival of several* E. coli *strains carrying* λ cI857 *prophage.* Bacteria were heated at 42°C just before X-irradiation with the indicated doses. The survival is the ratio of surviving cells after X-irradiation to unirradiated surviving cells heated at 42°C for the same length of time. The bacterial strains used were AB2470 *recB21,* DM456, and DM803 *lex-1.* Symbols: o, strains lysogenic for λcI857 *red⁺gam⁺*; △, AB2470 lysogenic for λcI857 *red3 gam⁺*; □, DM456 lysogenic for λcI857 *red⁺gam210.* The cI857 mutation in λ causes the λ repressor to be thermolabile at 42°C.

gators have obtained results that are qualitatively similar to those obtained in *E. coli* (Cleaver and Thomas, 1969; Rupp et al., 1970; Lehmann, 1972; Meyn and Humphrey, 1971). It was found by Zipser (1973) that strand exchanges were below the level of detection. Thus, if exchanges occur at all, they are confined to much smaller lengths in mammalian cells than in *E. coli*. Furthermore, Buhl et al. (1973) have suggested that photoproducts other than pyrimidine dimers may be causing the gaps in the new strands synthesized in UV-irradiated mammalian cells.

RADIORESISTANCE INDUCED BY λ BACTERIOPHAGE

Since Rec⁻ *E. coli* strains are more sensitive to ionizing radiation than are Rec⁺ strains, recombination enzymes are thought to be involved in the repair of X-irradiated DNA. This interrelation between repair and recombination is further confirmed by recent studies of a phenomenon called phage-induced radioresistance (Trgovcevic and Rupp, 1974). It has been shown that introduction of λ bacteriophage gene products into a cell by transient induction of a λ lysogen results in a substantial increase in radioresistance (Fig. 1). Genetic studies, summarized in Table 1, demonstrate the existence of at least two

Table 1. Effect of Bacterial and Bacteriophage Genotype on
Lambda-Induced Radioresistance[a]

Bacterial genotype	X-ray survival without transient induction	Effect of transient induction on X-ray resistance		
		λ *red⁺ gam⁺*	λ *red⁻ gam⁺*	λ *red⁺ gam⁻*
recB	0.04% (10 krads)	+ +	−	+ +
recB recC sbcA	0.1% (20 krads)	+ +	−	
recB recC xonA	0.03% (20 krads)	+	−	
recB recC sbcB recF	0.07% (8 krads)	±	−	
recB recC recF[b]	0.1% (8 krads)	−		
recB recC sbcB	0.1% (20 krads)	−		
recA[c]		−		
lex	0.04% (10 krads)	−		
polA	0.02% (6 krads)	+ +	+ +	±
recF	0.2% (10 krads)	+	+	
"Wild type" (DM 456)	1.4% (20 krads)	+	+	−

[a]−,No effect of transient induction on X-ray survival (dose reduction factor=0.9–1.1); ±, slight effect (dose reduction factor=1.1–1.3); +, slight effect (dose reduction factor=1.3–2.0); ++ strong effect (dose reduction factor >2.0).
[b]In some experiments, a slight increase in survival after transient induction was observed.
[c]Trgovcevic, Petranovic, and Zgaga (unpublished).

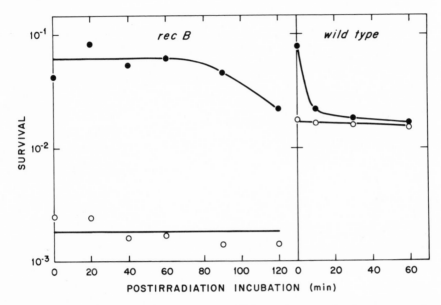

Fig. 2. Effect of postirradiation incubation at 30°C on red-*promoted and* gam-*promoted radioresistance.* At the times indicated, surviving cells were either scored directly without any additional treatment (○) or first pulse-heated for 5 min at 42°C and then scored (●). In the left panel, strain AB2470 carrying the *recB21* mutation and lysogenized with λcI857 was irradiated with 10 krads. In the right panel, the wild-type strain DM456, also lysogenic for λcI857, was irradiated with 20 krads.

separate classes of radioresistance, as well as certain strains that do not show phage-induced radioresistance.

The first type is observed in bacteria that are mutated in the *recB* gene and depends on the presence of the λ *red*⁺ genes (Fig. 1). It is thought that the λ *red*⁺ products, λ exonuclease and γ-protein, compensate for the missing RecB,C nuclease in the repair of X-ray damage. The maximum radioresistance is obtained even when the λ *red*⁺ products are supplied 1 hr after irradiation (Fig. 2). Although the mechanism of the *red*⁺-dependent radioresistance is uncertain, the affected lesions are apparently not DNA strand breaks, because the extent of DNA strand rejoining is not altered by the *red*⁺ products.

The second type of phage-induced radioresistance requires the *gam*⁺ product of λ and is observed in wild-type and *polA* strains. The λ *gam*⁺ gene product must be present immediately after irradiation to exert its full effect. In its presence, DNA breakdown is decreased, and a greater fraction of the DNA is converted back to a high molecular weight. Since it is known from studies by Sakaki et al. (1973) that γ-protein inhibits the RecB,C nuclease, it may act in this instance by inhibiting the attack of the RecB,C nuclease on particular susceptible DNA structures that are present immediately after irradiation. If the

DNA breakdown can be prevented at this early critical stage, it can then be a substrate for repair enzymes. Thus, the net effect of the action of the γ-protein would seem to be to prolong the time during which the X-irradiated DNA can be a substrate for the repair enzymes.

ACKNOWLEDGMENT

This work was supported by United States Public Health Service Grant CA 06519 and Grant GM 18587.

REFERENCES

Buhl, S., Setlow, R. and Regan, J. (1973). *Biophys. J.* **13**, 1265.
Cleaver, J. and Thomas, G. (1969). *Biochem. Biophys. Res. Commun.* **36**, 203.
Ganesan, A. (1974). *J. Mol. Biol.* **87**, 103.
Ganesan, A. and Smith, K. (1972). *J. Bacteriol.* **111**, 575.
Iyer, V. N. and Rupp, W. D. (1971). *Biochim. Biophys. Acta* **228**, 117.
Leclerc, J. and Setlow, J. (1972). *J. Bacteriol.* **110**, 930.
Lehmann, A. (1974). *J. Mol. Biol.* **66**, 319.
Ley, R. (1973). *Photochem. Photobiol.* **18**, 87.
Meyn, R. and Humphrey, R. (1971). *Biophys. J.* **11**, 295.
Radman, M. (1974). In *Molecular and Environmental Aspects of Mutagenesis* (in press).
Rupp, W. D. (1972). *Biophys. Soc. Abstr.* **12**, 148a.
Rupp, W. D. and Howard-Flanders, P. (1968). *J. Mol. Biol.* **31**, 291.
Rupp, W. D., Zipser, E., vonEssen, C., Reno, D., Prosnitz, L. and Howard-Flanders, P. (1970). In *Dose Rate Effects in Radiobiology as Applied to Radiotherapy* p. 1. Brookhaven Nat. Lab. Publication No. 50203 (C–57).
Rupp, W. D., Wilde, C., Reno, D. and Howard-Flanders, P. (1971). *J. Mol. Biol.* **61**, 25.
Sakaki, Y., Karu, A., Linn, S. and Echols, H. (1973). *Proc. Nat. Acad. Sci. U.S.A.* **70**, 2215.
Sigal, N. and Alberts, B. (1972). *J. Mol. Biol.* **71**, 789.
Smith, K. and Meun, D. (1970). *J. Mol. Biol.* **51**, 459.
Trgovcevic, Z. and Rupp, W. D. (1974). *Proc. Nat. Acad. Sci. U.S.A.* **71**, 503.
Zipser, E. (1973). Ph. D. Thesis, Yale University.

41

Ultraviolet-Light-Induced Incorporation of Bromodeoxyuridine into Parental DNA of an Excision-Defective Mutant of *Escherichia coli*

Ronald D. Ley

Division of Biological and Medical Research
Argonne National Laboratory
Argonne, Illinois 60439

ABSTRACT

Bromodeoxyuridine-containing regions approximately 1.5×10^4 nucleotides in length, and at intervals equivalent to the pyrimidine dimer content of the DNA, have been observed in the parental DNA of an excision-defective strain of *Escherichia coli* exposed to 10 ergs mm^{-2} at 254 nm followed by prolonged incubation in the presence of bromodeoxyuridine.

According to the model for excision-repair, pyrimidine dimer excision is prerequisite for repair replication. Therefore, *uvr* mutants of *Escherichia coli* that are unable to excise pyrimidine dimers would not be expected to exhibit ultraviolet-light-induced repair replication [operationally defined as incorporation of bromodeoxyuridine (BrdUrd) into parental DNA]; and, in previous reports, a *uvrA6* mutant was observed to undergo little (Ley and Setlow, 1972) or no (Cooper and Hanawalt, 1971) UV-induced repair replication. However, under conditions of a low dose of 254-nm light (10 ergs mm^{-2}) and prolonged postirradiation incubation in the presence of BrdUrd, I have observed extensive incorporation of BrdUrd into the parental DNA of a *uvr*$^-$ mutant of *E. coli* C (Ley, 1973). The presence of regions containing bromouracil (BrUra) in parental DNA was observed by the BrUra-photolysis technique (Regan et al., 1971) as described below.

Cells were uniformly labeled in their DNA with tritiated thymidine, exposed to 10 ergs mm^{-2} at 254 nm, and then incubated for 200 min in growth medium containing nonradioactive BrdUrd instead of thymidine. During the incubation, approximately 75% of the parental DNA had undergone normal semiconservative replication. The incubated cells were then exposed to 313-nm light and lysed in alkali, and the released DNA was sedimented through an alkaline sucrose gradient. If BrdUrd is incorporated into parental DNA during postirradiation incubation, the BrUra-containing regions can be selectively photolyzed by 313-nm radiation. Hence, irradiation at 313 nm reduces the sedimentation rate of the BrUra-containing DNA in an alkaline sucrose gradient.

Molecular weights calculated from the sedimentation profiles were used to determine the number of 313-nm-induced single-strand breaks in DNA of cells exposed to 0 or 10 ergs mm^{-2} at 254 nm and incubated with either thymidine or BrdUrd prior to exposure to 313-nm light. These data are presented in Fig. 1 as the number of breaks per 10^8 daltons plotted against exposure to 313-nm radiation. From these plots it is apparent that the initial 254-nm exposure, the BrdUrd treatment, and the inability of the cell to excise pyrimidine dimers are necessary to obtain the maximum increase in sensitivity of cellular DNA to

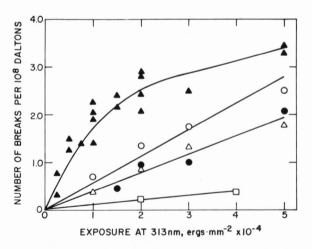

Fig. 1. The induction of single-strand breaks (or alkali-labile bonds) in tritium-labeled parental DNA by 313-nm irradiation of uvr⁻ and uvr⁺ strains of E. coli C which have been exposed to 0 or 10 ergs mm⁻² at 254 nm and subsequently incubated for 200 min with BrdUrd or dThd. Uvr⁻, 0 (△) and 10 (▲) ergs mm⁻² plus BrdUrd. Uvr⁻, 10 ergs mm⁻² plus dThd (□). Uvr⁺, 0 (●) and 10 (○) ergs mm⁻² plus BrdUrd.

313-nm light (closed triangles). Obviously these results cannot be explained in terms of the excision mode of repair, since the incorporation of BrdUrd observed under these conditions was more apparent in the *uvr⁻*(closed triangles) than in the *uvr⁺* (open circles) strain. The parental DNA of unirradiated cells incubated with BrdUrd was also sensitized, though to a lesser degree than with irradiated cells, to 313-nm radiation (open triangles and closed circles). The sensitization observed with unirradiated cells presumably does not result from incorporation of BrdUrd into parental DNA, but it appears that a small fraction (~1%) of the photolytic events that occur in the BrUra-substituted daughter strand result in a break in the base-paired parental strand of DNA (Beattie, 1972). If this nonspecific sensitization of parental DNA is substracted from the 313-nm dose-response curve for *uvr⁻* cells exposed to 10 ergs mm⁻² at 254 nm, the degradation due to random breakage of the parental DNA is eliminated, and the remaining sensitivity is due presumably to UV-induced BrUra-containing regions. From such a corrected dose-response curve one can calculate that 10 ergs mm⁻² at 254 nm followed by incubation with BrdUrd results in approximately one BrUra-containing region per 55×10^6 daltons of parental DNA. Since this exposure also results in one pyrimidine dimer per 45×10^6 daltons (Rupp and Howard-Flanders, 1968), roughly one BrUra-substituted region occurs for every pyrimidine dimer. The average size of the BrUra-containing region also may be estimated from how readily the regions are photolyzed upon exposure to 313-nm light. We calculate that each region contains, on the average, 3.7×10^3 BrUra residues. Thus, following exposure to 10 ergs mm⁻² at 254 nm and subsequent incubation with BrdUrd, the parental DNA of the *uvr⁻* strain of *E. coli* C contains BrUra-substituted regions approximately 1.5×10^4 nucleotides in length (4.5×10^6 daltons) which occur at intervals of approximately 55×10^6 daltons.

Since the observed incorporation of BrdUrd does not result from excision-repair, it is reasonable to assume that it may result from postreplication (recombination) repair. Thus, BrUra-containing regions could occur from the resynthesis of single-strand gaps in parental DNA which result from genetic exchanges as proposed by Rupp et al. (1971). Alternatively, the BrUra observed in the parental DNA might have arisen by either repair synthesis or a reciprocal recombination event involving parental and daughter strands of DNA from sister duplexes. The latter interpretation is in accord with the model for postreplication repair proposed by Ganesan (this volume).

If the incorporation of BrdUrd observed in the *uvr⁻* strain is a result of postreplication repair, only that parental DNA which has served as a template for semiconservative replication is the presence of BrdUrd should be sensitized to 313-nm light. We have found this to be the case. When UV-irradiated cells were incubated with BrdUrd for various priods of time, the fraction of parental DNA converted to a hybird density was equivalent to the fraction of DNA sensitized to 313-nm radiation (Ley, 1973).

It would appear, therefore, that BrdUrd is incorporated into parental DNA as a result of postreplication repair, whereby cells are able to circumvent pyrimidine dimers in the absence of excision-repair. The BrUra-containing regions occurs for each dimer in parental DNA which has undergone semiconservative replication. In contrast, the excision of a pyrimidine dimer in *E. coli* results, for the most part, in a BrUra-containing region 20–40 nucleotides in length (Ley and Setlow, 1972). However, some long regions of repair synthesis may also result from excision-repair (Cooper and Hanawalt, 1972).

ACKNOWLEDGMENT

This work was supported by the U.S. Atomic Energy Commission.

REFERENCES

Beattie, K. L. (1972). *Biophys. J.* **12**, 1573–1582.
Cooper, P. K. and Hanawalt, P. C. (1971). *Photochem. Photobiol.* **13**, 83–85.
Cooper, P. K. and Hanawalt, P. C. (1972). *J. Mol. Biol.* **67**, 1–10.
Ley, R. D. (1973). *Photochem. Photobiol.* **18**, 87–95.
Ley, R. D. and Setlow, R. B. (1972). *Biophys. J.* **12**, 420–431.
Regan, J. D., Setlow, R. B. and Ley, R. D. (1971). *Proc. Nat. Acad. Sci. U.S.A.* **68**, 708–712.
Rupp, W. D. and Howard-Flanders, P. (1968). *J. Mol. Biol.* **31**, 291–304.
Rupp, W. D., Wilde, C. E., III, Reno, D. L. and Howard-Flanders, P. (1971). *J. Mol. Biol.* **61**, 25–44.

42

Distribution of Pyrimidine Dimers During Postreplication Repair in UV-Irradiated Excision-Deficient Cells of *Escherichia coli* K12

Ann Ganesan

Department of Biological Sciences
Stanford University
Stanford, California 94305

ABSTRACT

During postreplication repair in excision-deficient mutants of *Escherichia coli* K-12, pyrimidine dimers are gradually lost from UV-irradiated DNA. Our data indicate that dimers are transferred, by a process which may involve genetic exchange, into daughter strands made after irradiation. Dimer transfer appears to continue through several rounds of replication, resulting in the gradual dilution of dimers into successive generations of DNA molecules.

We have used an enzymatic assay for pyrimidine dimers (Ganesan, 1973, and this volume) to follow the distribution of dimers during postreplication repair (Howard-Flanders et al., 1968) in UV-irradiated cells of *Escherichia coli*. The enzyme, T4 endonuclease V, produces single-strand breaks in DNA containing dimers. The number of breaks produced in a DNA sample by the enzyme, and from this the number of dimers originally present, was estimated from the relative sedimentation rate of the treated DNA in alkaline sucrose gradients compared to that of untreated control DNA. We found that excision-deficient (*uvrA6* or *uvrB5*) cells gradually lost dimers from UV-irradiated DNA under conditions permitting postreplication repair (Fig. 1). The dimer loss did not appear to result from residual excision or photoreactivation, since it could be prevented by starving the cells for required amino acids for 2 hr before irradiation and continuing the starvation during and after irradiation. These conditions permitted excision to occur in excision-proficient control cultures (Ganesan, 1973).

We detected dimers in daughter DNA synthesized *after* irradiation when excision-deficient cells were incubated under conditions permitting postreplication repair (Fig. 2). Thus, it appeared that dimers could be transferred from irradiated parental DNA into unirradiated daughter strands. Our data were

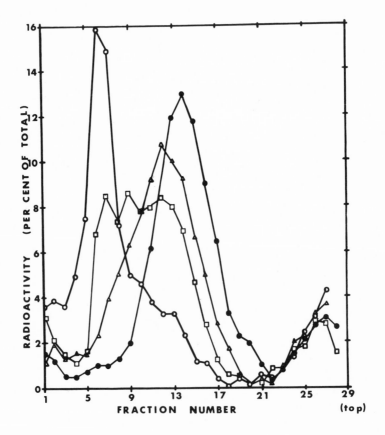

Fig. 1. Loss of dimers from UV-irradiated DNA of excision-deficient cells of E. coli. AB2499 (*uvrB5*) labeled with [³H] thymine was irradiated with 20 ergs/mm² at 254 nm (survival = 84%). Samples were removed at the indicated times thereafter, treated sith the T4 endonuclease V, and lysed, and their DNA was analyzed on alkaline sucrose gradients. Sample removed immediately after irradiation (●); at 60 min (△); at 120 min (□). Control sample removed immediately after irradiation, but not treated with nuclease (○).

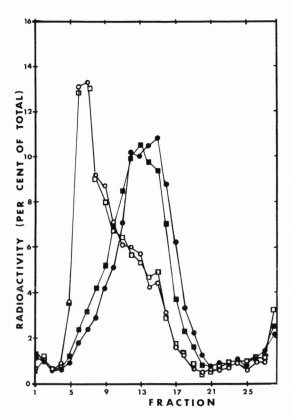

Fig. 2. Dimers in DNA synthesized in excision-deficient cells after UV irradiation. AB2499 (*uvrB5*) labeled with [^{14} C] thymine was irradiated with 40 ergs/mm² at 254 nm (survival = 50%), then incubated with the T4 endo-nuclease V and lysed, and the DNA was analyzed on alkaline sucrose gradients. DNA made before irradiation (•); DNA made after irradiation (■). One portion was analyzed without endonuclease treatment. DNA made before irradiation (○); DNA made after irradiation (□).

compatible with the idea that dimers were gradually diluted into successive generations of DNA molecules made after irradiation. The rate of loss of dimers from irradiated DNA was proportional to the rate of DNA synthesis, about one-half of the dimers being lost at each DNA doubling (Ganesan, 1974), while the frequency of dimers in daughter strands was comparable to the frequency of dimers in parental DNA in every sample examined (e.g., Fig. 2). Results of density transfer experiments had previously indicated that genetic exchanges between parental DNA made prior to irradiation and daughter strands synthesized after irradiation occurred during postreplication repair (Rupp et al., 1971). Our results support this idea and provide evidence that dimers may be included in the regions exchanged. This implies that intact, dimer-free DNA molecules are not necessarily generated by the gap filling which occurs during postreplication repair.

REFERENCES

Ganesan, A. K. (1973). *Proc. Nat. Acad. Sci. U.S.A.* **70**, 2753–2756.
Ganesan, A. K. (1974). *J. Mol. Biol.* **87**, 103–119.
Howard-Flanders, P., Rupp, W. D., Wilkins, B. M. and Cole, R. S. (1968). *Cold Spring Harbor Symp. Quant. Biol.* **33**, 195–205.
Rupp, W. D., Wilde, C. E., III, Reno, D. L. and Howard-Flanders, P. (1971). *J. Mol. Biol.* **61**, 25–44.

43

Experiments on the Filling of Daughter-Strand Gaps During Postreplication Repair

B. A. Bridges and S. G. Sedgwick

MRC Cell Mutation Unit
University of Sussex
Falmer, Brighton BN1 9QG, England

When *Escherichia coli* deficient in excision-repair replicate their DNA after UV irradiation, the newly synthesized strands contain gaps, presumed to be opposite pyrimidine dimers, which are rejoined during subsequent incubation (Rupp and Howard-Flanders, 1968), by a process believed to be recombinational in nature (Rupp et al., 1971). The detailed mechanism of postreplication repair is, however, far from clear. We have attempted to answer two somewhat different questions: (1) whether or not the daughter-strand gap persists because of the continued presence of the dimer on the opposite strand, and (2) which if any of the three DNA polymerases are required for filling of daughter-strand gaps.

EFFECT OF PHOTOREACTIVATION ON GAP FILLING

We have tried to determine whether the daughter-strand gaps are simple, possibly with 5'-P and 3'-OH ends so that they could be repaired rapidly by repair replication with DNA polymerase I and polynucleotide ligase if the dimer on the opposite strand were removed by photoreactivation. We used the excision-deficient strain WP2 *uvrA* (Hill) and a number of derivatives: CM611 (*uvrA exrA*), WP100 (*uvrA recA*), and WP12 (*uvrA exrA polA1*), the latter two kindly supplied by Dr. E. M. Witkin.

Logarithmic-phase bacteria were UV-irradiated and grown for a period in the presence of [*3H-methyl*]thymidine, thymine, and deoxyadenosine to label newly synthesized DNA. The labeling periods were adjusted so that equal amounts of label were incorporated in cultures synthesizing DNA at different rates, and extended over at least one-sixth of a replication cycle to minimize "end-error artifact" (Lehmann and Ormerod, 1969). The bacteria were then washed and resuspended (except where otherwise stated) in phage buffer. The filling of gaps was followed by measuring the change in size of the labeled DNA by alkaline sucrose centrifugation during incubation at room temperature (approximately $22°C$) either in the dark or in the presence of photoreactivating light from two Phillips "blacklight" tubes.

With WP2 *uvrA* incubated in growth medium, repair of daughter-strand gaps was so rapid that it was not possible to detect an effect of photoreactivating light with any confidence. When the bacteria were incubated in phosphate buffer *after* gap formation, repair was faster in the light than in the dark, suggesting that removal of pyrimidine dimers permits a more rapid repair of gaps, with no requirement for growth medium. With CM611 (*uvrA exrA*) there was little or no gap filling in buffer in the dark after doses in excess of 40 ergs mm^{-2}, but in the presence of light, repair was complete within 30 min. Under comparable conditions the DNA-polymerase-I-deficient WP12 (*uvrA exrA polA1*) did not show gap filling after doses in excess of 40 ergs mm^{-2} either in the dark or in the light. In WP100 (*uvrA recA*) there was no daughter-strand gap filling after any dose either in the dark or in the light, and there was significant degradation of newly synthesized DNA.

We conclude that, in the dark, gap filling can occur relatively slowly in phosphate buffer at room temperature, presumably allowing the conclusion that continued semiconservative DNA replication is not necessary for the repair of gaps already formed. No gap filling occurs in the absence of the $recA^+$ gene product, but in *exrA* strains gap filling can occur at doses below 40 ergs mm^{-2}. When the pyrimidine dimers presumed to be opposite daughter-strand gaps are removed by photoreactivation, gap filling is stimulated and has characteristics that differ from other reported mechanisms for the repair of DNA breaks or gaps. The process occurs rapidly in buffer at room temperature and requires both the $recA^+$ and $polA^+$ genes but not the $exrA^+$ gene.

We further conclude that daughter-strand gaps are not committed to recombination repair because of some structural feature of the "ends" of the gap. The gaps are kept "open" by the presence of a pyrimidine dimer on the opposite strand; removal of this dimer premits closure by the action of DNA polymerase I and (presumably) polynucleotide ligase. This result may also be taken as further circumstantial evidence that a large proportion of daughter-strand gaps are opposite pyrimidine dimers and not opposite nonphotoreversible lesions. The requirement of the $recA^+$ gene for light-stimulated gap filling suggests that its gene product may be needed at a very early stage, perhaps to stabilize the gap. In contrast, the requirement for the $exrA^+$ gene may be relieved by photoreactivation and presumably occurs at a later stage.

Full details of these experiments have been published elsewhere (Bridges and Sedgwick, 1974).

REQUIREMENT FOR EITHER DNA POLYMERASE I OR DNA POLYMERASE III

We have studied the involvement of the various DNA polymerases by utilizing bacteria carrying *polA* (DNA polymerase I) and *dnaE* (temperature-sensitive DNA polymerase III) mutations. The strains used were BT1026 (*dnaE polA1 thy*), 1026-1 (*dnaE* PolA⁺ *thy*), and BT1126 (*dnaE polA1 thy*), which carries a different *dnaE* mutation from BT1026. All these strains are deficient in endonuclease I but proficient in excision-repair (in contrast to the strains used in the work with photoreactivating light). The techniques were similar in principle to those described above, except that gap filling was followed in growth medium rather than in phosphate buffer, and always in the absence of light.

Bacteria singly mutant at either *polA* or *dnaE* (at restrictive temperature) could fill daughter-strand gaps, although one strain of *polA* bacteria was slower to make normal-sized DNA not only after UV but also without irradiation. We conclude that bacteria deficient in either DNA polymerase I or DNA polymerase III are able to carry out postreplication gap filling. The result with *dnaE* bacteria at restrictive temperature confirms the conclusion above, that continued semi-conservative DNA replication is not necessary for the repair of daughter-strand gaps already in existence at the time DNA replication is halted.

Double-mutant bacteria (*polA dnaE*) filled daughter-strand gaps at 33°C but not at 44°C (the restrictive temperature), where newly synthesized DNA was degraded. In the absence of either polymerase I or III activity, therefore, daughter-strand gaps are not filled. If there is any residual repair (perhaps involving polymerase II), it must be very limited in extent and does not give rise to DNA of normal size.

These results show that polymerase I and III can substitute for one another in daughter-strand gap filling as in repair replication after excision (Youngs and Smith, 1973). It should be pointed out that these experiments were obtained with excision-proficient strains, and it is at least in principle possible that their postreplication repair processes may differ from those in excision-deficient strains.

Full details of these experiments will be published elsewhere (Sedgwick and Bridges, 1974).

ACKNOWLEDGMENT

S. G. S. was supported by a Medical Research Council Scholarship.

REFERENCES

Bridges, B. A. and Sedgwick, S. G. (1974). *J. Bacteriol.* **117**, 1077–1081.

Lehmann, A. R. and Ormerod, M. G. (1969). *Nature* **221**, 1053–1056.

Rupp, W. D. and Howard-Flanders, P. (1968). *J. Mol. Biol.* **31**, 291–304.

Rupp, W. D., Wilde, E. G., Reno, D. L. and Howard-Flanders, P. (1971). *J. Mol. Biol.* **61**, 25–44

Sedgwick, S. G. and Bridges, B. A. (1974). *Nature* **249**, 348–349.

Youngs, D. A. and Smith, K. C. (1973). *Nature New Biol.* **244**, 241–242.

44

Postreplication Repair Gap Filling in an *Escherichia coli* Strain Deficient in *dnaB* Gene Product

Robert Carey Johnson

Department of Microbiology
Medical University of South Carolina
Charleston, South Carolina 29401

ABSTRACT

Gaps in daughter-strand DNA synthesized after exposure of *Escherichia coli* E279 to ultraviolet light are filled during reincubation at 30°C for 20 min. *Escherichia coli* E279 is phenotypically DnaB⁻ when incubated at 43°C. Cells incubated at 43°C were tested for their ability to complete postreplication repair gap filling. It is concluded that the *dnaB* gene product is essential for postreplication repair gap filling and that the inhibition seen is not initially the result of degradation.

Postreplication repair and excision-repair are two dark-repair processes by which ultraviolet-irradiated bacteria can remove or bypass pyrimidine dimers in their DNA. When bacteria are exposed to UV light, DNA synthesis slows but does not stop (Bowerstock and Moses, 1973). There is evidence that gaps are formed across from the pyrimidine dimers when they pass through the replication point (Rupp and Howard-Flanders, 1968). Rupp et al. (1971) have shown that recombinational exchanges from the parental DNA are involved in filling gaps. Ley and Setlow (1972) have suggested that *de novo* DNA synthesis is also involved in gap filling. Recently, polymerases I, II, and III have been shown to have roles in postreplication repair gap filling (Tait et al., 1974; Sedgwick and Bridges, 1974).

In my laboratory I have attempted to establish that the *dnaB* gene product has a role in postreplication repair gap filling. The *dnaB* gene product demonstrates a ribonuclease activity and possibly has a role in the initiation of DNA synthesis from single-strand primer DNA (Wickner et al., 1974). In this study, postreplication repair gap filling was studied in an *Escherichia coli* mutant with a thermolabile *dnaB* protein.

MATERIALS AND METHODS

E. coli strain E279 (*thr⁻, leu⁻, thi⁻, thy⁻, lac⁻, str^r, tonA, suII^+, dna B279*) was provided by J. A. Wechsler. Procedures for growth, ultraviolet irradiation, pulse-labeling, and centrifugation are the same as those previously published (Rupp and Howard-Flanders, 1971). Postreplication incubation was at 30°C or 43°C for 0, 20, or 70 min as indicated for each experiment. The shift from 30°C to 43°C took place in less than 15 sec. Recovery from the gradient was between 75 and 95%, with no correlation with particular strain or sample.

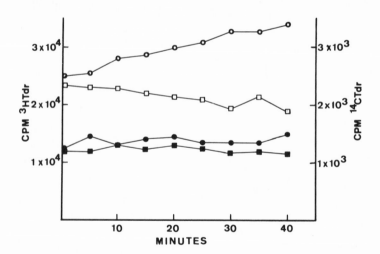

Fig. 1. *Acid-precipitable DNA from* E. coli *E279* dnaB *cells uniformly labeled with [¹⁴C]dThd, irradiated with UV (60 ergs/mm² at 254 nm), labeled with [³H]dThd for 10 min, and then reincubated with cold thymine for 40 min.* Open symbols represent ³H label (newly replicated DNA) and closed symbols represent ¹⁴C label (parental DNA). Circles represent reincubation at 30°C and squares reincubation at the nonpermissive temperature (43°C).

RESULTS AND DISCUSSION

E. coli E279 is phenotypically DnaB$^+$ when incubated 30°C and DnaB$^-$ when incubated at 43°C. This strain was used to study the role of the *dnaB* gene product in postreplication repair gap filling of DNA synthesized on UV-irradiated parental strands. E279 cells previously labeled with [^{14}C-*methyl*] thymidine were exposed to 60 ergs mm^{-2}, pulse-labeled with [^{3}H-*methyl*] thymidine, and reincubated in 2 μg/ml thymine at either 30°C or 43°C. The amount of acid-precipitable DNA present during this reincubation of the irradiated cells at the two different temperatures was then measured. These measurements were used to determine what if any effects the UV irradiation had with respect to the processes of DNA degradation and DNA synthesis in a strain with a thermolabile *dnaB* protein.

During a 40-min incubation period, no loss of acid-precipitable DNA was observed for either the parental or the postirradiation pulse-labeled DNA incubated at 30°C (Fig. 1). Some loss of acid-precipitable [^{3}H-*methyl*] thymidine-labeled DNA was observed (approximately 15%) after a 40-min incubation at 43°C. As expected, following a shift from 30°C to 43°C incubation, there was no indication of continued DNA synthesis even in this UV-irradiated cell culture.

Sedimentation profiles in alkaline sucrose gradients of DNA synthesized on irradiated parental DNA showed that, for cells incubated at 30°C, only 20 min was needed for the DNA to reach normal size (Fig. 2A, B). In other experiments (not shown), DNA from an isogenic strain (CR34, DnaB$^+$ phenotype at both 30°C and 43°C) reached normal size in a 20-min incubation at 43°C as well as at 30°C. In contrast, when cells of the strain E279 were reincubated at 43°C for 20 min there was no increase of DNA to normal size. Inhibition of the increase to normal size is not concomitant with a loss in total counts relative to the total counts of the sample exposed to UV irradiation only (Fig. 2C). Additional incubation time at 43°C produced little if any conversion of DNA to normal size but resulted in approximately 50% degradation of the lower-molecular-weight DNA (Fig. 2C).

Previous studies have shown that a double mutant BT 1026 (phenotype PolA$^-$ DnaE$^-$ when incubated at 44°C) is unable to complete postreplication repair gap filling and degrades DNA extensively with incubation at 44°C for 20 or 80 min (Sedgwick and Bridges, 1974). In my study a similar result was observed with a mutation in the *dnaB* gene only, except that little degradation was apparent for as long as 40 min. Thus the most significant observation concerns the inability of strain E279, phenotypically DnaB$^-$ when incubated at 43°C, to complete postreplication repair gap filling in 20 min. This fact is significant since (1) 20 min is the minimum time required for bacteria phenotypically DnaB$^+$ to complete gap filling, (2) no degradation is apparent within 20 min, and (3) possible residual activity of the *dnaB* gene product is probably not significant after only 20 min of incubation.

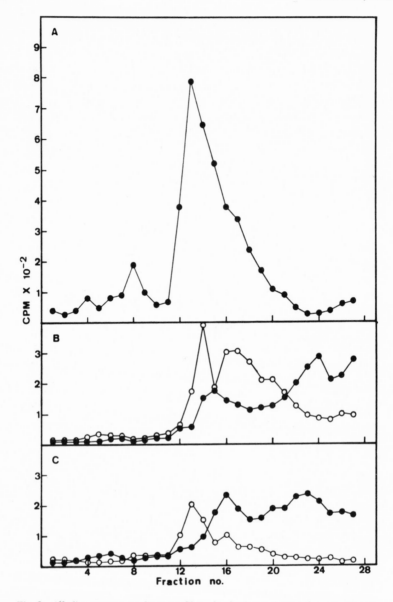

Fig. 2. Alkaline sucrose gradient profiles of pulse-labeled DNA from irradiated and unirradiated E. coli E279. (A) No UV irradiation, 5 min [³H] thymidine pulse. (B) •, 60 ergs mm⁻² at 254 nm, 10-min pulse of [³H] thymidine; ○, 60 ergs mm⁻² at 254 nm, 10-min pulse of [³H] thymidine, reincubation with 2 μg/ml thymine for 20 min at 43°C. (C) • 60 ergs mm⁻² at 254 nm, 10-min [³H] thymidine pulse, reincubation with 2 μg/ml thymine for 20 min at 43°C; ○ 60 ergs mm⁻² at 254 nm, 10-min pulse of [³H] thymidine, reincubation with 2 μg/ml thymine at 43°C for 70 min.

In conclusion, these results support the hypothesis that the *dnaB* gene product is essential for postreplication repair gap·filling. Extensive degradation does not appear to be the cause of inhibition of gap filling. A functioning *dnaB* gene product may be an early prerequisite for postreplication repair gap filling, needed to initiate DNA synthesis from primer DNA strands. It should be emphasized that a role for DNA synthesis in postreplication repair gap filling does not conflict conceptually with the role of strand-exchange repair previously demonstrated (Rupp et al., 1971).

ACKNOWLEDGMENTS

This research was supported by the National Institute of General Medical Sciences GM 18504. I thank Drs. Robb Moses, W. Dean Rupp, and Bruce Alberts for helpful discussions. The excellent technical assistance of E. Terry is gratefully acknowledged.

REFERENCES

Bowerstock, D. and Moses, R. E. (1973). *J. Biol. Chem.* **248**, 7449–7455.
Ley, R. and Setlow, R. B. (1972). *Biophys. Soc. Abstr.* **12**, 151a.
Rupp, W. D. and Howard-Flanders, P. (1968). *J. Mol. Biol.* **31**, 291–304.
Rupp, W. D. and Howard-Flanders, P. (1971). *Methods Enzymol.* **21**, 237–244.
Rupp, W. D., Wilde, C. E., III, Reno, D. L. and Howard-Flanders, P. (1971). *J. Mol. Biol.* **61**, 25–44.
Sedgwick, S. C. and Bridges, B. S. (1974). *Nature* **249**, 348–349.
Tait, R. C., Harris, A. L. and Smith, D. W. (1974). *Proc. Nat. Acad. Sci. U.S.A.* **71**, 675–679.
Wickner, S., Wright, M. and Hurwitz, J. (1974). *Proc. Nat. Acad. Sci. U.S.A.* **71**, 783–787.

45

Involvement of *uvrD, exrA,* and *recB* Genes in the Control of the Postreplicational Repair Process

David A. Youngs, Emmanuel Van der Schueren, and Kendric C. Smith

Department of Radiology
Stanford University School of Medicine
Stanford, California 94305

ABSTRACT

Ultraviolet radiation survival studies support the hypothesis that the *uvrD, exrA* and *recB* mutations inhibit separate branches of the postreplicational repair process.

The *uvrA* and *uvrB* strains appear to lack the UV-specific endonuclease which is required for the initial incision step of the excision-repair process (Braun, this volume). Thus, any mutation which sensitizes these strains to UV radiation might be assumed to interfere with the other known dark-repair process, post-replicational repair.

Several mutations sensitize *uvrA* or *uvrB* strains of *Escherichia coli* K12 to killing by ultraviolet (UV) radiation. These include *uvrD* (Ogawa et al., 1968), *exrA* (Mattern et al., 1966), *recB, recC* (Ganesan and Smith, 1970), *recF* (Horii and Clark, 1973), and *recA* (Howard-Flanders and Boyce, 1966). A partial deficiency in postreplicational repair (as assayed by the repair of breaks in DNA synthesized after irradiation) has been demonstrated for the *uvrB recF* (R. H. Rothman and A. J. Clark, unpublished results), *uvrB exrA* (Youngs and Smith,

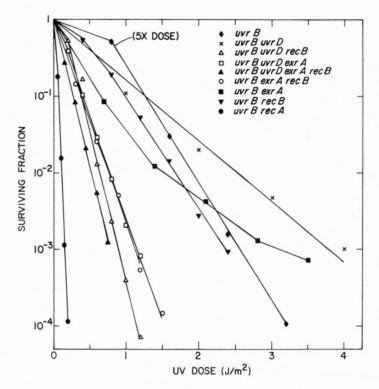

Fig. 1. Survival of uvrB mutants after UV irradiation. Experimental procedures and characteristics of the parent strain DY145 have been published
(Youngs and Smith, 1973). Noble agar was used here instead of agar-agar
No. 3. Details of the strains and their derivation will be described elsewhere.

1973), and *uvrB recB* (Youngs and Smith, unpublished results)[1] strains. There
appears to be little or no postreplicational repair in the *uvrB recA* strain (Smith
and Meun, 1970).

We have examined the interaction of the *uvrD, exrA,* and *recB* mutations in
uvrB strains by comparing the UV sensitivities of the multiple mutants. From
the results shown in Fig. 1 it is apparent that a *uvrB* strain carrying any two of
these mutations is more sensitive than either singly mutant parent. In addition, a
uvrB strain carrying all three additional mutations was slightly more sensitive
than the strains carrying *uvrB* and any two of the mutations. However, the most

---

[1] The earlier results (Smith and Meun, 1970) indicated that a *recB21* strain was not deficient
in postreplicational repair after a UV dose of 6.3 J m^{-2}. However, more recent results
indicate that the *recB21* mutation does result in a deficiency in postreplicational repair
after higher doses, both in *uvr$^+$* and *uvrB5* strains (Youngs and Smith, unpublished results).

sensitive multiple mutant, *uvrB uvrD exrA recB,* was still not as sensitive as the *uvrB recA* strain.

These results suggest that each of the mutations *uvrD, exrA,* and *recB* acts on a branch of the postreplicational repair process which is at least partially independent of those inhibited by the other mutations. Since the *uvrB uvrD exrA recB* strain was not as UV sensitive as the *uvrB recA* strain, at least one additional branch of postreplicational repair, possibly dependent on *recF* (Rothman et al., this volume), must be postulated.

ACKNOWLEDGMENTS

This work was supported by Public Health Service grant CA–02896 and Research Program Project grant CA–10372 from the National Cancer Institute. E. V. D. S. is a fellow of the NFWO (Belgium).

REFERENCES

Ganesan, A. K. and Smith, K. C. (1970). *J. Bacteriol.* **102**, 404–410.
Horii, Z. and Clark, A. J. (1973). *J. Mol. Biol.* **80**, 327–344.
Howard-Flanders, P. and Boyce, R. P. (1966). *Radiat Res.* **Suppl. 6**, 156–184.
Mattern, I. E., Zwenk, H. and Rörsch, A. (1966). *Mutat. Res.* **3**, 374–380.
Ogawa, H., Shimada, K. and Tomizawa, J. (1968). *Mol. Gen. Genet.* **101**, 227–244.
Smith, K. C. and Meun, D. H. C. (1970). *J. Mol. Biol.* **51**, 459–472.
Youngs, D. A. and Smith, K. C. (1973). *J. Bacteriol.* **116**, 175–182.

46

Replication and Expression of Constructed Plasmid Chimeras in Transformed *Escherichia coli*— A Review

Stanley N. Cohen and Annie C. Y. Chang

Department of Medicine
Stanford University School of Medicine
Stanford, California 94305

ABSTRACT

*Eco*RI restriction-endonuclease-generated fragments of bacterial plasmids isolated from *Staphylococcus aureus* or *Escherichia coli*, or of amplified DNA coding for the 18S and 28S ribosomal RNA of *Xenopus laevis*, have been linked to the pSC101 plasmid replicon and introduced into *E. coli* by transformation. The constructed plasmid chimeras can be cloned as stable replicons in *E. coli*, where they synthesize RNA and/or protein products specified by their component genes.

Treatment of *Escherichia coli* with calcium chloride renders this bacterial species competent for transfection by purified bacteriophage DNA (Mandel and Higa, 1970) and for transformation by plasmid (Cohen et al., 1972) or *E. coli* chromosomal (Cosloy and Oishi, 1973; Wackernagel, 1973) DNA. Transformation of *E. coli* by plasmid DNA, which unlike chromosomal DNA does not require a recipient strain defective in the *recBC* nuclease, has proved to be a useful tool for investigating the genetic and molecular properties of discrete plasmid species, and for introducing nonconjugative plasmids into specific *E. coli* hosts in the absence of transducing phage or conjugally proficient transfer plasmids (Cohen et al., 1972; Cohen and Chang, 1973; van Embden and Cohen, 1973; Lederberg et al., 1973; Guerry et al., 1974; Oishi and Cosloy, 1974).

The requirements for transformation of *E. coli* by plasmid DNA (Cohen et al., 1972) are shown in Table 1. Transformation is accomplished by plasmids of various incompatibility groups, and by closed-circular, open-nicked circular, and catenated forms of plasmid DNA; however, denaturation or sonication of the DNA destroys its transforming ability. Transformation efficiency is unaffected by treatment of plasmid DNA with ribonuclease, Pronase, or phenol but is abolished by treatment with pancreatic deoxyribonuclease if carried out prior to the 42°C heat-pulse step responsible for DNA uptake (Cohen et al., 1972). Transformed bacteria acquire a closed-circular autonomously replicating DNA species having the characteristics of the parent plasmid (Fig. 1), and plasmid DNA isolated from such transformants can itself be used to transform other bacteria.

Earlier experiments have shown that some DNA fragments that have been generated from larger plasmids by shearing (Cohen and Chang, 1973) or by

Table 1. Requirements for Transformation by R-Factor DNA[a]

DNA species	Transformants/μgDNA
R6–5 (F-like)	
Closed circular	9.2×10^4
+DNase (before)	<0.3
+DNase (after)	7.7×10^4
+RNase (before)	9.6×10^4
+Pronase (before)	8.1×10^4
Phenol extraction	9.9×10^4
Isolated from transformed bacteria	7.0×10^4
Catenated	3.2×10^4
Open-circular	5.6×10^4
Denatured	<0.3
Sonicated	<0.3
No DNA	[b]
No bacteria	<0.3
R64–11	
Closed circular	9.4×10^3
No DNA	[b]
No bacteria	<0.3

[a]Conditions for isolation of plasmid DNA and for transformation under the various conditions listed in the table have been described elsewhere (Cohen et al., 1972). Transformation efficiency was determined after a 2-hr incubation in antibiotic-free medium, to allow full expression of drug resistance. The terms "before" and "after" refer to the period of incubation of plasmid DNA and $CaCl_2$-treated cells at 42°C.

[b]No colonies were observed when 10^9 bacteria were assayed in the absence of DNA.

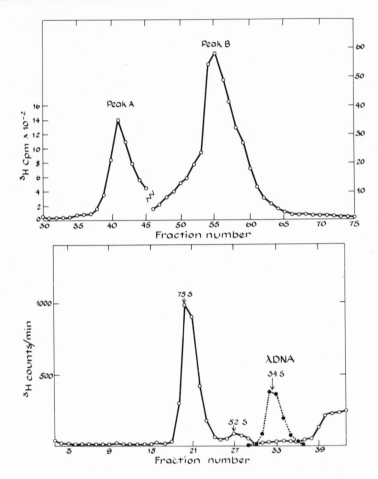

Fig. 1. *Centrifugation analysis of R-factor DNA isolated from transformed cells.* (Top) Tritium-labeled DNA isolated from transformed cells by a detergent-lysis procedure was centrifuged to equilibrium in ethidium bromide. Fractions were collected and assayed for radioactivity. (Bottom) After removal of ethidium bromide from the covalently closed circular DNA contained in peak A (fractions 40–43), this DNA was analyzed in a 5–20% linear sucrose gradient in the presence of ^{14}C-labeled DNA marker. The DNA species observed have the molecular properties of the transforming plasmid. Experimental details have been described elsewhere (Cohen et al., 1972). Note the change in scale in the top graph.

treatment with the *Eco*RI restriction endonuclease (Cohen et al., 1973) can be introduced by transformation into *E. coli*, where they can recircularize and become functional plasmid replicons. One such plasmid, pSC101, (Cohen and Chang, 1973; Cohen et al., 1973) that was formed after transformation of *E. coli* by a sheared fragment of the larger antibiotic-resistance plasmid R6–5 (Silver and

Cohen, 1972; Sharp et al., 1973) carries genetic information necessary for its own replication and for expression of resistance to tetracycline. This plasmid has proved to be of considerable use in the construction of plasmid chimeras *in vitro* because of the location of its single cleavage site for the *Eco*RI endonuclease.

The *Eco*RI restriction endonuclease (Hedgepeth et al., 1972) cleaves double-stranded DNA so as to produce short, overlapping single-strand ends. On a random basis, the nucleotide sequence cleaved is expected to occur once in every 4000–16000 nucleotide pairs (Hedgepeth et al., 1972); thus, most *Eco*RI-generated DNA fragments contain one or more intact genes. The nucleotide sequences cleaved by the enzyme are unique and self-complementary (Hedgepeth et al., 1972; Mertz and Davis, 1972; Sgaramella, 1972), so that DNA fragments produced can associate by hydrogen bonding with either end of any other *Eco*RI-generated fragment. After hydrogen bonding, the 3′-hydroxyl and 5′-phosphate ends can be joined by DNA ligase (Mertz and Davis, 1972; Sgaramella, 1972). Thus the enzyme appeared to be useful for the construction of DNA molecules having segments originating from diverse sources. Molecular chimeras produced by the joining of different *Eco*RI-generated DNA fragments could potentially be introduced into bacterial strains by transformation, provided that at least one of the segments carries a capability for replication and selection in transformed bacteria. As noted above, the pSC101 plasmid was especially useful for this purpose, since insertion of a DNA segment at its single *Eco*RI cleavage site does not interfere with either its replication functions or expression of its tetracycline-resistance genes. Our initial plasmid-construction experiments involved the linkage of this replicon to a fragment of another "synthetic" antibiotic-resistance plasmid, pSC102.

The pSC102 plasmid, which carries resistance to kanamycin, neomycin, and

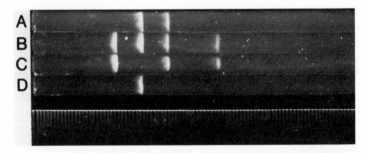

Fig. 2. Agarose gel electrophoresis of EcoRI digests of plasmids. DNA fragments subjected to electrophoresis in agarose gels as described elsewhere (Cohen et al., 1973) were stained with ethidium bromide, and the fluorescing DNA bands were photographed under long-wavelength ultraviolet light. *A,* pSC105 DNA; *B,* mixture of pSC101 and pSC102 DNA; *C,* pSC102 DNA; *D,* pSC101 DNA.

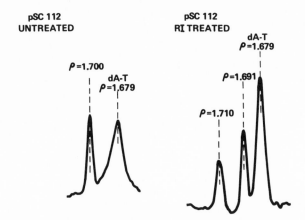

Fig. 3. Analytical ultracentrifugation of the E. coli–S. aureus plasmid chimera, pSC102. Densitometer tracings are shown. (Left) Untreated plasmid chimera showing a buoyant density intermediate between *E. coli* and *Staphylococcus*. (Right) Cleaved plasmid showing the component *Eco*RI generated fragments. The experimental conditions have been described elsewhere (Chang and Cohen, 1974).

sulfonamide, was formed by reassociation of several *Eco*RI-generated fragments of the larger plasmid, R6–5 (Cohen et al., 1973). Agarose gel electrophoresis (Fig. 2) demonstrates that the pSC102 plasmid is cleaved into its three component fragments by the *Eco*RI restriction endonuclease. The molecular weight of 17.4×10^6 for the sum of the three fragments is consistent with the size of the pSC102 plasmid determined by electron microscopy and sucrose gradient sedimentation. The pSC101 plasmid is cleaved by the *Eco*RI restriction endonuclease into a single linear DNA fragment (Fig. 2*D*) as noted above.

A mixture of *Eco*RI-cleaved pSC101 and pSC102 plasmid DNA was treated with DNA ligase, and the ligated molecules were used to transform *E. coli.* Transformants that were resistant to both tetracycline and kanamycin contained a new plasmid (pSC105) (Fig. 2*A*), which contained fragment II of pSC102 (i.e. the fragment that carries kanamycin resistance) plus the *Eco*RI-generated fragment that comprises the entire pSC101 tetracycline-resistance plasmid.

This procedure for the *in vitro* construction of recombinant plasmid molecules has also proved useful for the formation of plasmids that include DNA derived from different bacterial species. Such experiments have been carried out with a penicillinase-producing plasmid of *Staphylococcus aureus* [pI258, molecular weight $\sim 20 \times 10^6$ (Rush et al., 1969; Novick and Bouanchaud, 1971; Lindberg and Novick, 1973)]. This plasmid appeared to be especially appropriate for interspecies genome construction, since its properties have been well defined and it carries several different genetic determinants that were potentially detectable in *E. coli*. Moreover, agarose gel electrophoresis indicated that this plasmid is cleaved into four easily identifiable fragments by the *Eco*RI restriction endonuclease.

Molecular chimeras containing both staphylococcal and *E. coli* DNA were constructed by ligation of a mixture of *Eco*RI-cleaved pSC101 and pI258 DNA and were used to transform a restrictionless strain of *E. coli* (C600 $r_k^- m_k^-$). Selection was carried out for transformants that expressed the penicillin-resistance determinant carried by the staphylococcal plasmid, and plasmid DNA isolated from penicillin-resistant transformants was characterized. CsCl gradient analysis of an *E. coli–Staphylococcus* plasmid chimera isolated from one such transformant (pSC112) is shown in Fig. 3. The buoyant density of the intact plasmid chimera ($\rho = 1.700$ g/cm^3) is intermediate between the buoyant density of the *E. coli* plasmid (ρ - 1.710) and the density of the staphylococcal plasmid ($\rho = 1.691$). Upon treatment with the *Eco*RI endonuclease, the pSC112 plasmid is cleaved into separate fragments having the buoyant-density characteristics of its component DNA species. Further study of this plasmid chimera by agarose gel electrophoresis and electron microscope heteroduplex analysis (Chang and Cohen, 1974) confirmed that it contains DNA sequences derived from both *E. coli* and *Staphylococcus*. The staphylococcal genes linked to the pSC101 replicon were shown to be transferable among different *E. coli* strains by a conjugally proficient transfer plasmid indigenous to the Enterobacteriaceae.

In the experiments described thus far, genetic determinants carried by DNA fragments linked to the pSC101 plasmid replicon were utilized to select for transformants that contain plasmid chimeras. More recent experiments reported by Morrow et al. (1974) involve the *in vitro* construction of plasmid chimeras containing both prokaryotic and eukaryotic DNA, and the recovery of recombinant DNA molecules from transformed *E. coli* in the *absence* of selection for genetic properties expressed by the eukaryotic DNA. The amplified ribosomal DNA (rDNA) coding for 18S and 28S ribosomal RNA of *Xenopus laevis* was used as a source of eukaryotic DNA for these experiments, since this DNA has been well characterized and can be isolated in quantity (Dawid et al., 1970; Birnstiel et al., 1971). Moreover, the repeat unit of *X. laevis* rDNA contains a site susceptible to cleavage by the *Eco*RI endonuclease, resulting in the production of DNA fragments of characteristic size (primarily fragments of 3.0, 3.9, and 4.2×10^6 molecular weight that can be linked to the pSC101 plasmid (Morrow et al., 1974).

A mixture of *Eco*RI-cleaved pSC101 DNA and *X. laevis* rDNA was ligated and used to transform *E. coli* strain C600 $r_k^- m_k^-$. Tetracycline-resistant transformants were selected, and the plasmid DNA isolated from 55 separate clones was analyzed by agarose gel electrophoresis, cesium chloride gradient centrifugation, and/or electron microscopy to determine the presence of *X. laevis* rDNA linked to the pSC101 plasmid replicon. The results of these experiments are

Table 2. *X. laevis*–*E. coli* Recombinant Plasmids[a]

Plasmid DNA	Molecular weight of EcoRI plasmid fragments estimated by gel electrophoresis (× 10^{-6})	Molecular weight from contour length (× 10^{-6})	Buoyant density of intact plasmid in CsCl (g/cm^3)	Buoyant density of EcoRI-generated fragments in CsCl (g/cm^3)
4	5.8, 4.2, 3.0	13.6	1.721	1.710, 1.729
7	5.8, 4.2	—	—	—
12, CD 20, CD 45, CD 47, CD 51	5.8, 3.0	—	—	—
14	5.8, 4.2, 3.0	—	—	—
18	5.8, 3.0	9.2	1.720	1.710, 1.728
30	5.8, 3.9	10.0	1.719	1.710, 1.730
35	5.8, 3.9, 3.0	—	—	—
42	5.8, 4.2	10.6	1.720	1.710, 1.730
C101	5.8	6.0	1.710	

[a]The procedures employed for plasmid DNA isolation, agarose gel electrophoresis, cesium chloride gradient centrifugation, and calculation of molecular weight and buoyant density of DNA have been described (Morrow et al., 1974).

summarized in Table 2. Thirteen of the 55 tetracycline-resistant clones selected contained one or more *Eco*RI-generated fragment having the same size as the fragments produced by cleavage of *X. laevis* rDNA. Moreover, the plasmid chimeras isolated from *E. coli* were shown to contain DNA having a buoyant density characteristic of the high-(G+C) nucleotide base composition of *X. laevis* rDNA. The observed variation in size of the *Eco*RI-generated *X. laevis* rDNA fragments contained in plasmid chimeras was consistent with the observed size heterogeneity of fragments generated by *Eco*RI cleavage of the amplified *X. laevis* rDNA repeat unit (Morrow et al., 1974).

Electron microscope analysis (Fig. 4) of a heteroduplex formed between *X. laevis* rDNA and one of the plasmid chimeras (CD42) listed in Table 2 demonstrates that this plasmid contains DNA nucleotide sequences present in *X. laevis* rDNA. Moreover, in this and other heteroduplexes, two separate plasmid-DNA molecules were seen to form duplex regions with a single strand of *X. laevis* rDNA, consistent with the observation (Wensink and Brown, 1971; Hourcade et al., 1973) that the rDNA sequences of *X. laevis* are tandemly repeated.

Plasmid chimeras containing both *E. coli* and *X. laevis* DNA replicate stably in the bacterial host as part of the pSC101 plasmid replicon and can be recovered from transformed *E. coli* by procedures commonly employed for the

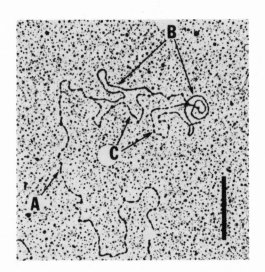

Fig. 4. Electron photomicrograph of heteroduplex of X. laevis rDNA and two separate molecules of a tetracycline-resistance plasmid chimera (CD42) isolated from E. coli. A, Single strand of *X. laevis* rDNA; *B,* double-stranded regions of homology between plasmid CD42 and *X. laevis* rDNA; *C,* single-stranded regions corresponding in length to the DNA segment of the CD42 plasmid that was derived from pSC101. Experimental conditions were as indicated by Morrow et al. (1974).

isolation of bacterial plasmids. Moreover, ^{3}H-labeled RNA isolated from bacteria harboring these plasmids hybridizes *in vitro* to amplified *X. laevis* rDNA isolated directly from the eukaryotic organism (Morrow et al., 1974), indicating that transcription of this eukaryotic DNA occurs in its prokaryotic host.

The experiments reviewed here demonstrate that plasmids constructed *in vitro* and containing genes from diverse sources can be introduced by transformation into *E. coli*, where they exist as stable replicons. The procedures described offer a general approach which utilizes bacterial plasmids for the cloning of a variety of prokaryotic and eukaryotic DNA species in *E. coli*.

ACKNOWLEDGMENTS

This article reviews certain experiments that have been done collaboratively with H. W. Boyer, H. M. Goodman, R. B. Helling, L. Hsu, and J. F. Morrow, as cited in the text. The studies carried out in the laboratory of the author were supported by grant AI08619 from the National Institutes of Health and by grant GB30581 from the National Science Foundation.

REFERENCES

Birnstiel, M. L., Chipchase, M. and Speirs, J. (1971). *Progr. Nucl. Acid Res. Mol. Biol.* 11, 351–389.

Chang, A. C. Y. and Cohen, S. N. (1974). *Proc. Nat. Acad. Sci. U.S.A.* 71, 1030–1034.

Cohen, S. N. and Chang, A. C. Y. (1973). *Proc. Nat. Acad. Sci. U.S.A.* 70, 1293–1297.

Cohen, S. N., Chang, A. C. Y. and Hsu, L. (1972). *Proc. Nat. Acad. Sci. U.S.A.* 69, 2110–2114.

Cohen, S. N., Chang, A. C. Y., Boyer, H. W. and Helling, R. B. (1973). *Proc. Nat. Acad. Sci. U.S.A.* 70, 3240–3244.

Cosloy, S. D. and Oishi, M. (1973). *Proc. Nat. Acad. Sci. U.S.A.* 70, 84–87.

Dawid, I. B., Brown, D. D. and Reeder, R. H. (1970). *J. Mol. Biol.* 51, 341–360.

Guerry, P., van Embden, J. and Falkow, S. (1974). *J. Bacteriol.* 117, 619–630.

Hedgepeth, J., Goodman, H. M. and Boyer, H. W. (1972). *Proc. Nat. Acad. Sci. U.S.A.* 69, 3448–3452.

Hourcade, D., Dressler, D. and Wolfson, J. (1973). *Proc. Nat. Acad. Sci. U.S.A.* 70, 2926–2930.

Lederberg, E. M., Brothers, L. L. and Cohen, S. N. (1973). *Genetics* 74, 5152 (abstract).

Lindberg, M. and Novick, R. P. (1973). *J. Bacteriol.* 115, 139–145.

Mandel, M. and Higa, A. (1970). *J. Mol. Biol.* 53, 159–162.

Mertz, J. E. and Davis, R. W. (1972). *Proc. Nat. Acad. Sci. U.S.A.* 69, 3370–3374.

Morrow, J. F., Cohen, S. N., Chang, A. C. Y., Boyer, H. W., Goodman, H. M. and Helling, R. B. (1974). *Proc. Nat. Acad. Sci. U.S.A.* 71, 1743–1747.

Novick, R. P. and Bouanchaud, D. (1971). *Ann. N.Y. Acad. Sci.* 182, 279–294.

Oishi, M. and Cosloy, S. D. (1974). *Nature* 248, 112–116.

Rush, M. G., Gordon, C. N., Novick, R. N. and Warner, R. C. (1969). *Proc. Nat. Acad. Sci. U.S.A.* 63, 1304–1310.

Sgaramella, V. (1972). *Proc. Nat. Acad. Sci. U.S.A.* **69**, 3389–3393.
Sharp, P. A., Cohen, S. N. and Davidson, N. (1973). *J. Mol. Biol.* **75**, 235–255.
Silver, R. P. and Cohen, S. N. (1972). *J. Bacteriol.* **110**, 1082–1088.
van Embden, J. and Cohen, S. N. (1973). *J. Bacteriol.* **116**, 699–709.
Wackernagel, W. (1973). *Biochem. Biophys. Res. Commun.* **51**, 306–311.
Wensink, P. C. and Brown, D. D. (1971). *J. Mol. Biol.* **60**, 235–247.

VI

Relationships Among Repair, Mutagenesis, and Survival

47

Relationships Among Repair, Mutagenesis, and Survival: Overview

Evelyn M. Witkin

Department of Biological Sciences
Douglas College, Rutgers University
New Brunswick, New Jersey 08903

In genetically damaged cells, the pathways of survival, mutagenesis, and repair are inseparable. An irradiated or chemically treated cell, having sustained potentially lethal damage to its DNA, may or may not survive. Surviving, it may or may not have kept intact its store of genetic information. The fate of such a cell depends in part upon the nature and degree of primary DNA damage, in part upon the DNA repair systems available to the cell which are capable of neutralizing the damage, and to an important extent also upon the internal and external environmental factors which determine how effectively repair mechanisms can operate. A surviving cell may have undergone a mutation either because an unrepaired DNA lesion has generated a replication error or because an error-prone repair system has changed the base sequence in the course of restoring a viable DNA structure (see Witkin, 1969; Bridges, 1969; Kondo, 1973; Doudney, 1974).

It is clear, by now, that a detailed understanding of mutagenesis must be based upon progress in the enzymology and physiology of DNA repair. It may not be equally obvious, though, that the study of induced mutation can provide insights into the nature of DNA repair processes. For instance, the existence of an enzymatic system of dark repair, capable of neutralizing potentially mutagenic UV lesions in bacterial DNA was inferred, and many of its properties (e.g. its kinetics and its inhibition by caffeine and acriflavine) were described (Witkin, 1961; Leib, 1961) several years before the demonstration that pyrimidine dimers are subject to enzymatic excision. In another instance, analysis of postirradiation effects on UV-induced mutations in *Escherichia coli* led to the conclusion that excision gaps may be repaired by either of two mechanisms, one of which is $recA^+ lex^+$-dependent[1] (Nishioka and Doudney, 1969). Biochemical validation of this conclusion came much later (Cooper and Hanawalt, 1972).

A new possibility, not yet supported by direct biochemical evidence, has been emerging from recent work on UV mutagenesis in bacteria: that UV mutability may depend upon an *inducible* repair system, expressed in response to the "SOS" signal of DNA damage, rather than upon repair enzymes constitutively synthesized in undamaged cells. This idea is part of the "SOS repair" hypothesis (see Radman, Witkin, Doudney, and Devoret et al., this volume), which suggests that DNA damage serves as a regulatory signal triggering the coordinate induction of a group of functions all promoting the survival of genetically damaged cells, or of their integrated episomes, and that one of these functions is an error-prone repair system responsible for UV mutability. Radman's paper summarizes the evidence which suggested the hypothesis, while my own contribution and that of Devoret describe experiments which lend it experimental support. Doudney's article offers a reinterpretation of his earlier "two-lesion" hypothesis of UV mutagenesis, which incorporates the idea that induction of an error-prone repair system is one of the prerequisites for mutation induction. The papers by Blanco et al. and by Mount et al. (this volume) concern the *lex* and *recA* genes, both of which are centrally involved in the SOS hypothesis, since the expression of all of the postulated inducible functions in response to DNA damage requires the $recA^+ lex^+$ genotype.

As background for the group of papers dealing either directly or indirectly with the SOS hypothesis, Fig. 1 summarizes my current understanding of how *E. coli* strains with various repair capabilities cope with pyrimidine dimers, and how UV-induced mutations may be generated by some of these mechanisms. A pyrimidine dimer, produced by UV in the DNA of a Uvr^+ strain, is usually excised (branch 1), although a small fraction of the dimers produced at higher doses escape excision and remain in the DNA through at least one DNA replication (branch 2). [In Uvr^- (excision-deficient) strains, all pyrimidine dimers formed pass through a replication fork.] The gap formed by excision of a pyrimidine dimer must be repaired by an error-free mechanism (branch 3) in *lex* and *recA* strains, since these strains perform efficient excision-repair without producing any UV-induced mutations (see Witkin, 1969). In wild-type strains, indirect evidence suggests that excision gaps are occasionally repaired by an error-prone system (branch 4), which is $recA^+ lex^+$-dependent, resulting in a low

---

[1] The designation "*lex*" refers to the *lexA* locus, closely linked to *malB*, which is assumed to be the same as the *exrA* locus. However, Blanco et al. (this volume) have described a *lexB* locus, closely linked to *recA*, and have shown that *lexB* mutations duplicate many of the pleiotropic phenotypic effects of *lexA* mutations. It is very probable, therefore, that $recA^+ lex^+$-Dependent functions require both the $lexA^+$ and $lexB^+$ gene products, although this has not been demonstrated for all of the functions.

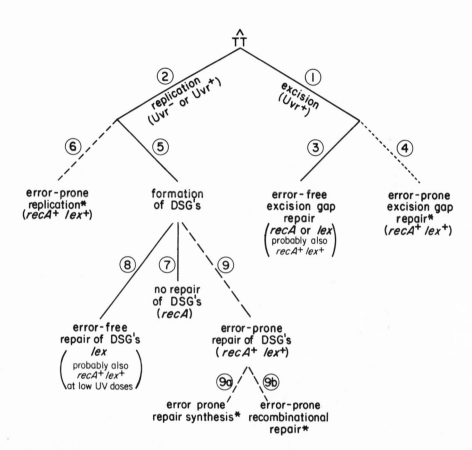

Fig. 1. Repair of UV damage in E. coli. strains of various genotypes and its possible relation to UV mutagenesis. TT, pyrimidine dimers; DSG, daughter-strand gap. Branches indicated by solid lines are those known to operate in strains of the genotypes or phenotypes indicated in parentheses. Branches indicated by broken lines (dashed or dotted) are hypothetical, but could theoretically operate in strains of the genotypes indicated in parentheses. Branch 4 (dotted line), while not known with certainty to occur, is supported by indirect evidence. Branches marked with asterisks (*) are possible sources of UV-induced mutations. See text for full explanation and references.

frequency of UV-induced mutations having excision gaps as the site of error (Nishioka and Doudney, 1969). It is likely, however, that most of the excision gaps in wild-type strains, as well as in *recA* and *lex* strains, are repaired by an error-free mechanism, probably utilizing DNA polymerase 1 (Kanner and Hanawalt, 1970; Boyle et al. 1970; Cooper and Hanawalt, 1972).

A pyrimidine dimer encountered by a replication fork usually (perhaps invariably) causes the production of a daughter-strand gap, which then requires some form of postreplication repair as a prerequisite for survival (Rupp and Howard-Flanders, 1968) (branch 5). The possibility has not been ruled out, however, that a small fraction of the unexcised pyrimidine dimers may permit continuous progress of semiconservative replication, a situation which would almost certainly generate errors with high probability. If such error-prone replication past a pyrimidine dimer does occur, it must require the *recA*$^+$ *lex*$^+$ genotype, since UV-induced mutations are totally absent in *recA* or *lex* strains. It is unlikely, therefore, that error-prone semiconservative replication past a pyrimidine dimer could be effected by any of the three known DNA polymerases of *E. coli,* at least in their unmodified constitutive form, since *recA* and *lex* cells are not known to be deficient in the activity of any of these enzymes. This type of replication, however, could depend upon a fourth inducible DNA polymerase, synthesized in response to interruption of DNA replication (see Radman, Witkin, and Devoret, this volume), or upon the modification of one of the constitutive polymerases by an inducible factor which relaxes its template requirements. Since the number of daughter-strand gaps produced is about equal to the number of pyrimidine dimers in the template strand of DNA (Rupp and Howard-Flanders, 1968), only a very small fraction of the dimers (if any) could permit continuous replication. However, if the probability of error in such replication is close to 100% (as might be expected), it is conceivable that all UV mutability might originate in this way.

Daughter-strand gaps, produced by all or most unexcised pyrimidine dimers in both Uvr$^+$ and Uvr$^-$ strains, are not repaired in *recA* strains (Smith and Meun, 1970) (branch 7). They must be repaired by an error-free mechanism in *lex* strains (branch 8), since these strains perform postreplication repair (Young and Smith 1973) and can sometimes be nearly as UV-resistant as *lex*$^+$ strains without producing any UV-induced mutations (Witkin, 1967). The same kind of error-free postreplication repair which must occur in *lex* strains may also occur in *recA*$^+$ *lex*$^+$ strains, at least at low doses. However, since *recA*$^+$ *lex*$^+$ strains are UV-mutable, they may also possess a *recA*$^+$ *lex*$^+$-dependent system of post replication repair (branch 9), operating conjointly or competitively with the error-free system seen in *lex* strains, or replacing it. Constitutive presence of an error-free system of postreplication repair operating at low UV doses must be assumed by the SOS hypothesis, since doses below the level required for efficient induction of the inducible error-prone SOS repair system produce up to about 30 pyrimidine dimers (Witkin, this volume) and should therefore be lethal unless a constitutive repair system capable of repairing at least that many daughter-strand gaps is present. Since the recombination ability of *lex* strains is essentially normal (Mount et al., 1972), it is possible that the error-free post-replication repair system (branch 8) is the recombinational mechanism which has been detected biochemically (Rupp et al., 1971).

If all unexcised pyrimidine dimers cause the formation of daughter-strand gaps at replication, then error-prone repair of such gaps (branch 9) would be the most probable source of UV-induced mutations, at least in Uvr⁻ strains. Two possible types of error-prone postreplication repair can be envisaged (Witkin, 1967). One possibility is an error-prone type of repair synthesis (branch 9*a*), utilizing a type of DNA polymerase capable of patching gaps despite the presence of a pyrimidine dimer opposite a portion of the gap. Another possibility (branch 9*b*) is that recombinational repair of such gaps (which may be error-free in *lex* strains) is rendered more efficient but less accurate by the participation of a $recA^+$ and lex^+-dependent component. The distinction between branches 9*a* and 9*b* may be artificial, if error-prone repair synthesis occurs in conjunction with recombinational exchanges. Recombinational repair may be error-free when a strand exchange replaces a dimer-containing region of the DNA but error-prone when the strand exchange leaves a single-stranded region, requiring repair synthesis, which still contains a pyrimidine dimer. The UV sensitivity of *lex* strains, in that case, might indicate the fraction of recombinational postreplication repair events occurring in these strains which result in viable repair without requiring the conjoint activity of a $recA^+$ lex^+-dependent error-prone patching mechanism.

Figure 1, in summary, shows four possible modes of origin of UV-induced mutations in *E. coli:* error-prone repair of excision gaps in Uvr⁺ strains (branch 3); error-prone postreplication repair of daughter-strand gaps opposite unexcised pyrimidine dimers via repair synthesis (branch 9*a*), or via recombinational repair (branch 9*b*), and error-prone semiconservative replication past an unexcised pyrimidine dimer (branch 6). In Uvr⁻ strains, the first mechanism (branch 3) is ruled out, but any one or more of the other three could, and at least one of these three *must* operate, in order to account for UV-induced mutations in excision-defective strains. All four of the possible mutagenic mechanisms require the $recA^+$ lex^+ genotype, and any or all could depend upon one or more inducible gene products, as postulated by the SOS repair hypothesis. It is quite possible that all four mechanisms contribute to UV mutability in wild-type strains. For example, if the SOS repair system is a DNA polymerase with relaxed template requirements [e.g., a nucleotidyl transferase capable of adding bases at ramdom to the end of a single DNA strand (Witkin, 1967)], it might be synthesized in response to UV damage before a round of postirradiation DNA synthesis has been completed. In that case, especially if the SOS repair system is responsible for the "aberrant" reinitiation of DNA synthesis that occurs after UV irradiation at the chromosomal origin (see Radman, this volume), it could cause errors in semiconservative replication by inserting wrong bases opposite unexcised pyrimidine dimers in previously unreplicated regions of the DNA, as well as errors in repair synthesis associated with postreplication repair of daughter-strand gaps, with or without a recombinational exchange.

The absence of any biochemical evidence for an error-prone DNA polymerase in extracts of *E. coli* may be due to its inducible nature: such activity should be sought in cells first damaged by UV (or some other inhibitor of DNA synthesis effective as an inducer of prophage), and then allowed to synthesize protein.

The possibility that an inducible repair system is responsible for UV-induced mutations requires renewed caution in the interpretation of the photoreversibility of induced mutations. Mutations which are photoreversible by the direct enzymatic splitting of pyrimidine dimers clearly requiare pyrimidine dimers for their induction, but this does not mean that the premutational lesion itself is a pyrimidine dimer. Pyrimidine dimers may be necessary for induction of the mutagenic repair system, and therefore for all mutagenesis resulting from its error-prone activity, whatever the nature of the lesions prompting the error at the actual site of the mutation. This could be one kind of "indirect enzymatic photoreactivation" (Witkin, 1969).

A wholly different approach to the study of mutagenesis as it relates to repair is exemplified by the two papers on yeast included in this chapter. In the article by Prakash, radiation-sensitive strains of yeast are used to show that functional repair systems are required for certain kinds of chemical mutagenesis. The paper by Lawrence et al. traces the specificity of UV mutagenesis observed at a particular site in an *ochre* codon to the operation of a particular nonexcision repair pathway. These studies offer exciting prospects for the analysis of specificity in mutation induction and repair, based as they are on the hard information provided by amino acid sequence analysis of wild-type and mutant gene products.

Another sector of the front represented in this chapter is covered by papers which to not deal directly with mutagenesis, but with the profound effects that loss of an enzyme such as DNA polymerase I (Smirnov et al., this volume) or Exonuclease V (Capaldo and Barbour, this volume), often regarded primarily as repair enzymes, may have on the economy of the undamaged cell. These papers underline the overlap between many enzymatic functions required for the "three *R*'s" of molecular genetics: repair, replication, and recombination.

REFERENCES

Boyle, J. M., Patterson, M. C. and Setlow, R. B. (1970). *Nature* **226**, 708–710.
Bridges, B. A. (1969). *Ann. Rev. Nucl. Sci.* **19**, 139–178.
Cooper, P. K. and Hanawalt, P. C. (1972). *J. Mol. Biol.* **67**, 1–10.

Doudney, C. O. (1975). In *Photochemistry and Photobiology of Nucleic Acids,* ed. S. Wang, Academic Press, New York, *in press.*

Kanner, L. and Hanawalt, P. (1970). *Biochem, Biophys. Res. Commun.* **39,** 149–155.

Kondo, S. (1973). *Genetics Suppl.* **73,** 109–122.

Lieb, M. (1961). *Z. Vererbungs l.* **92,** 416–429.

Mount, D. W., Low, K. B. and Edmiston, S. J. (1972). *J. Bacteriol.* **112,** 886–893.

Nishioka, H. and Doudney, C. O. (1969). *Mutat. Res.* **8,** 215–228.

Rupp, W. D. and Howard-Flanders, P. (1968). *J. Mol. Biol.* **31,** 291–304.

Rupp, W. D., Wilde, C. E., III, Reno, D. L. and Howard-Flanders, P. (1971). *J. Mol. Biol.* **61,** 25–44.

Smith, K. C. and Meun, D. H. C. (1970). *J. Mol. Biol.* **51,** 459–472.

Witkin, E. M. (1961). *J. Cell. Comp. Physiol.* **58** (Suppl. 1), 135–144.

Witkin, E. M. (1967). *Brookhaven Symp. Biol.* **20,** 17–55.

Witkin, E. M. (1969). *Ann. Rev. Microbiol.* **23,** 487–514.

Youngs, D. A. and Smith, K. C. (1973). *J. Bacteriol.* **116,** 175–182.

48

SOS Repair Hypothesis: Phenomenology of an Inducible DNA Repair Which Is Accompanied by Mutagenesis

Miroslav Radman

Laboratoire de Biophysique et Radiobiologie
Université libre de Bruxelles
1640 Rhode St Genèse
Belgium

ABSTRACT

A hypothesis was proposed several years ago that *Escherichia coli* possesses an inducible DNA repair system ("SOS repair") which is also responsible for induced mutagenesis. Some characteristics of the SOS repair are (1) it is induced or activated following damage to DNA, (2) it requires *de novo* protein synthesis, (3) it requires several genetic functions of which the best-studied are $recA^+$ and lex^+ of *E. coli*, and (4) the physiological and genetic requirements for the expression of SOS repair are suspiciously similar to those necessary for the prophage induction. The SOS repair hypothesis has already served as the working hypothesis for many experiments, some of which are briefly reviewed. Also, some speculations are presented to stimulate further discussions and experimental tests.

How do lesions in the DNA bring about cell death? By inactivating vital genes? This is an unlikely possibility, since even one pyrimidine dimer can cause cell death in some *Escherichia coli* mutants. The primary cause of the biological inactivation of DNA, i.e. the first "lethal hit," is most probably due to a loss of ability to replicate damaged DNA. The fact that the radiation sensitivity of bacteriophage λ is proportional to the length of its DNA, other physiological factors being unaffected, is in agreement with this proposal (Radman, 1971). In addition, biochemical studies of DNA metabolism following radiation damage show a reduction of rate blockage of DNA synthesis (Setlow, 1967; Bowersock and Moses, 1973) even in the absence of DNA breakdown.

E. coli cells can overcome the lethal and mutagenic effects of a substantial number of pyrimidine dimers (Howard-Flanders, 1968a,b; Witkin, 1969) in several ways: (1) dimers can be split in situ (photoreactivation), (2) they can be excised from the DNA (excision-resynthesis repair), (3) two or more copies of damaged DNA can yield a surviving molecule through multiple exchanges (recombinational repair in the strict sense), such as during multiplicity reactivation (Luria, 1947), prophage reactivation (Jacob and Wollman, 1955; Blanco and Devoret, 1973), and cross-reactivation (Luria, 1952), and (4) some dimers can apparently leak through the replication process. It is this fourth case that we least understand and which interests us most, since it seems to be responsible for mutagenesis (Witkin, 1969).

Is any of these DNA repair processes inducible, or are all repair enzymes produced constitutively? It has not been possible to answer this question directly, since the usual criteria of inducibility, such as dependence on de novo protein synthesis, were practically impossible to employ in systems where viability and mutagenesis were detected in clones originating from a single cell. However, if we use a phage assay as the test for the presence of a certain DNA repair process in E. coli, this difficulty can be overcome if the presumed repair system is not selective for either host or phage DNA. The first experiments of this kind were performed 20 years ago by Jean Weigle (1953). The phenomenon to be discussed is so-called UV reactivation. Since this is easily confused with other phenomena and designations such as host-cell reactivation, uvr, UV^R, etc., I propose here to call it Weigle reactivation or W reactivation. The term UV reactivation is inappropriate also, because it can be provoked by a whole variety of mutagenic treatments (Weigle, 1953; Hart and Ellison, 1970) in addition to UV.

W REACTIVATION (KNOWN AS UV REACTIVATION)

Weigle's original observation was that the survival of UV-irradiated bacteriophage λ was greatly increased when the host cells were irradiated as well (Weigle, 1953). He also found that photoreactivation of the host cells prior to infection by phage would eliminate this effect, showing that the presence of pyrimidine dimers in the host cell DNA was necessary for the manifestation of W reactivation. Among reactivated phage, Weigle found an increased frequency of mutants.

Two types of theories to account for the mechanism of W reactivation have been entertained until recently: (1) increased efficiency of excision-repair and (2) recombination between homologous parts of the UV-irradiated host DNA and the phage DNA. Recent careful examinations could eliminate both proposals (Devoret, Blanco, George, and Radman, this volume). Under optimal conditions three different excision-repair-deficient *E. coli* mutants show as strong W reactivation as does *E. coli* wild type (Radman and Devoret, 1971). Thus the *uvr*-controlled excision-repair is not likely to be essential for W reactivation. This has been supported biochemically as well (Boyle and Setlow, 1970). On the other hand, Blanco and Devoret (1973) provided arguments against the normal legitimate recombination as the mechanism of W reactivation. Thus it seems likely that W reactivation is a consequence of a yet undefined DNA repair system in *E. coli.*

DOES W REACTIVATION OPERATE ON HOST DNA?

There are no relevant experimental data which provide an answer to this question. Heuristically speaking, it would seem unlikely that cells would develop a DNA repair system operating on viral DNA only. We shall try here to deduce the existence of some new bacterial DNA repair system(s), probably involving W reactivation, from the known discrepancies between the lethality of pyrimidine dimers in bacterial DNA and in phage λ DNA. One lethal hit in *E. coli* is caused by $1-3 \times 10^3$ pyramidine dimers (see Howard-Flanders, 1968*a*), while phage λ in the same bacterial host requires only about 40 dimers for killing (Radman et al., 1970). Maurice Fox (personal communication) first noticed that under

Table 1. Lethality of Pyrimidine Dimers in *E. coli* DNA and in λ DNA

Organism tested	Functional repair systems[a]	Lethal effect per pyrimidine dimer
E. coli wild type	*E, R, W*	~ 10^{-3}
E. coli lex⁻	*E, R*	~ 0.02
λ on *E. coli* wild type	*E, R*	~ 0.025
λ on UV'd *E. coli lex⁻*	*E, R*	~ 0.02
λ on UV'd *E. coli* wild-type	*E, R, W*	~ 10^{-3}–10^{-2}
λ *red⁻* on *E. coli recA⁻uvrA⁻*	None	~ 0.5
E. coli recA⁻uvrA⁻	None	~ 0.8

[a]*E* is excision-repair, *R* is recombination repair, *W* is W reactivation. Relevant data were taken from Howard-Flanders (1968*b*), Moody et al. (1973), Radman et al. (1970), and M. Fox (unpublished), M. Defais and P. Fauquet (unpublished), and M. Radman (unpublished).

optimal conditions for W reactivation λ requires up to 10^3 dimers per lethal hit. Table 1 shows the general relationship between pyrimidine dimer lethalities in *E. coli* DNA and in phage λ DNA under genetical and physiological conditions allowing or restricting W reactivation. Of three *E. coli* mutants known to abolish W reactivation (*recA*, *lex*, and *zab*), we have chosen *lex*, which is recombination proficient.

It is clear that the discrepancy between pyrimidine dimer lethality in *E. coli* and λ DNA is diminished when W reactivation of *E. coli* DNA is taken into account. This same general relationship holds under genetic conditions where excision-repair is not functional.

I would like to suggest a provisory scheme for dark-repair processes operating in *E. coli* following UV irradiation:

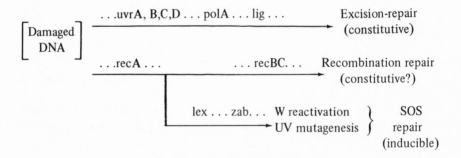

This scheme suggests that the UV sensitivity of a $recA^-$ single mutant should be similar to UV sensitivities of the double mutants $recA^- recBC^-$, $recA^- lex^-$, and $recBC^- lex^-$. This is true (Moody et al., 1973; Willetts and Clark, 1969).

The fact that $recA^-$ and lex^- mutations abolish W reactivation and UV mutagenesis of phage λ (Miura and Tomizawa, 1968; Defais et al., 1971) as well as a repair and UV mutagenesis of *E. coli* (Witkin, 1969) is not enough to assume that the same repair pathway is involved in these phenomena. Yet, for reasons discussed above and for the sake of simplicity, we shall assume that W reactivation does operate on bacterial DNA as indicated in the scheme.

EXPERIMENTS SUGGESTING A CORRELATION BETWEEN PHOPHAGE INDUCTION, DNA REPAIR, AND UV MUTAGENESIS

We previously found striking similarities between the genetic and physiological requirements for W reactivation, UV mutagenesis, and prophage induction of bacteriophage λ (Defais et al., 1971; also see Figure 2). This led us to propose common pathways for these three phenomena. A number of predictions were made, some of which have been tested experimentally.

W Reactivation in *E. coli tif* Mutant in the Absence of UV Irradiation

The *E. coli* T44 mutant induces prophage λ at elevated temperature (Goldthwait and Jacob, 1964). This mutant, now called *tif,* was recently shown to *W*-reactivate UV-irradiated λ phage following a temporary shift to elevated temperature without UV irradiation of the cell (Castellazzi et al., 1972*a,b*). Irradiation of this mutant host and incubation at low temperature yielded the same extent of *W* reactivation and UV mutagenesis of irradiated λ as following a shift to elevated temperature without UV treatment of this host. Thus, in the *tif* mutant, temperature shift mimics UV irradiation of the cell.

Increased UV Mutability of *E. coli tif⁻* Following Temperature Shift

The *tif⁻* mutant has been tested by Witkin for bacterial mutagenesis in another genetic background. The results (Witkin, 1974) can be summarized as follows: (1) UV mutability of *E. coli uvr⁻ tif⁻* at very low UV doses (100% survival) is increased by a factor of 10 following a shift to 42°C for 45 min, (2) the usual dose-squared shape of the UV-induced mutation frequency curve as a function of dose becomes linear with *tif⁻* at 42°C (*tif⁺* at 42°C and *tif⁻* at 30°C show the dose-squared relationship), and (3) these effects are abolished if chloramphenicol, pantoyl lactone, or high concentrations of cytosine and guanine are present during incubation at elevated temperature, just as in the case of prophage induction (Goldthwait and Jacob, 1964).

Witkin's work (Witkin, 1974; Witkin and George, 1973) provides the first evidence for the action on bacterial DNA of a mutagenic repair related to prophage induction and *W* reactivation.

Increased Mutability of *E. coli dnaB* Mutant Following Temperature Shift

DNA synthesis in *E. coli dnaB* mutants ceases immediately following a temperature shift to 42°C (Kohiyama, 1968), and, if the bacterium carries a prophage λ, lysogenic induction is provoked by the temperature shift (Noack and Klaus, 1972). Witkin (this volume) found that the temperature shift to 42°C also increases the UV mutagenesis of an *E. coli dnaB* mutant, just as in the case of the *tif* mutant.

Indirect Induction of *W* Reactivation

If prophage induction and *W* reactivation are strongly correlated, then we hoped to observe indirect *W* reactivation under conditions of indirect prophage induction. This prediction was fullfilled: when an F^- recipient *E. coli* bacterium receives $F-lac^+$ or Hfr DNA from a UV-irradiated donor, the survival and mutation frequency of UV-damaged phage λ in the F^- recipient are substantially increased (George et al., 1974). This experiment elegantly demonstrates that the signal which turns on *W* reactivation of λ DNA in *E. coli* can be a relatively small fragment of damaged DNA. The question remains whether the fragment of the transferred Hfr or F–*lac* DNA carries some inducible gene(s) or if any fragment of damaged DNA somehow induces the recipient's chromosomal function(s).

SOS REPAIR HYPOTHESIS

I would like to propose a general working hypothesis to account for phenomena and experiments briefly reviewed in the preceding paragraphs. Chronologically, this hypothesis preceded most of the reviewed experiments for which it has already served as the working hypothesis.

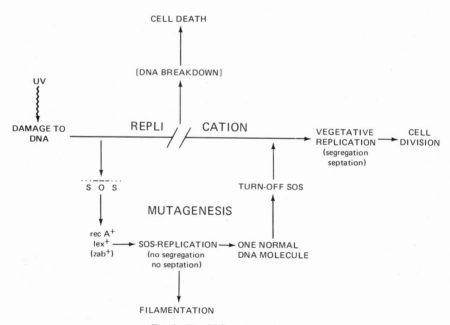

Fig. 1. The SOS repair pathway.

Table 2. Pleiotropic Effects of *recA* and *lex* Mutations[a]

Phenotype	*recA*	*lex*	Wild type
UV sensitivity	+++	++	+
X-ray sensitivity	+++	+++	+
DNA breakdown after UV	+++	++	+
Induction of prophage	−	−	+
W reactivation of	−	−	+
UV mutagenesis of	−	−	+
Filamentous growth (*lon*)	−	−	+
UV mutagenesis of E. coli	−	−	+
"Aberrant" DNA reinitiations	−	?	+
Reversion of *tif* phenotype	+	+	−
Recombination ability	−	+	+

[a]For references see text and Castellazzi et al. (1972*a,b*), Defais et al. (1971), Howard-Flanders (1968*a*), and Witkin (1969).

The principal idea is that *E. coli* possesses a DNA repair system which is repressed under normal physiological conditions but which can be induced by a variety of DNA lesions. Because of its "response" to DNA-damaging treatments, we call this hypothetical repair "SOS repair." The "danger" signal which induces SOS repair is probably a temporary blockage of the normal DNA replication and possibly just the presence of DNA lesions in the cell. During the action of SOS repair, mutation frequency is increased. The simplest assumption is that the SOS repair mechanism is error-prone; on the other hand, mutagenesis may be just a secondary consequence of the physiological conditions under which SOS repair operates. In order for SOS repair to function it should require some specific genetic elements, the inducing signal, and *de novo* protein synthesis (see Fig. 1).

GENETIC ELEMENTS REQUIRED FOR SOS REPAIR

Two well-known mutations of *E. coli* are deficient in SOS repair: *recA* and *lex*. The recently isolated *zab* mutant (Castellazzi et al., 1972*b*) and *lexB* mutant (Blanco et al., this volume) are likely to be deficient in SOS repair as well. It is characteristic of both *recA* and *lex* mutants that they exhibit pleiotropic effects listed (Table 2). This complicates the interpretation of any data based on comparisons between wild-type and *recA⁻* and *lex⁻* derivatives. It is likely that the *recA, lex, zab, tif* genetic system is a part of a complex control circuit (Castellazzi et al., 1972*a,b*).

PHYSIOLOGICAL CONDITIONS REQUIRED FOR SOS REPAIR

Treatments such as UV, X-rays, mitomycin, thymineless condition, and alkylating agents which cause a temporary blockage of the normal vegetative replication seem to induce "aberrant" reinitiations of new replication forks (reviewed by Bridges, 1972). This is true even when the abrupt blockage of DNA replication is produced by temperature shift of *dnaB* mutants in the absence of the above-listed mutagens (Worcel, 1970). In all of these cases $rec\widetilde{A}^+$ genotype and protein synthesis are necessary for reinitiations of new replication forks. When DNA replication cannot resume due to genetic deficiency (Table 2) or to inhibition of protein synthesis (Drakulić and Errera, 1959), extensive DNA breakdown takes place.

All of the listed conditions which lead to inhibition and subsequent reinitiation of DNA replication also induce *W* reactivation of phage λ (Weigle, 1953; Hart and Ellison, 1970), mutagenesis (Witkin, 1969), and prophage induction (Noack and Klaus, 1972). The connection between these three phenomena is also seen from their kinetics versus UV dose to cells (Fig. 2; Defais et al., 1971).

Optimal doses for induction of these three phenomena are overlapping (Defais et al., 1971). Figure 2 shows that we need about tenfold lower UV dose

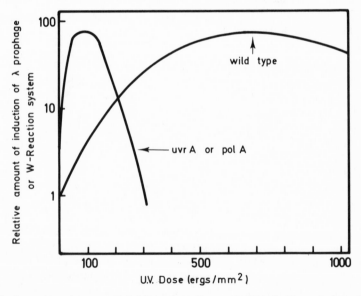

Fig. 2. General appearance of the dose dependence of UV induction of λ prophage, W reactivation, and UV mutagenesis of phage λ in E. coli with and without functional excision-repair. Data compiled from Defais et al. (1971), Radman and Devoret (1971), and Monk and Gross (1971).

to induce these phenomena in excision-repair-deficient (uvr^-) mutants than in uvr^+ wild type, suggesting that the inducing lesions are efficiently repaired by excision-repair. Photoreversibility of the three above-mentioned phenomena proves that the inducing lesions in this particular case are pyrimidine dimers.

WHICH FUNCTIONS ARE INDUCED BY UV IRRADIATION?

We do not know the full answer to this question, but we do know that the lex^+ genotype is expressed only following inducing treatment of the cell. The evidence follows from assays with UV-irradiated phage λ, which plates with identical efficiencies on lex^+ and lex^- hosts (Defais, Fauquet, and Radman, unpublished). Yet, when host cells are treated with inducing UV doses, a strong increase in survival and mutation frequency occurs only in E. coli lex^+ (W reactivation), but not in lex^-. We know that photoreactivation or inhibition of protein synthesis (Ono and Shimazu, 1966) will eliminate the latter effect. The $recA^-$ mutation shows an important effect on λ survival even without inducing treatment of the cell (Radman et al., 1970), but the difference between $recA^+$ and $recA^-$ genotypes is greatly increased following inducing UV doses (W reactivation). This suggests that the $recA$ gene product is active under normal physiological conditions but that its activity either increases or becomes modified following mutagenic treatment.

The first biochemical observation which may be directly correlated to the SOS repair is that the Ole Westergaard (1970), who found that following UV irradiation, thymine starvation, or ethidium bromide treatment of Tetrahymena cells, a DNA polymerase activity is induced which depends on de novo protein synthesis.

SPECULATIONS ABOUT THE MECHANISM AND CONSEQUENCES OF SOS REPAIR

Since SOS repair also acts on single-stranded phage DNA (Tessman and Ozaki, 1960), we would like to speculate that it acts during and/or after DNA replication. What do we know about replication of UV-damaged DNA? Does the normal replication fork stop as it encounters a dimer? Setlow's data (1967) indicate that DNA synthesis does stop; it recovers in UV-resistant E. coli B/r but is very poorly recovered in E. coli B_s-1, which carries both uvr and lex mutations. Poddar and Sinsheimer (1971) found that in an in vitro replication system synthesis of phage φX174 DNA stops at the first lesion encountered.

On the other hand, Rupp and Howard-Flanders (1968) found that following a UV dose of 60 ergs/mm^2 E. coli cells carrying the $uvrA$ mutation only incorporate thymidine into their DNA but at a very slow rate. The normal rate of synthesis is recovered after several hours. They have interpreted their data in terms of a symmetric semiconservative replication producing daughter molecules with gaps opposite each dimer in the parental strands. If we accept their scheme, we can speculate that SOS repair is necessary either for continuation of replication, i.e. bypassing of dimers, or for the gap-filling, or for both.

Regardless of molecular mechanisms, we know of two modes of DNA replication in *E. coli*: Cairns-type (θ) replication and rolling-circle (σ) replication (for review, see volume 33 of the *Cold Spring Harbor Symp. Quant. Biol.*). In addition to their topological differences, the two replication modes have very different physiologies which may be of some interest for the SOS repair phenomenon. The θ replication is typical for the vegetative replication of *E. coli* DNA: it is a "constitutive," symmetrical process which leads to segregation, septation, and cell division; it is an accurate, mutation-proof process which is inhibited by DNA lesions. On the other hand, the σ replication takes place during bacterial mating: it is an inducible, asymmetrical process which does not lead to segregation, septation, and cell division; it is mutation-prone for acridine mutagenesis (Sesnowitz-Horn and Adelberg, 1968) and is extremely resistant to UV damage to DNA (see Greenberg et al., 1970; George et al., 1974). UV doses which cause blockage and degradation of vegetative DNA do not greatly affect conjugational replication (Greenberg et al., 1970). Vegetative replication and conjugational replication can operate separately: some mutants blocking vegetative replication do not affect conjugational replication (Bresler et al., 1968; Marinus and Adelberg, 1970), and vice versa (Kingsbury and Helinski, 1973). It is interesting that the conditions under which both prophage λ and SOS repair are induced lead to filamentation of bacteria (Table 2, Fig. 1) as the consequence of continued DNA synthesis without septation and cell division. An interesting question is whether, under these filamentation conditions, DNA replication is of the rolling-circle type.

Finally, it is difficult to escape the impression of a superficial similarity between the SOS induction of bacteria and oncogenic transformation in mammalian cell cultures. In both cases an active cellular response to the inducing agent is required for "transformation" to occur. Such an example is the behavior of the *E. coli tif* mutant, which when "depressed" at 42°C becomes a sick cell, forming very long filaments and becoming genetically unstable (George, personal communication). Furthermore, it has been shown that carcinogens are potent mutagens in bacteria (Ames et al., 1973); and vice versa, some well-known mutagenic agents appear to be potent carcinogens (see e.g. Setlow, this volume; Ikenaga et al., this volume). All these carcinogenic DNA lesions are repairable by the prereplicative excision-repair system of bacteria and mammalian cells (Ames et al., 1973; Ikenaga et al., this volume), and another, postreplicative, DNA repair is likely to be responsible for the induced mutagenesis. Recently, Zajdela and Latarjet (1973) have shown that caffeine, a known inhibitor of the post-replicative DNA repair in animal cells (Lehmann, 1972), suppresses UV-induced tumors in mice.

ACKNOWLEDGMENTS

I thank Dr. Evelyn Witkin for many delightful discussions and for taking the SOS hypothesis out of forgetfulness by doing several crucial experimental tests. It is due to discussions with her and with R. Devoret, J. George, M. Blanco, P. Fauquet-Caillet, M. Defais, R. Cordone, and M. Fox, all of whom also provided much of their unpublished results, that the SOS hypothesis appears in the present form.

This work is supported by contract Euratom-ULB 099–72–1BIAB and by an agreement between the Belgian Government and Université libre de Bruxelles concerning priority action for collective basic research.

This paper is a modified version of a previously published paper (Radman, 1974).

NOTE ADDED IN PROOF

A *recA* and *lex* dependent temporary turn-off of respiration in *E. coli* treated by UV or ionizing radiation (Swenson and Schenley, 1974, *Int. J. Radiat. Biol.* **25**, 51–60) should be added to the list of pleiotropic effects of these mutations (Table II).

Our recent experiments have revealed several important characteristics of the SOS repair in *E. coli:* (1) the kinetics of expression of the SOS repair, as tested by phage λ and φX174 assays: both, the SOS repair and mutagenic activities show a peak after about 30 min of postirradiation incubation at 37°C. The SOS activities disappear after about 2 h of incubation at 37°C (Fox, Defais, Caillet-Fauquet and Radman, in preparation); (2) the presence of 100 μg/ml chloramphenicol during the first 30 min of postirradiation incubation abolishes the SOS repair and mutagenic activities (Defais, Caillet-Fauquet and Radman, in preparation); (3) cyclic AMP antagonizes the SOS repair (Defais, Caillet-Fauquet and Radman, in preparation); (4) there is no apparent induced lesion-specific endonuclease activity under conditions of SOS induction (Radman, in preparation); (5) phage φX174 with a single stranded DNA genome can be repaired in dark only by the induced SOS repair. The biochemical experiments performed under conditions of SOS repair and mutagenesis have shown a stimulation of DNA replication (which is otherwise blocked by the lesion) by the same increase factor as is the increase in survival (Caillet-Fauquet, Defais and Radman, in preparation). This is the first biochemical evidence for the existence of the hypothesized "SOS-replication" proposed in this paper and earlier (Radman, cf. Witkin and George, 1973, and Radman, 1974).

REFERENCES

Ames, B. N., Durston, W. E., Yamasaki, E. and Lee, F. D. (1973). *Proc. Nat. Acad. Sci. U.S.A.* **70**, 2281–2285.

Blanco, M. and Devoret, R. (1973). *Mutat. Res.* **17**, 293–305.

Bowersock, D. and Moses, R. E. (1973). *J. Biol. Chem.* **248**, 7449–7455.

Boyle, J. M. and Setlow, R. B. (1970). *J. Mol. Biol.* **51**, 131–144.

Bressler, S. E., Lanzov, V. A. and Lujaniec-Blinkova, A. A. (1968). *Mol. Gen. Genet.* **102**, 169–284.

Bridges, B. A. (1972). *Nature New Biol.* **240**, 52.

Castellazzi, M., George, J. and Buttin, G. (1972*a*). *Mol. Gen. Genet.* **119**, 139–152.

Castellazzi, M., George, J. and Buttin, G. (1972*b*). *Mol. Gen. Genet.* **119**, 153–174.

Defais, M., Fauquet, P., Radman, M. and Errera, M. (1971). *Virology* **43**, 495–503.

Drakulić, M. and Errera, M. (1959). *Biochim. Biophys. Acta,* **31**, 459–463.

George, J., Devoret, R. and Radman, M. (1974). *Proc. Nat. Acad. Sci. U.S.A.* **71**, 144–147.

Goldthwait, D. and Jacob, F. (1964). *C.R. Acad. Sci. Paris* **259**, 661–664.

Greenberg, J., Green, M. H. L. and Bar-Nun, N. (1970). *Mol. Gen. Genet.* **107**, 209–214.

Hart, M. G. R. and Ellison, J. (1970). *J. Gen. Virol.* **8**, 197–208.

Howard-Flanders, P. (1968*a*). *Ann. Rev. Biochem.* **31**, 175–200.

Howard-Flanders, P. (1968*b*). *Advan. Biol. Med. Phys.* **12**, 299–317.

Jacob, F. and Wollman, E. L. (1955). *Ann. Inst. Pasteur* **88**, 724–749.

Kingsbury, D. T. and Helinski, D. R. (1973). *Genetics* **74**, 17–31.

Kohiyama, M. (1968). *Cold Spring Harbor Symp. Quant. Biol.* **33**, 317–331.

Lehmann, A. R. (1972). *Eur. J. Biochem.* **31**, 438–445.

Luria, S. E. (1952). *J. Cell. Comp. Physiol.* **39**, 119–123.

Marinus, M. G. and Adelberg, E. A. (1970). *J. Bacteriol.* **104**, 1266–1272.

Miura, A. and Tomizawa, J. (1968). *Mol. Gen. Genet.* **103**, 1–10.

Moody, E. E. M., Low, K. B. and Mount, D. W. (1973). *Mol. Gen. Genet.* **121**, 197–205.

Monk, M. and Gross, J. (1971). *Mol. Gen. Genet.* **110**, 299–306.

Noack, D. and Klaus, S. (1972). *Mol. Gen. Genet.* **115**, 216–224.

Ono, J. and Shimazu, Y. (1966). *Virology,* **29**, 295–302.

Poddar, R. K. and Sinsheimer, R. L. (1971). *Biophys. J.* **11**, 355–369.

Radman, M. (1971). *Nature New Biol.* **230**, 277–278.

Radman, M. (1974). In *Molecular and Environmental Aspects of Mutagenesis* (Prakash, L., Sherman, F., Miller, M. W., Lawrence, C. W. and, Taber, H. W., eds.), C. C. Thomas, Publ., Springfield, Illinois, pp. 128–142.

Radman, M. and Devoret, R. (1971). *Virology* **43**, 504–506.

Radman, M., Cordone, L., Krsmanovic-Simic, D. and Errera, M. (1970). *J. Mol. Biol.* **49**, 203–212.

Rupp, W. D. and Howard-Flanders, P. (1968). *J. Mol. Biol.* **31**, 291–304.

Sesnowitz-Horn, S. and Adelberg, E. A. (1968). *Cold Spring Harbor Symp. Quant. Biol.* **33**, 393–402.

Setlow, R. B. (1967). *Brookhaven Symp. Biol.* **20**, 1.

Tessman, E. S. and Ozaki, T. (1960). *Virology* **12**, 431–449.

Weigle, J. J. (1953). *Proc. Nat. Acad. Sci. U.S.A.* **39**, 628–636.

Westergaard, O. (1970). *Biochim. Biophys. Acta* **213**, 36–44.

Willets, N. S. and Clark, A. J. (1969). *J. Bacterol.* 97, 231–239.
Witkin, E. M. (1969). *Ann. Rev. Microbiol.* 23, 487–513.
Witkin, E. M. (1974). *Proc. Nat. Acad. Sci. U.S.A.* 71, 1930–1934.
Witkin, E. M. and George, D. L. (1973). *Genetics* 73, 91–108.
Worcel, A. (1970). *J. Mol. Biol.* 52, 371–386.
Zajdela, M. and Latarjet, R. (1973). *C.R. Acad. Sci. Paris* 277, 1073–1076.

49

Thermal Enhancement of Ultraviolet Mutability in a *dnaB uvrA* Derivative of *Escherichia coli* B/r: Evidence for Inducible Error-Prone Repair

Evelyn M. Witkin

Department of Biological Sciences
Douglass College, Rutgers University
New Brunswick, New Jersey 08903

ABSTRACT

DNA damage triggers coordinate expression of a cluster of diverse functions in *Escherichia coli,* including prophage induction, filamentous growth, and "aberrant" reinitiation of DNA replication at the chromosomal origin. The "SOS repair" hypothesis proposes that one of these coordinately inducible functions is an error-prone system of DNA repair ("SOS repair") which is responsible for ultraviolet mutagenesis. In *dnaB* strains, incubation at 42°C stops DNA synthesis and induces lambda prophage and should, therefore, also induce the postulated error-prone repair activity. Thermal posttreatment of a *dnaB uvrA* derivative of *E. coli* B/r is found to enhance the yield of ultraviolet-light-induced mutations as much as 50-fold, while having no such effect in the *dnaB*⁺ parent strain. The results support the SOS repair hypothesis. The possibility is discussed that the inducible repair system is a mutagenic DNA polymerase.

In wild-type strains of *Escherichia coli,* DNA damage elicits the simultaneous expression of a diverse array of functions, including prophage induction, filamentous growth and W reactivation,[1] which are not ordinarily expressed in undamaged cells. Mutations in the *recA* or *lex* loci copleiotropically eliminate or drastically reduce the capacity to express these functions. The "SOS repair" hypothesis (Defais et al., 1971; Witkin and George, 1973; Devoret et al., this volume; Radman, 1974, and this volume; George et al., 1974; Witkin, 1974*a,b*) postulates that many *recA⁺lex⁺*-dependent functions are inducible, that their induction is triggered coordinately by treatments which interrupt DNA replication, and that one of these functions is an error-prone repair system responsible for UV mutagenesis. Recent work in my laboratory has centered on several mutant strains which have provided experimental support for the SOS repair hypothesis.

Figure 1 illustrates one possible interpretation of the coordinate expression of *recA⁺ lex⁺*-dependent functions ("reclex" functions) in *E. coli.* Inhibition of DNA synthesis, in *recA⁺ lex⁺* strains, is assumed to initiate a "reclex" induction pathway, which culminates in the synthesis of a "reclex" effector (or group of effectors) capable of inactivating a specific class of repressors, exemplified by λ repressor, thereby simultaneously derepressing operons controlled by repressors of the susceptible type. In the case of λ prophage induction, proteolytic cleavage of the λ repressor has been demonstrated directly (Roberts and Roberts, 1975). Other functions are included as possible members of the postulated inducible "reclex" cluster if they exhibit the same requirements for their expression as lambda prophage, namely, the *recA⁺ lex⁺* genotype and the occurrence of *de novo* protein synthesis during a period of interrupted DNA synthesis (see Witkin, 1974*b*; Radman, this volume). By this criterion, recent evidence implicates the induction of a product which limits the activity of exonuclease V (Marsden et al., 1974), another which switches off respiratory metabolism following damage to DNA (Swenson and Schenley, 1974), and still another responsible for type III repair of X-ray damage (Youngs and Smith, 1973), as likely candidates for membership in the "reclex" cluster of inducible functions. Functions belonging to this cluster presumably serve no purpose in healthy cells but are useful in promoting the survival of genetically damaged cells, or of the integrated episomes they may harbor. They may therefore have evolved similar repressors, all responsive to the same effector or group of effectors, the synthesis of which is normally initiated by interruption of DNA replication. This model implies that at least some mutations which alter the conditions required for induction of λ prophage should cause parallel changes in the expression of other "reclex" functions postulated to respond to the same regulatory signal. I shall discuss work on three mutant strains of *E. coli*, each of which exhibits anomalous induction of λ prophage under specific conditions, in which we have shown that UV mutagenesis is also anomalous, in the particular

---

[1] Abbreviations used: W reactivation and W mutagenesis, Weigle reactivation and Weigle mutagenesis. These terms are used in place of the older confusing terms UV reactivation and Uv mutagenesis, to describe the phenomenon first described by Weigle (1953).

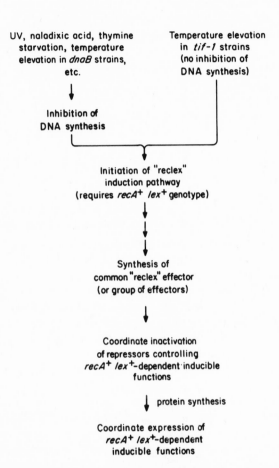

Fig. 1. An interpretation of the coordinate expression of recA⁺ lex⁺-dependent functions in E. coli. These "reclex" functions include prophage induction, filamentous growth, W reactivation, and W mutagenesis of λ phage, UV mutagenesis and most of X-ray mutagenesis, control of DNA breakdown by exonuclease V, type III repair of X-ray-induced single-strand breaks in DNA, the switching off of respiration after UV irradiation, and probably also "aberrant" reinitiation of DNA synthesis at the chromosomal origin after DNA damage. For references, see text, Witkin (1974*b*), and Radman (this volume).

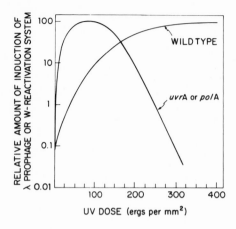

Fig. 2. Induction of λ prophage and of the W-reactivation system in E. coli K12 at various doses of UV. Composite data taken from Monk et al. (1971), Defais et al. (1971), and Fauquet and Defais (1972).

way predicted by the SOS repair hypothesis for each of the strains. Results with two of these strains (*polA* and *uvrA tif*) have been published elsewhere, and will be summarized briefly here. New results with a *uvrA dnaB* mutant will be described in greater detail.

Mutant strains lacking DNA polymerase I (*polA* mutants) exhibit a striking change in the UV dose range required to induce λ prophage, (Monk et al., 1971), and to effect *W* reactivation and *W* mutagenesis of UV-irradiated λ phage (Fauquet and Defais, 1972). Figure 2 shows that *polA* strains, although they are Uvr$^+$ and excise pyrimidine dimers normally (Boyle et al., 1970), express these functions in the dose range that is characteristic of excision-deficient *uvrA* strains, in contrast to the much higher dose range in which wild-type strains express these functions. If the effective trigger for UV induction of "reclex" functions is not a pyrimidine dimer *per se,* but a persistent gap in the DNA, the cause of the shifted induction dose range in *polA* strains may be its slow rate of closing excision gaps (Kanner and Hanawalt, 1970). Whatever the reason for the tenfold downshift in the induction dose range in *polA* strains, the SOS hypothesis predicts a comparable shift toward lower UV doses for all *recA$^+$ lex$^+$*-dependent functions postulated to respond to DNA damage in coordination with prophage induction. J. Donch (personal communication) has found a similar shift in the dose range required to elicit filamentous growth in a *lon polA* strain. Assuming that UV mutagenesis depends upon an inducible function (SOS repair) sharing the same induction requirements as other "reclex" functions, the *polA* strain should exhibit anomalous induction of the SOS repair function in the same low range of UV doses in which this strain expresses prophage induction. We have found (Witkin and George, 1973; Witkin, 1974*b*) that UV mutability in this dose range is significantly elevated in a *polA* strain.

A more telling test of the SOS hypothesis involves the *tif* mutation, originally described in *E. coli* K12 strain T44 (Kirby et al., 1967). The *tif-1* allele results in anomalous induction of both λ prophage and filamentous growth at elevated temperatures (42°C). Castellazzi et al. (1972*a*) have shown that *W* reactivation and *W* mutagenesis of UV-irradiated lambda phage are also thermally inducible in *tif* strains, suggesting that the *tif* mutation may permit thermal activation of the "reclex" induction pathway, resulting in coordinate thermal inducibility of the entire cluster. However, activation of the "reclex" induction pathway, whether initiated by DNA damage (as in normal strains) or by temperature elevation (as in *tif* mutants), requires the *recA⁺ lex⁺* genotype, since Castellazzi et al. (1972*b*) have shown that the *tif* phenotype is suppressed by mutations in the *recA, lex,* or *zab* loci. [The *zab* phenotype is similar to the *lexB* phenotype described by Blanco et al. (this volume)]. Since the *tif-1* mutation causes thermal inducibility of at least four "reclex" functions, the SOS hypothesis predicts that the postulated inducible function responsible for UV mutagenesis, SOS repair, should also be anomalously expressed in *tif* strains at high temperatures. Not only has the predicted thermal enhancement of UV mutability been demonstrated in a *uvrA tif-1* derivative of *E. coli* B/r (Witkin, 1974*a*), but this enhancement was shown to be promoted or reversed by treatments known to exert parallel effects on prophage induction in *tif* lysogens, consistent with the hypothesis of a common induction pathway. Furthermore, the *tif-1* mutation causes mutator activity in unirradiated cells grown at 42°C (Witkin, 1974*a*), suggesting that the inducible function responsible for UV mutagenesis can also introduce errors into undamaged DNA, although its mutagenic effect is greatly amplified by the presence in the DNA of even a very small number of pyrimidine dimers.

Another provocative result of the *tif* experiments was the linear rise of induced mutation frequency with increasing UV dose observed in the *tif* strain given thermal posttreatment, in contrast to the usual relation observed at low UV doses, in which the mutation frequency rises as the square of the dose. The mutation frequency in the Tif⁺ strains at 42°C or 30°C, as well as that in the *tif* mutant at 30°C, showed the typical "dose-squared" response of mutation frequency with increasing UV dose. We had earlier proposed (Witkin and George, 1973) that the normal "dose-squared" response reflects a requirement for two independent radiation-induced events: (1) the production of a mutagenic lesion (e.g. pyrimidine dimer) in the appropriate gene and (2) the induction of the SOS repair system. In thermally posttreated *tif* populations, all the cells are presumably induced and contain the mutagenic repair system, and the rising mutation frequency reflects the linear increase of pyrimidine dimers with increasing UV dose. In normal strains, there is a shift from "two-hit" mutation induction kinetics to "one-hit" kinetics as the dose approaches the level required to elicit maximal induction of prophage or *W* reactivation.

My most recent experiments concern mutant strains that are temperature-sensitive for DNA synthesis owing to a *dnaB* mutation. In such strains, elevation of the temperature to 42°C causes "instant stop" of DNA chain elongation

(Wechsler and Gross, 1971) and triggers prophage induction (Noack and Klaus, 1972). Assuming coordinate inducibility of prophage and the postulated error-prone SOS repair, *dnaB* strains (like *tif* strains, but for a quite different reason) should exhibit thermal enhancement of UV mutability at very low doses of UV, and should also produce mutations linearly with increasing dose, rather than as the square of the dose, when UV irradiation is followed by a period of incubation at 42°C. As explained elsewhere (Witkin, 1974a), thermal induction of the postulated SOS function must be assayed in UV-irradiated cells, since its activity (by definition) is expressed via the induction of mutations in DNA containing pyrimidine dimers. However, all but the lowest doses of UV are, in themselves, efficient inducers of "reclex" functions, and should thus be expected to induce SOS repair independently of any thermal induction. As can be seen in Fig. 2, extremely low doses of UV (e.g. doses below 10 ergs/mm^2 in *uvrA* strains) induce prophage and *W* reactivation in a negligibly small fraction of the population, and thus should introduce pyrimidine dimers into the DNA (about 6 dimers/erg/mm^2) without also inducing "reclex" functions to a significant extent. Irradiation with these low doses of UV, therefore, followed by a period of incubation at 42°C, provides a suitable assay for thermal induction of the postulated error-prone SOS repair activity, which should result in thermal enhancement of UV mutability in this dose range.

The *dnaB-43* allele from a K12 *dnaB* strain (Bonhoeffer, 1966) was transferred by P1 transduction into strain WP2$_s$, a *trp uvrA* derivative of *E. coli* B/r. The *dnaB* transductant of strain WP2$_s$, strain WP80, was found to be indistinguishable from WP2$_s$ in UV sensitivity and UV mutability at 30°C. At 42°C, strain WP80 shows immediate arrest of DNA synthesis, but recovers colony-forming ability completely if the incubation at 42°C does not exceed about 100 min. A spontaneous DnaB$^+$ revertant, selected as a colony formed at 42°C, was used interchangeably with the parent strain WP2$_s$ as a DnaB$^+$ control strain. Methods and media used in these experiments were those described elsewhere in detail (Witkin, 1974a), except that posttreatment at 42°C was for 90 min, and 5% SEM agar [minimal "E" agar supplemented 5% (v/v) with nutrient broth] was used as the plating medium. Saline suspensions of bacteria, exposed to doses of UV ranging from 0 to 10 ergs/mm^2, were spread on the surface of agar plates, incubated for the first 90 min either at 30°C or 42°C, and then incubated for 2 days (survival plates) or for 3 days (mutation plates) at 30°C. The mutations scored were those from Trp$^-$ to Trp$^+$, and were primarily ochre suppressors.

Figure 3 shows results of six experiments, and data from part of a typical experiment are shown in Table 1. The yield of UV-induced Trp$^+$ mutations is significantly enhanced in the *dnaB* strain by posttreatment at 42°C, about 30–50 times more mutations appearing at the lowest dose used than in either the same strain kept at 30°C after irradiation, or in the DnaB$^+$ strains with or without the thermal posttreatment. The magnitude of the enhancement diminishes with increasing dose, and no thermal enhancement of UV mutability

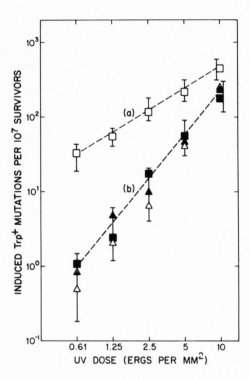

Fig. 3. Effect of postirradiation temperature elevation on the UV mutability of strains WP2$_S$ (dnaB$^+$) and WP80 (dnaB). ▲, WP2$_S$ at 30°C for 3 days; △, WP2$_S$ at 42°C for 90 min, then at 30°C for 3 days; ■, WP80 at 30°C for 3 days; □, WP80 at 42°C for 90 min, then at 30°C for 3 days. Points are averages and vertical lines show range of variability of data from six experiments. A DnaB$^+$ revertant of strain WP80 gave results indistinguishable from those shown for strain WP2$_S$. Broken line (a) shows theoretical "one-hit" curve (slope = 1); broken line (b) shows theoretical "two-hit" curve (slope = 2).

is observed at doses of 20 ergs/mm^2 or higher (not shown). The thermally enhanced yield of Trp$^+$ mutations induced in the *dnaB* strain increases linearly with increasing UV dose, while the Trp$^+$ mutation frequency rises as the square of the dose in the DnaB$^+$ strains, with or without thermal posttreatment, and in the *dnaB* strain incubated only at 30°C. The broken line (a) in Fig. 3 is the linear extrapolate toward lower UV doses made from the Trp$^+$ mutation frequency observed at 20 ergs/mm^2, the dose at which the mutation frequency curve begins to rise linearly with dose in strain WP2$_S$. Thus, thermal posttreatment of the *dnaB* mutant results in an induced Trp$^+$ mutation frequency that rises linearly with UV dose *throughout the entire range of doses in which mutations can be scored.* These results are similar to those described for *tif* strains (Witkin, 1974a), except that the yields of UV-induced mutations obtained in the *tif* experiments were somewhat lower than those shown here for *dnaB*, and the magnitude of the observed thermal enhancement of UV mutability was not as great. The reason for this difference is that a higher amount of nutrient broth supplementation (5%) was added to the minimal medium used to score Trp$^+$ mutations in the *dnaB* experiments, only 1.25% nutrient broth having been added to the minimal agar in the published experiments on the *tif* strain. When 5% SEM is used to score Trp$^+$ mutations in the *tif* strain WP44$_S$, the results essentially duplicate those shown in Fig. 3 for the *dnaB* strain WP80. Increasing

Table 1. The Effect of Temperature Elevation After UV Irradiation on the
Frequency of Induced Trp$^+$ Mutations in *dnaB uvrA* Strain WP80

	Temperature of incubation during first 90 min after UV irradiation[a]					
	30°C			42°C		
UV dose (ergs/ mm^2)	Trp$^+$ colonies per plate	Survival (%)	Induced Trp$^+$ mutations per 10^7 survivors	Trp$^+$ colonies per plate	Survival (%)	Induced Trp$^+$ mutations per 10^7 survivors
0	16, 10, 11	100	–	44, 43, 45	100	–
0.61	16, 13, 14	100	1.0	110, 90, 118	100	31
1.25	23, 24, 18	100	4.7	157, 131, 182	100	56.3
2.5	42, 49, 55	100	18.2	275, 262, 251	100	109.3
5.0	118, 189, 166	100	72.7	368, 360, 410	83	201.2
10.0	525, 492, 501	74	333.6	402, 420, 415	40	460.4

[a]Temperature after first 90 min was 30°C for all plates. Number of bacteria plated: 2 X 10^7 per plate.

the level of nutrient broth enrichment in the selective minimal medium is known to result in higher yields of UV-induced mutations in Uvr$^-$ strains (Green et al., 1972), but we have found that the yields obtained on 5% SEM cannot be further increased by additional nutrient broth enrichment, and probably represent truly maximal yields.

Thermal posttreatment of *dnaB* cells thus enhances the yield of UV-induced mutations at UV doses too low to cause significant UV induction of "reclex" functions, and converts the mutation frequency vs. dose response from a "two-hit" to a "one-hit" curve, duplicating the effect first described in *tif* strains. However, these strikingly similar effects undoubtedly have a very different basis in the two strains. In *tif* strains, incubation at 42°C results in induction of "reclex" functions without any apparent damage to DNA, and without any inhibition of DNA replication (Castellazzi et al., 1972a). The *tif* mutation somehow renders the "reclex" induction pathway susceptible to activation by heat, bypassing the usual requirement for DNA damage as the initial trigger. Incubation at 42°C in the *dnaB* strain, however, interrupts DNA replication and thus provides the standard signal recognized by normal strains as the trigger for induction of "reclex" functions. Both strains, each exhibiting anomalous thermal induction of lambda prophage for a different reason, also exhibit the thermal enhancement of UV mutability and the linear mutation induction kinetics predicted by the SOS hypothesis. Together, they provide substantial evidence that UV mutagenesis depends upon the induction of an error-prone function (SOS repair) not normally present in undamaged *E. coli* cells, and sharing the induction requirements of prophage induction and filamentous growth.

In Table 1, data for unirradiated *dnaB* cells, incubated for 90 min at 42°C, show a significant increase in the frequency of spontaneous Trp^+ mutations, compared to the 30°C controls. No such increase is observed in $DnaB^+$ strains. Increased spontaneous mutability at 42°C was also seen in the *tif* strain WP44$_s$ (Witkin, 1974*a*), and was interpreted as an indication that the inducible error-prone function responsible for UV mutagenesis is also capable of causing errors in DNA not containing pyrimidine dimers, albeit with a much lower probability. The mutagenic effect of incubation at 42°C in unirradiated *dnaB* cells, as seen in Table 1, supports this conclusion. However, in the *tif* strain, DNA replication continues at 42°C, while the temperature elevation arrests DNA replication sharply in the *dnaB* strain. Prolonged incubation at 42°C, in *tif* strains, thus permits many replications of DNA under conditions of continuous expression of the SOS function, providing a reasonable explanation for the mutator-like effect of the *tif-1* allele under these conditions. In the *dnaB* strain, however, the thermal treatment which promotes induction of the error-prone function does not permit DNA replication to continue while "reclex" functions are derepressed. High spontaneous mutability, in the *dnaB* strain, does not reach mutator levels because it is largely limited to the first DNA replication following reduction of the temperature to permissive levels (Witkin, unpublished results), implying that SOS activity persists briefly once the signal for induction is quenched.

The known DNA polymerases in *E. coli* appear to be constitutive components of normal cells, and exhibit extrordinary fidelity to template instructions in DNA replication. The error-prone repair activity inducible as part of the "reclex" cluster could be (1) a fourth DNA polymerase, not normally present in undamaged cells, having relaxed template requirements and the ability to insert bases opposite such noncoding lesions as pyrimidine dimers; (2) a factor which modifies one of the constitutive DNA polymerases so as to confer upon it the ability to replicate DNA past a noncoding lesion; or (3) a component of recombinational repair which increases its efficiency but reduces its accuracy. A constitutive DNA polymerase with relatively poor fidelity would hardly find favor by natural selection. In a damaged cell, the ability to induce synthesis of such a DNA polymerase in response to the "SOS" signal of arrested DNA replication could make the difference between life and death. A heightened frequency of mutations, under such circumstances, may well be an acceptable price to pay for survival.

ACKNOWLEDGMENTS

This work was supported by research grant (AI-10778) from the National Institute of Allergy and Infectious Diseases. I thank Ingbritt Wermundsen for expert technical assistance.

REFERENCES

Bonhoeffer, F. (1966). *Z. Vererbungs l.* **98**, 141–149.

Boyle, J. M., Paterson, M. C. and Setlow, R. B. (1970). *Nature* **226**, 708–710.

Castellazzi, M., George, J. and Buttin, G. (1972*a*). *Mol. Gen. Genet.* **119**, 139–152.

Castellazzi, M., George, J. and Buttin, G. (1972*b*). *Mol. Gen. Genet.* **119**, 153–174.

Defais, M., Fauquet, P., Radman, M. and Errera, M. (1971). *Virology* **43**, 495–503.

Fauquet, P. and Defais, M. (1972). *Mutat. Res.* **15**, 353–355.

George, J., Devoret, R. and Radman, M. (1974). *Proc. Nat. Acad. Sci. U.S.A.* **71**, 144–147.

Green, M., Rothwell, M. A. and Bridges, B. A. (1972). *Mutat. Res.* **16**, 225–234.

Kanner, L. and Hanawalt, P. (1970). *Biochem. Biophys. Res. Commun.* **39**, 149–155.

Kirby, E. P., Jacob, F. and Goldthwait, D. A. (1967). *Proc. Nat. Acad. Sci. U.S.A.* **58**, 1903–1910.

Marsden, H. S., Pollard, E. C., Ginoza, W. and Randall, E. P. (1974). *J. Bacteriol.* **118**, 465–470.

Monk, M., Peacey, M. and Gross, J. D. (1971). *J. Mol. Biol.* **58**, 623–630.

Noack, D. and Klaus, S. (1972). *Mol. Gen. Genet.* **115**, 216–224.

Radman, M. (1974). In *Molecular and Environmental Aspects of Mutagenesis* (Prakash, L., Sherman, F., Miller, M., Lawrence, C. and Tabor, H. W., eds.), Thomas, Springfield, Ill., pp. 128–142.

Roberts, J. W. and Roberts, C. W. (1975). *Proc. Nat. Acad. Sci. U.S.A.* **72**, 147–151.

Swenson, P. A. and Schenley, R. L. (1974). *Int. J. Radiat. Biol.* **25**, 51–60.

Wechsler, J. A. and Gross, J. D. (1971). *Mol. Gen. Genet.* **113**, 273–284.

Weigle, J. J. (1953). *Proc. Nat. Acad. Sci. U.S.A.* **39**, 628–636.

Witkin, E. M. (1974*a*). *Proc. Nat. Acad. Sci. U.S.A.* **71**, 1930–1934.

Witkin, E. M. (1974*b*). *Genetics* (suppl.) **78**, in press.

Witkin, E. M. and George, D. L. (1973). *Genetics* (suppl.), **73**, 91–108.

Youngs, D. A. and Smith, K. C. (1973). *J. Bacteriol.* **114**, 121–127.

50

lexB: A New Gene Governing Radiation Sensitivity and Lysogenic Induction in Escherichia coli K12

M. Blanco,[1] A. Levine, and R. Devoret[2]

Laboratoire d'Enzymologie, C.N.R.S.
91190 Gif-sur-Yvette, France

Bacterial mutants resistant to thymine deprivation were selected from *Escherichia coli* K12 $(\lambda)^{+}$. Mutants in which wild-type λ prophage had become noninducible after thymine starvation and/or UV irradiation were isolated (Devoret and Blanco, 1970). One of them, GY6130 (Devoret et al., 1972), has a phenotype very similar to that of AB2494 *lex-1* (Howard-Flanders and Boyce, 1966). It is highly sensitive to UV light and X-rays and degrades its DNA rather extensively after irradiation (see Figs. 1 and 2); furthermore, the mutant is recombination proficient and allows the growth of phage *λred gam*. The mutation responsible for this phenotype has been located by Hfr $\times$ F$^-$ crosses and P1 transduction; it lies near *recA* between *cysC* and *pheA*. Consequently, the mutation carried by GY6130 has been named *lexB30;* we propose that mutations mapping between *malB* and *metA* (Mount et al., 1972) be now designated *lexA*. Castellazzi et al. (1972) have described, in strains carrying mutation *tif* (Kirby et al., 1967), a mutation they called *zab* that suppresses the *tif* phenotype. Mutation *lexB30* has a phenotype and a genetic location very similar to *zab*. We do not know yet whether or not *lexB30* and *zab-53* are in the same cistron. It must be pointed out that, since *zab* is so close to *tif*, it has been

[1] Present address: Instituto de Investigaciones Citológicas, Amadeo de Saboya 4, Valencia 10, Spain.
[2] Present address: Laboratory of Molecular Biology, NIAMDD, Bldg. 2, Room 322, National Institutes of Health, Bethesda, Maryland 20014.

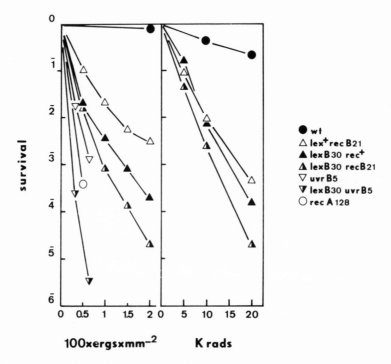

Fig. 1. UV irradiation: Cells from exponentially growing cultures were harvested by centrifugation and resuspended in M9 buffer at a concentration of 2×10^8 cells/ml. This suspension was then placed in a petri dish and irradiated with a 15-watt General Electric germicidal lamp giving a maximum output at 254 nm. Dose rates were measured with a Latarjet dosimeter. The thickness of the suspensions was 1 mm. *X-irradiation:* The general procedure was as indicated for UV irradiation except that bacteria was resuspended in Luria broth. The source of X-rays was a Machlett OEG60 tube. The dose rate was 4.8 krads/min. *Bacterial survival:* After UV or X-irradiation, bacteria were diluted and plated immediately on Luria agar medium. After UV irradiation, photoreactivation was avoided.

difficult to study *zab* out of a *tif* genetic background (Castellazzi, personal communication). In contrast, the study of the genotype and phenotype of *lexB30* is not subject to the same limitation.

Double mutants *lexB30 uvrB5* and *lexB30 recB21* are viable. The two double mutants are more sensitive to UV light than either parental single mutant (Fig. 1). Presence of the *recB21* defect prevents DNA degradation after UV irradiation (Fig. 2*b,c*). These results are similar to those obtained with double mutants carrying *lexA* (Moody et al., 1973; Mount and Kosel, 1973). It is interesting to note that mutation *uvrB5* does not prevent DNA degradation of the DNA after UV irradiation.

Host-cell reactivation and prophage-reactivation of phage λ takes place in *lexB30* bacteria whereas UV reactivation does not.

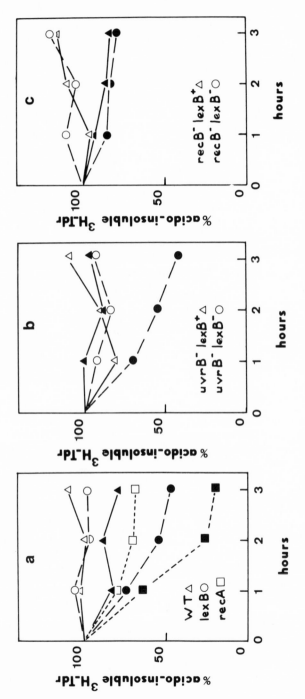

Fig. 2. Measurement of DNA breakdown. Degradation of DNA after UV irradiation was estimated by observing the loss of radioactivity precipitable by trichloroacetic (TCA) in the labeled cultures. Cultures grown overnight in M9 with 2 μg/ml thymidine were diluted 1:20 into the same fresh medium containing 4 μCi of [6-³H] thymidine and incubated at 37°C for 120–150 min. The labeled cells were collected by centrifugation and resuspended in warm M9 medium with 10 μg/ml cold thymidine. Incubation was continued for 20 min to "chase" any residual, unincorporated, [6-³H] thymidine. The cultures were then centrifuged and the bacteria were resuspended in M9 buffer at a concentration of 2 × 10⁷ cells/ml and irradiated with UV light. After irradiation the cultures were diluted 1:10 into fresh prewarmed M9 with 10 μg/ml thymidine and incubated at 37°C with aeration. At various times during the incubation period, triplicate samples (1 ml) were withdrawn and added to an equal volume of cold 10% TCA. The mixtures were kept in an ice bath for at least 2 hr. The resulting precipitates were collected on glass-fiber filters (Whatman GF/C), washed twice with cold 5% TCA, dried, and counted in a Packard Tri-Carb scintillation counter in Bray's liquid scintillator. Open and closed symbols are used for nonirradiated and irradiated samples, respectively.

The phenotypic similarities displayed by *lexA, zab,* and *lexB* bacteria pose many questions about the gene functions. However, the idea that *lex* genes might regulate *recA* (Moody et al., 1973) now appears untenable, since there are too many *lexA*-like genes. Instead, one might speculate on the fact that four mutations governing lysogenic induction—*tif, recA, zab, lexB*—map at the same chromosomal site.

REFERENCES

Castellazzi, M., George, J. and Buttin, G. (1972). *Mol. Gen. Genet.* 119, 153–174.

Devoret, R. and Blanco, M. (1970). *Mol. Gen. Genet.* 107, 272–280.

Devoret, R., Blanco, M. and Bailone, A. (1972). *Molecular Mechanisms of Genetics Processes,* Nauka, Moscow, pp. 321–330.

Howard-Flanders, P. and Boyce, R. P. (1966). *Radiat. Res.* 6 (suppl.), 156–184.

Kirby, E. P., Goldthwait, D. A. and Jacob, F. (1967). *Proc. Nat. Acad. Sci. U.S.A.* 58, 1903–1910.

Moody, E. E. M., Low, B. and Mount, D. W. (1973). *Mol. Gen. Genet.* 121, 197–205.

Mount, D. W. and Kosel, C. (1973). *Mol. Gen. Genet.* 120, 291–299.

Mount, D. W., Low, B. and Edmiston, S. J. (1972). *J. Bacteriol.* 112, 886–893.

51

Indirect Suppression of Radiation Sensitivity of a *recA⁻* Strain of *Escherichia coli* K12

David W. Mount, Anita C. Walker,[1] and Candace Kosel

Department of Microbiology
College of Medicine
University of Arizona
Tucson, Arizona 85724

ABSTRACT

It has been shown previously that the radiation sensitivity of LexA⁻ strains of *Escherichia coli* K-12 can be suppressed by thermosensitive mutations (designated *tsl*) that are closely linked to the *lexA* locus. These are thought to be intragenic suppressors that reduce the activity of the diffusible product that gives rise to the LexA⁻ phenotype (Mount et al., 1973). When a *recA* mutation is crossed into a suppressed *tsl⁻* strain, the extreme radiation sensitivity usually conferred by a *recA* mutation is considerably reduced without any detectable change in genetic recombination deficiency. Suppression of UV sensitivity depends upon the activity of the *uvrA⁺* product. We propose that at least part of the radiation sensitivity of a *recA⁻* strain is due to a DNA repair defect that is different from inability to perform genetic exchanges and depends upon the presence of the *lexA⁺* product. We hypothesize that the *lexA⁺* product is a repressor of the synthesis of repair enzymes. In *recA⁺* cells with DNA lesions, repressor is inactivated leading to enzyme induction but this does not occur in *recA⁻* cells. *tsl* mutations inactivate repressor leading to constitutive enzyme synthesis and bypassing the need for *recA⁺* product to inactivate the *lexA⁺* product.

[1] Present address: Roche Institute of Molecular Biology, Nutley, New Jersey.

RecA⁻ strains of *Escherichia coli* K12 are deficient in genetic recombination and in repair of a variety of DNA lesions including pyrimidine dimers produced by irradiation with UV light (for review see Clark, 1973). They also respond abnormally to UV irradiation in certain other respects that are reviewed by Radman (this volume). The prevailing view is that they are defective in closure of gaps left opposite pyrimidine dimers (Rupp and Howard-Flanders, 1968) when damaged DNA is replicated (Smith and Meun, 1970). Genetic exchanges between newly formed sister chromosomes each containing gaps could be one mechanism for the repair of these gaps (Rupp et al., 1971; Ganesan, this volume).

Essentially all the abnormalities in DNA repair found in *recA*⁻strains are also observed in *lexA*⁻ strains[2] of *E. coli* K12 (Smith and Meun, 1970; Sedgwick and Bridges, 1972; Bridges, 1973; Moody et al., 1973; Radman, this volume). Moody et al. (1973) have provided evidence that the *recA*⁺ and *lexA*⁺ products act in a common pathway of DNA repair. The properties of *lexA*⁻strains do differ from those of *recA*⁻ strains in two respects. First, *lexA*⁻ strains are recombination proficient, and, second, they are not as sensitive as *recA*⁻strains to agents that induce DNA lesions. These observations have led to the hypothesis that the *recA*⁺ product acts in two pathways of DNA repair; one that is also acted upon by the *lexA*⁺ product, and a second that is a DNA repair and genetic recombination pathway (Moody et al., 1973).

Other experiments have shown that the LexA⁻ phenotype results from the synthesis of a diffusible product (Mount et al., 1972; Castellazzi et al., 1972). UV-resistant derivatives in which the activity of this product may be reduced have been obtained from *lexA*⁻strains (Mount et al., 1973). These derivatives are defective in cell division and grow into multinucleate filaments at 42.5°C yet grow apparently normally and are as UV resistant as *lexA*⁺ strains at 30°C. The thermosensitive mutations, designated *tsl,* are tightly linked to the *lexA* locus, and their genetic complementation behavior is that expected if *tsl* and *lexA* mutations affect the same product (Mount et al., 1973). These studies have suggested that above normal activity of the *lexA* gene product leads to a UV-sensitive phenotype while below normal activity of this same product results in a UV-resistant, thermosensitive phenotype.

Since *lexA* and *recA* mutations appear to affect a common pathway of DNA repair, *tsl* mutations might be expected to influence the radiation sensitivity of a *recA*⁻ strain. The following experiments show that *tsl*⁻ can suppress the radiation sensitivity of a *recA*⁻ strain.

Strains are all derivatives of *E. coli* K12, and have been constructed by standard genetic methods from stock strains described previously; methods for

---

[2] We refer to the *lex* locus at 80.9 min on the standard *E. coli* K12 linkage map (Taylor and Trotter, 1972) as *lexA,* as suggested by Blanco et al. (this volume). We assume that the *exrA* locus in *E. coli* B is identical to *lexA* and will use *lexA* to refer to the *exrA* locus (Donch and Greenberg, 1974).

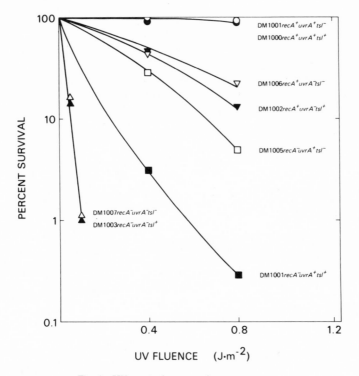

Fig. 1. UV survival curves of mutant strains.

measuring UV sensitivity and genetic recombination proficiency are given elsewhere (Mount et al., 1972, 1973; Moody et al., 1973).

UV survival curves of a set of strains with combinations of mutant or normal
tsl, recA, and *uvrA* alleles are shown in Fig. 1. For cultures grown and tested at
30°C, *tsl* increases the UV survivals of a *uvrA⁺ recA⁻* strain to a large extent but
has little effect on the UV survivals of *uvrA⁺ recA⁺, uvrA⁻recA⁺* and *uvrA⁻recA⁻*
strains.[3] These results indicate that *tsl⁻* partially suppresses UV sensitivity due to
a *recA* mutation, and that this suppression depends upon a functioning *uvrA⁺*
allele. A *tsl⁻* mutation also suppresses the UV sensitivity of a *lon⁻* strain
(Howard-Flanders et al., 1964; Adler and Hardigree, 1964) to about this same
extent.

tsl⁻ does not affect the genetic recombination deficiency of a *recA⁻* strain at
30°C, but a *tsl⁻ recA⁻* strain is more resistant to the alkylating agent MMS and
the cross-linking agent mitomycin C than a *tsl⁺ recA⁻* strain. Suppression by *tsl⁻*

[3] Whenever the *tsl⁻* allele is present in a given strain, *lex⁻* may also be present. Because they
are so tightly linked it is impossible to separate them in genetic crosses. *tsl* mutations could
also be deletions or other aberrations in the *lexA* gene (Mount et al., 1973).

thus influences one or more pathways of DNA repair without having any detectable influence on genetic recombination.

The above results recall earlier experiments of Ganesan and Smith (1968), who demonstrated that UV survivals of a *recA*⁻ strain can be increased by incubating UV-irradiated cells in buffer prior to inoculating them on to nutrient plates, or by inoculating them onto minimal medium plates rather than complete broth medium plates. These recovery processes are not seen in *uvr*⁻*recA*⁻ strains that are deficient in both excision and recombination repair. The treatments are proposed to increase recovery by delaying replication of DNA containing pyrimidine dimers and other photoproducts, thereby allowing more time for excision repair (Ganesan and Smith, 1968).

By comparison, the increased UV survivals conferred by *tsl*⁻ on *recA*⁻*uvrA*⁺, but not *recA*⁻ *uvrA*⁻, strains could be due to a slowing effect of *tsl*⁻ on the replication of UV-damaged DNA. Our experiments suggest that this is not the case. In the first place, *tsl*⁻ does not influence the growth rate of cells at 30°C. Second, when *recA*⁻ *tsl*⁺ and *recA*⁻ *tsl*⁻ strains are UV-irradiated with the same UV fluence, rates of DNA breakdown, DNA synthesis, and the time of restoration of cell division are indistinguishable. Finally, when UV-irradiated *tsl*⁻*recA*⁻ cells are incubated in buffer prior to inoculating them on to complete broth medium plates, survivals increase to approximately the same relative extent as those of *tsl*⁺ *recA*⁻ cells. This result would not be expected if *tsl*⁻-mediated UV resistance originated from the same mechanism as recovery in buffer. Consequently, suppression of radiation sensitivity of a *recA*⁻ strain by *tsl*⁻ cannot be explained by a delay in replication of damaged DNA.

A second possible explanation of the suppression that we have observed is that *tsl*⁻ *recA*⁻ cultures at 30°C may contain UV-resistant filaments. Two experiments indicate that this is probably not the mechanism of resistance. In the first place, microscopic examination of 30°C cultures of a *recA*⁻*tsl*⁻ strain reveals that the distribution of cell sizes is approximately the same as that of a *recA*⁻ *tsl*⁺ strain. Second, a culture of a *tsl*⁻ *recA*⁻ strain stops dividing and grows into long filaments at 42.5°C. These filaments which grow to be 50–100 times the length of normal *E. coli* rods following 2 hr incubation at 42.5°C retain their ability to form colonies on nutrient agar at 30°C and can be tested for their sensitivity to UV irradiation. Filaments of the *tsl*⁻ *recA*⁻ strain prepared in this manner are slightly more UV resistant than the rods in a 30°C culture but the difference is considerably smaller than the resistance conferred by *tsl*⁻ on *recA*⁻ cells. It therefore appears unlikely that the *tsl*⁻ *recA*⁻ rods at 30°C owe their resistance to physiological changes in growth and division.

The above results show that at least part of the radiation sensitivity of a *recA*⁻ strain cannot be attributed to genetic recombination deficiency, and suggest that *recA*⁻ cells are defective in a DNA repair pathway that is independent of genetic recombination. This observation supports the earlier suggestion of Moody et al. (1973) that *recA*⁻ cells are deficient in two pathways of DNA repair.

To account for these results, we present the following hypothesis: The $lexA^+$ product is a repressor that in unirradiated cells lowers the level of certain repair enzymes. In cells with DNA lesions, certain unknown biochemical changes lead to inactivation of repressor and enzyme induction. These enzymes are necessary for efficient filling of gaps left in replicated DNA opposite pyrimidine dimers or other DNA lesions. *lexA* mutations alter repressor so that it is not inactivated in cells with DNA lesions and *tsl* mutations lower the activity of repressor so that constitutive synthesis of the repair enzymes results. This constitutive synthesis of enzymes in a tsl^- strain results in a UV-resistant, thermosensitive phenotype.

The radiation sensitivity of a $recA^-$ strain results, in part, from failure to inactivate repressor in cells with inducing DNA lesions. This requirement of the $recA^+$ product is similar to its role in the destruction of phage repressor in induced lysogenic cells (Brooks and Clark, 1967). Since repair repressor is already inactive in tsl^- cells, this repressor-inactivating function of *recA* is not required, with the result that $tsl^- recA^-$ cells are more resistant than $tsl^+ recA^-$ cells. Failure of $tsl^- recA^-$ cells to show a fully UV-resistant $recA^+$ phenotype could be due to the requirement of the $recA^+$ product in a second pathway of DNA repair and genetic recombination as proposed by Moody et al. (1973). This second pathway is still deficient in $recA^- tsl^-$ strains and could be repair that occurs through sister strand exchanges (Rupp et al., 1971; Ganesan, this volume).

The hypothesis that repair enzymes are induced has been given earlier (Defais et al., 1971; Witkin and George, 1973; Radman, 1974) and is discussed in detail by Witkin (this volume) and by Radman (this volume). At the present time, we do not have enough information on the properties of $ts1^-$ strains to speculate whether or not the postulated *lexA* regulated enzymes could be identical to those of SOS repair (Radman, 1974).

Experiments with strains that are deficient in excision and recombination repair indicate that tsl^--promoted repair of UV damage depends upon active $uvrA^+$ product. tsl^- does not appear to affect rate of replication of UV-damaged DNA, but instead influences a repair process that depends upon the $uvrA^+$ product. It could either enhance the efficiency of excision repair or promote a repair process that depends upon the presence of endonucleolytic scissions in DNA next to pyrimidine dimers by the UV endonuclease known to be under control of the *uvrA* gene (Braun et al., this volume). A second possibility for the dependence of tsl^--promoted suppression of $recA^-$-UV sensitivity on the $uvrA^+$ product is that the system depends upon a critical number of DNA lesions to induce other repair enzymes (Radman, 1974, and this volume; Witkin, this volume). These could act in the same pathway of repair as tsl^--controlled enzymes, and the extreme UV sensitivity of the $recA^- uvrA^-$ strains, one to two pyrimidine dimers corresponding to a lethal hit (Howard-Flanders et al., 1969), would not permit an adequate number of DNA lesions to induce these other functions.

ACKNOWLEDGMENTS

This research was supported by NSF grant GB27910. Bill Howe, Richard Hull, and Marilyn Medweid are thanked for their comments.

REFERENCES

Adler, H. I. and Hardigree, A. A. (1964). *J. Bacteriol.* **87**, 220–226.
Bridges, B. A. (1973). *Genetics* **73** (suppl), 123–129.
Brooks, K. and Clark, A. J. (1967). *J. Virol.* **1**, 283–293.
Castellazzi, M., George, J. and Buttin, G. (1972). *Mol. Gen. Genet.* **119**, 153–174.
Clark, A. J. (1973). *Ann Rev. Genet.* **7**, 67–86.
Defais, M., Faquet, P. and Radman, M. (1971). *Virology* **43**, 495–503.
Donch, J. J. and Greenberg, J. (1974). *Mol. Gen. Genet.* **128**, 277–281.
Ganesan, A. K. and Smith, K. C. (1968). *Cold Spring Harbor Symp. Quant. Biol.* **33**, 235–242.
Howard-Flanders, P., Simson, E. and Theriot, L. (1964). *Genetics* **49**, 237–246.
Howard-Flanders, P., Theriot, L. and Stedeford, J. B. (1969). *J. Bacteriol.* **97**, 1134–1141.
Moody, E. E. M., Low, K. B. and Mount, D. W. (1973). *Mol. Gen. Genet.* **121**, 197–205.
Mount, D. W., Low, K. B. and Edmiston, S. J. (1972). *J. Bacteriol.* **112**, 886–893.
Mount, D. W., Walker, A. C. and Kosel, C. (1973). *J. Bacteriol.* **116**, 950–956.
Radman, M. (1974). In *Molecular and Environmental Aspects of Mutagenesis* (Miller, M., ed.), Thomas, Springfield (in press).
Rupp, W. D., and Howard-Flanders, P. (1968). *J. Mol. Biol.* **31**, 291–304.
Rupp, W. D., Wilde, C. E., Reno, D. L. and Howard-Flanders, P. (1971). *J. Mol. Biol.* **61**, 25–44.
Sedgwick, S. G. and Bridges, B. A. (1972). *Mol. Gen. Genet.* **119**, 93–102.
Smith, D. C. and Meun, D. H. C. (1970). *J. Mol. Biol.* **51**, 459–472.
Taylor, A. L. and Trotter, C. D. (1970). *Bacteriol. Rev.* **36**, 504–524.
Witkin, E. M. and George D. L. (1973). *Genetics* **73** (suppl.), 91–108.

52

The Two-Lesion Hypothesis for UV-Induced Mutation in Relation to Recovery of Capacity for DNA Replication

Charles O. Doudney[1]

Department of Genetics
Albert Einstein Medical Center
Philadelphia, Pennsylvania 19141

Evidence has accumulated (Doudney, 1961, 1963, 1965, 1966, 1968, 1969; Doudney and Young, 1962) that supports the hypothesis that in *Escherichia coli* two different UV-induced lesions are necessary for induction of mutations. This hypothesis was prompted by the dose-squared response to UV seen with mutation in bacteria. The evidence suggested that the two different lesions responsible for mutation are a pyrimidine dimer within the potentially mutated gene length together with a "DNA-synthesis blocking lesion" bringing a requirement for RNA and protein synthesis if recovery of DNA replication (and mutation) is to occur.

Recent evidence (DeFais et al., 1971; Witkin and George, 1973) has suggested the basis for the two lesions and the requirement for DNA replication inhibition in the mutation process. This evidence suggests that "error-prone repair" is an inducible function rather than constitutive and is induced by the inhibition of DNA replication (e.g. by UV), sharing this response with a number of other UV-inducible functions including prophage induction, filamentous growth, reactivation and mutagenesis of irradiated bacteriophage. The work of Witkin and George (1973) supports a hypothesis that two effects of UV are necessary for mutation to occur—i.e. damage presumably in the form of a pyrimidine dimer within the gene in question and damage to DNA replication

[1] Current Address: Division of Laboratories and Research, New York State Dept. of Health, Albany, New York 12201.

causing the induction of an error-prone repair system. RNA and protein synthesis must occur during the period of blocked DNA replication if induction is to take place.

Recent evidence supports the hypothesis that UV inactivates DNA replication by direct damage to the active replication site and that recovery of capacity for replication involves reinitiation of replication at the chromosomal fixed origin (Doudney, 1973). This reinitiation requires protein synthesis and is prevented. by chloramphenicol, an antibiotic which specifically blocks protein formation. Chloramphenicol loses its capacity to prevent DNA replication with postirradiation incubation and before initiation of DNA replication (see Doudney, 1965). It was shown that the decrease in rate of replication in the culture enforced by chloramphenicol is due to the inhibition of replicating units rather than a slowed rate of replication in each bacterium.

Figure 1 shows a correlation of loss of the ability for blockage of DNA replication by chloramphenicol with loss of ability to promote *mutation frequency decline* (Witkin, 1956). This study was on mutation to tryptophan independence in *E. coli* WP2, a strain widely used for mutation studies. If the event which prevents *mutation frequency decline* is required for induction of the "error-prone" repair mechanism as seems likely, then, in view of the above-described correlation, the initiation proteins for DNA replication and the protein required for induction of the repair system could be one and the same with mutation induction thus depending on reinitiation of DNA replication at the chromosomal fixed origin.

The two-lesion hypothesis for UV-induced mutation, taking the above into consideration, would include the induction of a lesion damaging the active DNA replication site followed by protein and RNA formation leading to restoration of DNA replication. These events in one way or another would lead to induction of the error-prone repair mechanism. The macromolecules involved in induction of the error-prone mechanism could be the initiation protein and RNA or/and other RNA and protein formed at about the same time as the initiation macromolecules. The other lesion necessary for UV-induced mutation would be a pyrimidine dimer within the length of the potentially mutated gene and subject to repair by the error-prone mechanism. The thermal enhancement of UV mutation in the *tif* strain of *E. coli* (see Witkin, this volume) presumably eliminates the necessity for UV-induction of the "DNA synthesis-blocking lesion."

ACKNOWLEDGMENTS

This research was supported by research grant No. CA 12133–03 from the National Cancer Institute of the U.S. Public Health Service and by U.S. Atomic Energy Commission contract AT (11–1)3095.

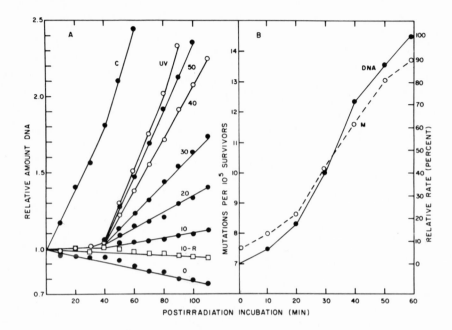

Fig. 1. (A) The ability of chloramphenicol added at specific intervals after UV exposure (225 ergs/mm²) to limit rate of DNA synthesis in E. coli strain WP2 B/r trp⁻ thy⁻ (Witkin, 1956) cultures with postirradiation incubation. The culture procedures and irradiation techniques have been described (Doudney, 1973). The culture was prelabeled before UV exposure with [¹⁴C]thymine from a small innoculum. The increase in labeled thymine incorporation after UV exposure was determined at 10-min intervals and used to calculate the relative amount of DNA. The symbol "C" represents the unirradiated control; "UV" designates the UV-exposed culture without postirradiation addition of chloramphenicol; the curves denoted by numbers represent the UV-exposed cultures to which chloramphenicol (160 µg/ml) was added at the time indicated by the numbers. The curve labeled "10-R" represents a culture incubated without added [¹⁴C]thymine after UV exposure (but with nonlabeled thymine present) and shows that when chloramphenicol is added after 10 min of exposure very little DNA breakdown occurs, contrary to the results obtained when chloramphenicol is added immediately after exposure. *(B) A comparison of relative rate of DNA replication after chloramphenicol addition at specific times with decrease in mutation frequency of trp⁺ reversions in E. coli strain WP2 (Witkin, 1956) enforced by chloramphenicol addition at the same times.* In order to follow decrease in mutation frequency the bacteria were incubated after chloramphenicol addition for 40 min and then plated out on appropriate agar medium. The mutation studies were carried out as described by Doudney (1968). The relative rate figures for DNA replication in chloramphenicol were calculated from the slopes in Fig. 1A relative to the slope of the UV-exposed culture incubated without chloramphenicol. After 60 min of incubation, no further fixation of mutation (to resistance to the chloramphenicol effect) occurred.

REFERENCES

DeFais, M., Fauquet, P. and Radman, M. (1971). *Virology* **43**, 495.

Doudney, C. O. (1961). *J. Cell. Comp. Physiol.* (suppl. 1) **58**, 145.

Doudney, C. O. (1963). In *Repair from Genetic Radiation Damage and Differential Radiosensitivity in Germ Cells* (Sobels, F., ed.), p. 125. Pergamon Press, New York.

Doudney, C. O. (1965). In *Cellular Radiation Biology,* p. 120. Williams and Wilkins, Baltimore.

Doudney, C. O. (1966). *Mutat. Res.* **3**, 280.

Doudney, C. O. (1968). *Mutat. Res.* **6**, 345.

Doudney, C. O. (1969). *Curr. Top. Microbiol.* **46**, 117.

Doudney, C. O. (1973). *Mutat. Res.* **17**, 1.

Doudney, C. O. and Young, C. (1962). *Genetics* **47**, 1125.

Witkin, E. M. (1956). *Cold Spring Harbor Symp. Quant. Biol.* **21**, 123.

Witkin, E. M. and George, D. L. (1973). *Genetics* **73** (suppl.), 91.

53

The Effect of Genes Controlling Radiation Sensitivity on Chemical Mutagenesis in Yeast

Louise Prakash

Department of Radiation Biology and Biophysics
University of Rochester School of Medicine and Dentistry
Rochester, New York 14642

A special set of mutants containing identified altered codons in the iso-1-cytochrome *c* gene of yeast were recently used as tester strains for determining mutagenic specificities (Prakash and Sherman, 1973). One of the mutants, *cyc*1-131, contains a GUG codon in place of the normal chain-initiation codon AUG and requires a G·C-to-A·T transition to yield the normal protein (Stewart et al., 1971). This mutant was reverted preferentially and with a high frequency by ethyl methanesulfonate (EMS), diethyl sulfate (DES), nitrous acid (HNO$_2$), *N*-methyl-*N*'-nitro-*N*-nitrosoguanidine (NTG), nitrosoimidazolidone (NIL), [^{3}H] uridine ([^{3}H]U), nitroquinoline oxide (NQO), and β-propioloactone (β-PL), and it was therefore concluded that these agents selectively induced G·C-to-A·T transitions (Prakash and Sherman, 1973; Prakash et al., 1974). In addition, ultraviolet light (UV), X-rays, nitrogen mustard (HN2), methyl methanesulfonate (MMS), and dimethyl sulfate (DMS) were found to revert all the tester strains with the same efficiency or without any dependence on simple types of base-pair changes, and it was concluded that these mutagens were nonspecific in the types of base-pair changes produced. Since the *cyc*1-131 tester reverted well with all 11 chemical mutagens and with ionizing and UV radiation, it was used in studies designed to determine the genetic control of mutation induction using a wide variety of mutagens (Prakash, unpublished results). This tester was coupled to over 20 different *rad* genes (the symbol for loci conferring radiation sensitivity in yeast), and diploids were constructed which were homozygous for *cyc*1-131 as well as a particular *rad* gene.

A mutation in any of the four loci which control excision of UV-induced pyrimidine dimers, namely *rad*1 (Unrau et al., 1971), *rad*2 (Resnick and Setlow, 1972; Parry and Parry, 1969), and *rad*3 and *rad*4 (Parry and Parry, 1969), results in increased reversion of *cyc*1-131 after low doses of either NQO or HNO_2. At NQO doses greater than 0.3 μg/ml for a 30-min treatment and HNO_2 doses

Table 1. Frequency of Induced $cyc1-131 \rightarrow CYC1$ Mutations (Expressed as Revertants/10^7 Survivors) with Various Mutagens in Normal and Radiation-Sensitive Diploid Yeast[a]

	Strain		
Mutagenic treatment	RAD	rad6	rad9
HNO_2 (0.35 mg/ml, 20 min)	142 (100%)	151 (78%)[b]	218 (100%) [344 (93%)][c]
NIL (0.4 mg/ml, 30 min)	1588 (89%)	541 (100%)[b]	470 (81%)[c]
EMS (0.5%, 5 hr)	1124 (80%)	0 (4.6%)	37 (100%)
DES (0.5%, 15 min)	226 (94%)	29 (57%)	9 (100%)
MMS (0.1%, 40 min)	5 (94%)	0 (1.8%)	<1 (87%) [0 (100%)][c]
DMS (0.05%, 12 min)	24 (89%)	0 (0.25%)	<1 (1.8%) [0 (3.6%)][c]
HN2 (0.2 mg/ml, 1 hr)	10 (80%)	0 (0.07%)	0 (55%)
NQO (1 μg/ml, 1 hr)	395 (78%)	0 (1.6%)	34 (34%)
NTG (40 μg/ml, 40 min)	1506 (76%)	0 (8.5%)	10 (20%)
β-PL (0.02%, 1 hr)	16 (100%)	0 (3%)	—[e]
[³H]U (10 days decay)	14 (98%)	0 (89%)	—[e]
UV (25 J/m²)[d]	15 (96%)	0 (0.12%)	8 (53%)

[a] Reversion frequencies are expressed as the number of revertants obtained after treating cells minus those that occurred spontaneously, except for treatments with [³H]U. Reversion frequencies for cells grown in the presence of [³H]U are expressed as the number of revertants obtained after 10 days of storage at 4°C, minus the number obtained after no storage. In most experiments, spontaneous reversion frequencies were less than 1 per 10^7 cells. An entry of 0 indicates that no revertants were found out of $2–8 \times 10^7$ total cells plated, or that the difference between the induced spontaneous reversion frequency was 0.
[b] Reversion rates decrease at higher doses.
[c] LC-9A.
[d] The UV survival and reversion data were kindly provided by Dr. C. Lawrence.
[e] No entry indicates that the strain was not tested with the mutagen.

greater than 1.5 mg/ml for a 20-min treatment, reversion frequencies in *rad*1, *rad*2, *rad*3, *rad*4, and the normal *RAD* strains are similar.

Preliminary results also indicate that the *rad*52 gene substantially reduces EMS-induced reversion of *cyc*1-131, while *rad*1, *rad*2, *rad*3, *rad*4, *rad*11, *rad*14, *rad*15, *rad*16, *rad*17, *rad*18, *rad*19, and *rad*22 seem to have no effect.

The *rad*6 and *rad*9 genes, on the other hand, greatly reduce the frequency of chemically induced reversion of *cyc*1-131. Mutations induced by EMS, DES, MMS, DMS, NQO, NTG, HN2, β-PL, and $[^3H]U$, as well as mutations induced by UV and ionizing radiation (Lawrence et al., 1974), were greatly diminished in strains homozygous for either the *rad*6 or *rad*9 gene (Table 1). Nitrous acid and NIL, on the other hand, were highly mutagenic in these two repair-deficient mutants, and at low doses these mutagens acted with about the same efficiency as in the normal *RAD* strain. At high doses of either nitrous acid or NIL, however, reversion frequencies were significantly reduced in the two *rad* mutants compared to normal *RAD* strains. Although both *rad*6 and *rad*9 are immutable to about the same extent, the *rad*9 strains tend to be less sensitive to the lethal effect of chemical mutagens than *rad*6 strains. It is concluded that in yeast, the induction of mutation by chemical agents requires a functional repair system.

ACKNOWLEDGMENTS

I greatly appreciate the technical assistance given by Miss Susan Mancuso. This work was supported in part by the U.S. Public Health Services research grant GM19261 and in part by the U.S. Atomic Energy Commission at the University of Rochester Atomic Energy Project, Rochester, New York, and has been assigned USAEC Report No. UR–3490–530.

REFERENCES

Lawrence, C. L., Stewart, J. W., Sherman, F. and Christensen, R. (1974). *J. Mol. Biol.* **85,** 137–162.
Parry, J. M. and Parry, E. M. (1969). *Mutat. Res.* **8,** 545–556.
Prakash, L. and Sherman, F. (1973). *J. Mol. Biol.* **79,** 65–82.
Prakash, L., Stewart, J. W. and Sherman, F. (1974). *J. Mol. Biol.* **85,** 51–65.
Resnick, M. A. and Setlow, J. K. (1972). *J. Bacteriol.* **109,** 979–986.
Stewart, J. W., Sherman, F., Shipman, N. A. and Jackson, M. (1971). *J. Biol. Chem.* **246,** 7429–7445.
Unrau, P., Wheatcroft, R. and Cox, B. S. (1971). *Mol. Gen. Genet.* **113,** 359–362.

54

Influence of Repair on the Specificity of Ultraviolet-Induced Reversion of an Ochre Allele of the Structural Gene for Iso-1- cytochrome *c*

Christopher W. Lawrence, John W. Stewart, Fred Sherman, and Roshan Christensen

Department of Radiation Biology and Biophysics
University of Rochester School of Medicine and Densitry
Rochester, New York 14642

ABSTRACT

The specific action of UV on the reversion of the ochre allele *cyc*1-9, in which 21 out of 23 revertants have been shown to arise from $A \cdot T$-to-$G \cdot C$ transitions at position one in the UAA codon, was found to depend on the function of the *RAD*6 gene, since *cyc*1-9 reversion occurred by a variety of single-base-pair substitutions in a strain carrying the *rad*6-1 allele.

It has been shown by amino acid replacements that UV-induced reversion of *cyc*1-9, an ochre allele of the structural gene for iso-1-cytochrome *c* (Stewart et al., 1972), occurs predominantly by $A \cdot T$-to-$G \cdot C$ transition at position one in the UAA codon (Stewart et al., 1972; Sherman and Stewart, 1974). Since all possible base-pair substitutions could be obtained using various other mutagens, such specificity is clearly a property of UV mutagenesis itself. Analogies with bacteria suggest that the specificity depends on error-prone DNA repair, and this hypothesis was examined by reverting the *cyc*1-9 allele in strains which carried one or other of the radiation-sensitive mutations *rad*1-2, *rad*6-1, or *rad*18-2 (Lawrence et al., 1974). It was found that specificity depended largely on the action of the *RAD*6 locus, since *cyc*1-9 reverted by $A \cdot T$-to-$C \cdot G$ transversions at positions one and two in the UAA codon, and by A·T-to-T·A transversion at position two, as well as by the usual A·T-to-G·C transition at position one, in a strain carrying the *rad*6-1 allele. The frequency of UV reversion in such UV- and X-ray-sensitive strains is much reduced, though significantly greater than the spontaneous frequency. The phenotypically similar *rad*18-2 mutation, which appears to block the same non-excision-repair pathway as *rad*6-1, also had some effect on the reversion specificity, but its action depended on the presence of other, unidentified, mutations. Specificity was completely unaltered, however, in an excision-defective strain carrying the *rad*1-2 allele. Induced reversion frequency of *cyc*1-9 was much higher than normal in this strain. Photoreactivation studies indicated that pyrimidine dimers were responsible for most of the revertants in all strains studied.

The origin of the reversions induced in *rad*6-1 strains was studied by means of double mutants carrying *rad*6-1 and *rev*3-1, a mutation isolated by Lemontt (1971) which, like *rad*6-1, greatly reduces though does not entirely abolish UV mutagenesis. Double mutant strains, which were no more sensitive than the most sensitive of the single mutants, failed to give any UV-induced mutations. It is concluded that UV mutagenesis in *rad*6 or *rev*3 strains arises from a residual trace of activity from the *RAD*6 or *REV*3 loci, possibly of an abnormal kind, or by the action of error-prone repair enzymes not normally involved in mutagenesis which depend on these functions.

ACKNOWLEDGMENTS

We greatly appreciate the technical assistance of Ms. S. Dennis, Mrs. E. Risen, and Mrs. N. Brockman. This work was supported by the U.S. Atomic Emergy Commission at the University of Rochester Atomic Energy Project, Rochester, N.Y., and by Public Health Service Research Grant GM12702 from the National Institutes of Health. It has been designated USAEC Report No. UR–3490–531.

REFERENCES

Lawrence, C. W., Stewart, J. W., Sherman, F. and Christensen, R. (1974). *J. Mol. Biol.* **85**, 137–162.
Lemontt, J. F. (1971). *Genetics* **68**, 21–33.
Sherman, F. and Stewart, J. W. (1974). *Genetics* **78**, 97–113.
Stewart, J. W., Sherman, F., Jackson, M., Thomas, F. L. X. and Shipman, N. (1972). *J. Mol. Biol.* **68**, 83–96.

55

The Role of DNA Polymerase I in Genetic Recombination and Viability of *Escherichia coli*

G. B. Smirnov

*The Gamaleya Institute for Epidemiology and Microbiology AMS USSR
Moscow, U.S.S.R.*

and

B. I. Sinzinis and A. S. Saenko

*Research Institute of Medical Radiology AMS USSR
Obninsk, Kaluga Region, U.S.S.R.*

ABSTRACT

The rate of formation of high-molecular-weight daughter DNA in the conditionally lethal double mutant *polA12 uvrE502,* incubated at nonpermissive temperature, was slower than that in the single *polA12* mutant. There exist at least two pathways determining viability of *Escherichia coli* cells: one of them is dependent on *polA⁺* and *recB⁺* genes, while another is *polA recB* independent but requires the *uvrE⁺* gene and can be blocked by exonuclease I. The RecF but not the RecBC pathway of genetic recombination was found to be absolutely dependent on the polymerizing activity of DNA polymerase I. The involvement of DNA polymerase I in genetic recombination in the *recB⁻ C⁻ sbsB⁻* strain and viability in the *uvrE⁻* or *recB⁻* strains suggest the existence of the common steps required for the accomplishing of the RecF pathway of recombination and for viability of *E. coli.*

The existence of the radiation-sensitive mutants of *Escherichia coli,* showing reduced recombination ability and viability (e.g. *recB⁻*), suggests the existence of common steps in the processes of normal growth, recombination, and repair of DNA. The aim of the present work was to study the role of DNA polymerase I (Pol I) in the processes required for viability and genetic recombination in *E. coli*.

Although the single *polA*⁻ mutants lacking polymerase activity of Pol I are viable, it is clear that Pol I performs some essential vital function(s), since the double mutants *polA*⁻*recA*⁻ (Gross et al., 1971), *polA*⁻*recB*⁻ (Monk and Kinross, 1972), *polA*⁻ *uvrE*⁻; (Mattern, 1971; Smirnov et al., 1973c), and *polA*⁻ *uvrB*⁻Δ (Shizuya and Dykhuizen, 1972) are inviable. In no case was the molecular nature of the essential function blocked by the mentioned combinations determined.

Pol I carries out the repair replication step of excision-repair in UV-irradiated cells (Kanner and Hanawalt, 1970; Cooper and Hanawalt, 1972). The *uvrE*⁺ allele is also required for the accomplishing of excision-repair and most probably controls some reaction of repair replication (Sinzinis et al., 1973; Smirnov et al., 1973a). The *uvrE*⁻ mutation leads also to the mutator phenotype, increasing the probability of $A \cdot T \to G \cdot C \to A \cdot T$ substitutions (Smirnov et al., 1973b). This fact suggests the involvement of the *uvrE*⁺ gene product in semiconservative DNA replication. These observations lead us to the assumption that inviability of the double mutants *polA*⁻ *uvrE*⁻ is due to their inability to carry out effective sealing of gaps or breaks in the DNA arising in the course of discontinuous replication (Smirnov et al., 1973c).

There was no accumulation of single-strand breaks in parent DNA of the conditionally lethal double mutant *polA12 uvrE*⁻ during its incubation at non-permissive temperature (Smirnov et al., 1973c), whereas the rate of formation of high-molecular-weight daughter DNA was significantly lower than that in the single *polA12* mutant (Fig. 1). Thus the formation of high-molecular-weight daughter DNA appears to be an essential vital function affected by the combination of mutations *polA*⁻ and *uvrE*⁻.

Reactions similar to resynthesis necessary to close single-strand gaps or breaks during repair or replication may be required for the integration of donor DNA during recombination. On this assumption we have performed the study of the role of Pol I and the *uvrE*⁺ gene product in genetic recombination. The extensive study of recombination pathways carried out by Clark and his co-workers has shown that there are at least three pathways of genetic recombination in *E. coli*—namely, the RecBC pathway dependent upon exonuclease V (Exo V), which is the main pathway operating in *recB*⁺*C*⁺ strains; the RecF pathway controlled by the *recF, recK*, and *recL* alleles and blocked by exonuclease I (Exo I); and the RecE pathway dependent on the exonuclease VIII (Horii and Clark, 1973; Kato et al., this volume). We have shown that *polA*⁻ and *uvrE*⁻ were not capable of producing recombination deficiency in the *recB*⁺ *C*⁺ *sbcB*⁺ genetic background in both transduction and conjugal crosses (Table 1). These results were in agreement with those of Gross and Gross (1969) and showed that Pol I and *uvrE*⁺ were almost unessential for the RecBC recombination pathway.

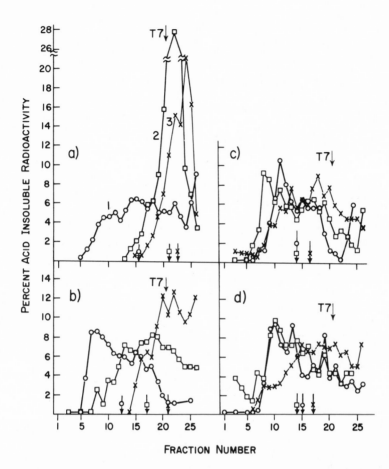

Fig. 1. Sedimentation of single-strand DNA synthesized at 45°C. Bacteria growing exponentially at 30°C to a density of 5 × 10⁷ cells/ml in tris-buffer-triptone medium were transferred in a 45°C water bath. After 5 min incubation the cells were pulse-labeled for 10 min with [³H] thymidine (50 μCi/ml; specific activity 9.1 Ci/mmole), then unlabeled thymidine was added to the final concentration of 500 μg/ml to stop the incorporation of [³H] thymidine, and incubation was continued at 45°C. 4-ml aliquots were taken at 0 (*a*), 20 (*b*), 40 (*c*), and 80 (*d*) min after the addition of unlabeled thymidine for measurement of sedimentation of single-strand [³H] DNA through alkaline sucrose gradient. The details of sedimentation analysis are described elsewhere (Saenko et al., 1972). ○-*polA⁺ uvrE⁺*; □-*polA12 uvrE⁺*; x-*polA12 uvrE502*. The sedimentation pattern of DNA from the single *uvrE⁻* mutant is not presented since it was indistinguishable from that of DNA from *polA⁺ uvrE⁺* cells. Arrows at the top of sedimentograms show the position of T7 phage DNA centrifuged under the same conditions. Direction of sedimentation is from right to left. Arrows under the sedimentograms indicate the first moments of the sedimentation distributions (Kaplan, 1966). All experiments were performed in five duplications.

Table 1. The Effect of $polA^-$ and $uvrE^-$ Mutations on the Recombination Ability in Transductional and Conjugal Crosses and on the Ability to Inherit F'104 Plasmid

Recipient strains[a]	In transductional crosses with phage P1 vir[c]	Deficiency indices[b]		
		In conjugal crosses with		
		AB259 Hfr H Thr⁺ Leu⁺ [Str^R]	JC 8563 Hfr P4x Thr⁺ Leu⁺ [Met⁺ Ilv⁺ Str^R]	KS258 F' 104 Thr⁺ Leu⁺ Pro⁺ [Thy⁺]
KS31 $polA6$	1.5 ($thyA^+$)	7.5		
KS26 $polA6$	1.1 ($thyA^+\ argA^+$)			
KS59 $uvrE502$		1.2		
KS20 $uvrE502$				
KS199 $recB21\ recC22\ sbcB15\ polA6$	1 ($proA^+$)	$>6 \times 10^4$	$>5 \times 10^3$	2
KS180 $recB21\ recC22\ sbcB15\ uvrE502$	5 ($proA^+$)	1.6×10^1	9.2×10^1	1.3×10^1

[a] Only relevant genotypes are presented.

[b] Average data from three or more experiments are presented. A deficiency index was calculated by dividing the frequency at which progeny were produced in a cross with a $polA^+$ or $uvrE^+$ recipient by the frequency at which progeny were produced in a cross of the same donor with the isogenic $polA^-$ or $uvrE^-$ recipient. In the cross wich KS20 recipient Hfr AB313 was used and Lys⁺ [Thr⁺ Ilv⁺] progeny were selected. 1-h matings were performed in all cases.

[c] Selective markers are shown in parentheses.

To study the role of *polA* and *uvrE* genes in the RecF pathway of recombination, the strains *polA⁻* (KS199) and *uvrE⁻* (KS180) also defective in Exo I and Exo V (*recB⁻C⁻ sbcB⁻*) were constructed, and their recombination abilities were determined. The data presented in Table 1 show that these strains have almost normal recombination ability necessary for the formation of transductants when compared with the isogenic *polA⁺ uvrE⁺* strain. However, the *polA⁻* mutation completely blocked the ability of the *recB⁻C⁻ sbsB⁻* strain to form progeny in conjugal crosses, while the *uvrE⁻* mutation affected this ability only partially (Table 1). The strain KS199 *recB⁻ C⁻ sbcB⁻ polA⁻* has normal ability to inherit plasmid F′104 DNA by repliconation, suggesting that the transfer of donor DNA into the cells of this recipient is unaffected. In contrast, the strain KS 180*recB⁻ C⁻sbcB⁻ uvrE⁻* was found to be significantly deficient in inheriting F′104 (Table 1). Earlier we found that only 5% of the cells of strain KS180 are viable, while the triple mutant *recB⁻C⁻sbcB⁻* has shown 100% viability (Smirnov, 1974). Thus the reduced yield of sexductants and recombinants obtained in the crosses with the recipient strain KS180 appears to be due to the loss of potential progeny resulting from pairing of donor cells with inviable recipient bacteria.

Thus we conclude that Pol I plays a significant role in the RecF pathway of recombination during conjugation, while the *uvrE⁻* mutation has little if any effect on the functioning of this pathway.

The RecF pathway of recombination seems to have steps similar to the pathway determining the viability of *recB⁻* strains, since both require Pol I. Inviability of the *recB⁻ polA⁻* strains and viability of the *recB⁻C⁻ sbcB⁻ polA⁻* strains suggest the existence of an alternative pathway for viability independent on Exo V and Pol I and blocked by Exo I. This pathway was found to be *uvrE* dependent (and named the UvrE pathway), since the strain *recB⁻C⁻sbcB⁻polA⁻ uvrE⁻* was inviable (Smirnov, unpublished). The UvrE pathway for viability is similar to the RecF pathway of recombination since both are blocked by Exo I. The difference between these pathways is that the RecF pathway but not the UvrE pathway is Pol I dependent.

ACKNOWLEDGMENTS

The authors are particularly grateful to Dr. A. G. Skavronskaya and Dr. A. J. Clark for stimulating interest to their work and helpful suggestions, and to Dr. A. J. Clark for kind gift of bacterial strains.

REFERENCES

Cooper, P. K. and Hanawalt, P. C. (1972). *Proc. Nat. Acad. Sci. U.S.A.* **65**, 1156.
Gross, J. D. and Gross, M. (1969). *Nature* **224**, 1166.

Gross, J. D., Grunstein, J. and Witkin, E. M. (1971). *J. Mol. Biol.* **58**, 631.

Horii, Z.-I. and Clark, A. J. (1973). *J. Mol. Biol.* **80**, 327.

Kanner, L. and Hanawalt, P. C. (1970). *Biochem. Biophys. Res. Commun.* **39**, 149.

Kaplan, H. S. (1966). *Proc. Nat. Acad. Sci. U.S.A.* **55**, 1442.

Mattern, I. E. (1971). In *First European Biophysical Congress, Baden, Austria,* p. 237.

Monk, M. and Kinross, J. (1972). *J. Bacteriol.* **109**, 971.

Saenko, A. S., Sinzinis, B. I. and Smirnov, G. B. (1972). *Mol. Biol. (USSR)* **6**, 655.

Shizuya, H. and Dykhuizen, D. (1972). *J. Bacteriol.* **112**, 676.

Sinzinis, B. I., Smirnov, G. B. and Saenko, A. S. (1973). *Biochem. Biophys. Res. Commun.* **53**, 309.

Smirnov, G. B. (1974). *Genetika (USSR)* **10**, 102.

Smirnov, G. B., Filkova, E. V. and Skavronskaya, A. G. (1973a) *J. Gen. Microbiol.* **76**, 407.

Smirnov, G. B., Filkova, E. V. and Skavronskaya, A. G. (1973b). *Mol. Gen. Genet.* **126**, 225.

Smirnov, G. B., Filkova, E. V. and Skavronskaya, A. G., Saenko, A. S. and Sinzinis, B. I. (1973c). *Mol. Gen. Genet.* **121**, 139.

56

The Role of the *rec* Genes in the Viability of *Escherichia coli* K12

Florence N. Capaldo[1] and Stephen D. Barbour

Department of Microbiology, School of Medicine
Case Western Reserve University
Cleveland, Ohio 44106

ABSTRACT

Mutations in the *recA, recB,* or *recC* genes significantly reduce the growth rate and viability of *Escherichia coli.* Cultures of *rec⁻* strains are composed of three populations of cells: viable cells, nonviable but residually dividing cells, and nonviable and nondividing cells. Nondividing cells can be separated from dividing cells by penicillin treatment and velocity sedimentation. Nondividing cells of all *rec⁻* strains are greatly reduced in their ability to synthesize DNA. *recB⁻ recC⁻* and *recA⁻ recB⁻ recC⁻* nondividing cells contain DNA. This DNA is synthesized in dividing cells and segregated into the nondividing cells. *recA⁻* nondividing cells contain little or no DNA. *recA⁻ recB⁻ recC⁻* nondividing cell DNA accumulates single-strand breaks.

Strains of *Escherichia coli* K12 with lesions in either the *recA, recB* or *recC* genes exhibit reduced growth rates and viabilities when compared with isogenic *rec⁺* strains (Capaldo-Kimball and Barbour, 1971; Capaldo et al., 1974; Haefner, 1968; Hertman, 1969). We are investigating the role of the *rec* gene products in normal growth. In this report we shall review the effects of *rec* mutations on

[1] Present address: Department of Biochemistry, School of Medicine, Stanford University, Stanford, California 94305.

viability, present a method for separating two classes of cells present in Rec⁻
cultures, and present the results of experiments designed to characterize the cells
in Rec⁻ cultures with regard to DNA content, synthesis and integrity.

EFFECTS OF *rec* MUTATIONS ON VIABILITY

The viability of a bacterial culture is commonly measured by determining
the fraction of total particles in the culture which are capable of forming visible
colonies when plated on solid medium. As can be seen in Table 1, when assayed
in this manner, strains carrying mutations in either the *recA, recB,* or *recC* genes
are significantly less viable than an isogenic *rec⁺* strain (Capaldo-Kimball and
Barbour, 1971). Restoration of recombination proficiency to *recB* and *recC*
mutants by the indirect suppressor, *sbcA* (Barbour et al., 1970), leads to a
significant restoration of viability (Capaldo-Kimball and Barbour, 1971).

Viability may also be measured directly in liquid medium by inoculating
single cells into each of a number of tubes and scoring the tubes for development
of visible turbidity. This technique minimizes any effects on viability which
might be caused by the manipulations associated with plating. As can be seen in
Table 1, the results of this assay are in substantial agreement with the results of
the plate assay for all strains (Capaldo et al., 1974).

Both assays, formation of a visible colony and development of a turbid
culture, require that a cell be capable of giving rise to at least 20 generations of
progeny in order to be scored as viable. Thus, exponentially growing Rec⁻

Table 1. Viability of Rec⁻ Strains

Relevant genotype	Tube assay[b]	Plate assay[c]
rec⁺	1.00	1.00
recA⁻	0.66	0.49
recB⁻ recC⁻	0.34	0.27
recA⁻ recB⁻ recC⁻	0.18	0.22
recB⁻ recC⁻ sbcA⁻[a]	0.80	0.75

[a]*sbcA⁻* is a mutation in a gene outside of *recB* and *recC* which
indirectly suppresses *recB⁻* and *recC⁻* mutations (Barbour et
al., 1970; Lloyd and Barbour, 1974).
[b]The data for the first four strains are taken from Capaldo et
al. (1974). The datum for the fifth strain is the unpublished
work of G. Ramsey.
[c]The data for the first three and the fifth strain are taken
from Capaldo and Barbour, (1971). The datum for the
fourth strain is taken from Capaldo et al. (1974).

Table 2. Classes of Cells in Rec⁻ Cultures[a]

Relevant genotype	Fraction of		
	Viable cells	Nondividing cells	Residually dividing cells
rec⁺	1.00	0	0
recA⁻	0.66	0.13	0.21
recB⁻ recC⁻	0.34	0.40	0.26
recA⁻ recB⁻ recC⁻	0.18	0.40	0.42

[a]These data are taken from Capaldo et al. (1974).

cultures contain a significant fraction of nonviable cells, that is, cells incapable of giving rise to 20 or more generations of progeny.

We next asked whether all of the nonviable cells are incapable of any division or whether there is a fraction of nonviable cells capable of undergoing some residual divisions. To answer this question, we took advantage of the fact that penicillin causes dividing cells to elongate, forming "snakes" which are 3–10 times as long as cells not treated with penicillin. Penicillin-induced snakes thus may be easily distinguished from normal cells using the light microscope. Hence, when an exponentially growing Rec⁻ culture is treated with penicillin, dividing cells should elongate whereas cells incapable of division should remain unchanged in size.

We therefore treated *rec⁺*, *recA⁻*, and *recB⁻ recC⁻* cultures with penicillin and followed cell elongation microscopically over a period of 2 hr. Initially, all three cultures contained less than 5% long cells. For each strain, the total number of particles, long and short, remained constant throughout the experiment. However, after exposure to penicillin for 2 hr, all of the cells in the *rec⁺* culture elongated, whereas only 60% of the *recB⁻ recC⁻* cells and 80% of the *recA⁻* cells elongated (Capaldo et al., 1974).

Thus, the viability assay on the one hand and penicillin treatment on the other establish two limit classes of cells in Rec⁻ cultures: that is, cells capable of giving rise to at least 20 generations of progeny, and cells incapable of a single division. The remainder of cells in Rec⁻ cultures are nonviable, but are capable of undergoing residual divisions. The fraction of cells in each class for several Rec⁻ strains is summarized in Table 2.

When a viable cell divides, it may give rise to a cell which is restricted in the number of further divisions it can undergo. Such a cell along with all of its progeny has been termed a "lethal sector" by Haefner (1968). Haefner and members of our laboratory have observed the formation of lethal sectors in Rec⁻

cultures grown on solid medium. This is direct evidence for the existence of the class of nonviable but residually dividing cells.

SEPARATION OF DIVIDING AND NONDIVIDING CELLS

To approach the problem of how the *rec* gene products are involved in the growth of *E. coli,* we developed a procedure for separating the nondividing cells from dividing cells in Rec⁻ cultures (Capaldo and Barbour, 1973). Using this technique, we examined the nondividing cells for specific defects which might be responsible for the nonviability.

To isolate the nondividing cells, we treat an exponentially growing culture with penicillin to cause the dividing cells to elongate. We then separate the two size classes of cells by velocity sedimentation through 5–20% preformed linear neutral sucrose gradients. The gradients are spun for 8 min at 3000 rpm. Under these conditions, dividing cells band in the lower half of the gradient, whereas nondividing cells band in the upper half of the gradient.

Figure 1*a* shows the sedimentation pattern of cells *not* treated with penicillin. Cells which have not been treated with penicillin sediment in a sharp band in the upper half of the gradient. When the culture is uniformly labeled with [^{14}C]leucine, the radioactivity, total particles, and viable cells cosediment. Similar results are obtained with Rec⁺ and Rec⁻ cultures.

Figures 1*b* and *c* show the sedimentation patterns of penicillin-treated *rec*⁺ and *recB*⁻ *recC*⁻ cultures. If a *rec*⁺ culture is uniformly labeled with [^{14}C]leucine and then treated with penicillin and centrifuged, the cells sediment in a broad band in the lower half of the gradient. Radioactivity and microscopically observable particles, most of which are long, cosediment, and most of the viable cells are also found in the broad band (Fig. 1*b*).

However, when a culture of a *recB*⁻ *recC*⁻ strain is uniformly labeled with [^{14}C]leucine and then treated with penicillin and centrifuged, a different sedimentation pattern is observed (Fig. 1*c*). There are two distinct, though overlapping, bands of radioactivity, one in the lower half of the gradient where penicillin-treated *rec*⁺ cells band, and the other in the upper half of the gradient where cells not treated with penicillin band. The faster-moving band is almost entirely composed of long cells, whereas the slower-moving band is largely composed of short cells. The distribution of viable cells throughout the gradient coincides with the distribution of long cells, indicating that the short cells are inviable. We wish to emphasize that, after sedimentation of the Rec⁻ culture, the region of the gradient having the lowest viability is that region where the cells showing the least visible effects of penicillin band. Thus, the lower band is composed of cells which before penicillin treatment were largely viable, whereas the upper band is composed of the nondividing cells. We have obtained qualitatively similar results with *recA*⁻ and *recA*⁻*recB*⁻*recC*⁻ strains.

Thus, we have developed a method for separating nondividing and dividing

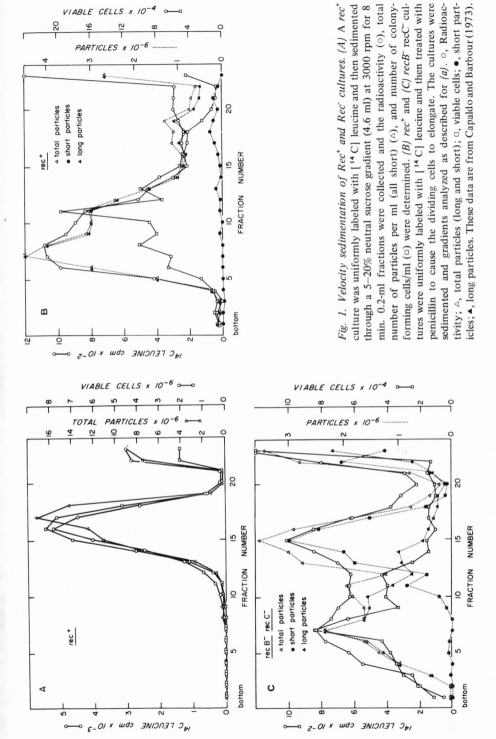

Fig. 1. Velocity sedimentation of Rec⁺ and Rec⁻ cultures. (A) A rec⁺ culture was uniformly labeled with [¹⁴C] leucine and then sedimented through a 5–20% neutral sucrose gradient (4.6 ml) at 3000 rpm for 8 min. 0.2-ml fractions were collected and the radioactivity (○), total number of particles per ml (all short) (△), and number of colony-forming cells/ml (□) were determined. (B) rec⁺ and (C) recB⁻ recC⁻ cultures were uniformly labeled with [¹⁴C] leucine and then treated with penicillin to cause the dividing cells to elongate. The cultures were sedimented and gradients analyzed as described for (a). ○, Radioactivity; △, total particles (long and short); □, viable cells; ●, short particles; ▲, long particles. These data are from Capaldo and Barbour (1973).

cells. This method is primarily an analytical rather than a preparative technique. Using this technique, one may carry out an experiment on an exponentially growing culture, and then separate the dividing and nondividing cells and analyze the results. This circumvents possible deleterious effects of the isolation procedure on cell metabolism, such as alterations in the cell membrane. Using this technique, we have examined a variety of parameters, including enzyme synthesis (Capaldo and Barbour, 1973) and the incorporation of radioactive precursors into macromolecules (Capaldo and Barbour, 1975).

DNA SYNTHESIS, CONTENT, AND INTEGRITY IN Rec⁻ CELLS

One approach to understanding the role of the *rec* genes in growth is to study the nondividing cells produced during the growth of Rec⁻ strains. Since several of the functions of the *rec* genes involve the repair of DNA which has been damaged by irradiation and chemicals, it is plausible that the *rec* gene products are required to repair DNA damage produced during normal growth processes such as transcription and replication. The nondividing cells, produced as a result of the absence of normal *rec* gene products, would be likely to reveal this damage. Using the method described above for isolating Rec⁻ nondividing cells, we have examined the content, synthesis and integrity of DNA in Rec⁻ nondividing and dividing cells.

To determine whether or not Rec⁻ nondividing cells synthesize DNA, we uniformly labeled Rec⁺ and Rec⁻ cultures with [³H] thymidine for one-tenth of a generation. The cultures were treated with penicillin and centrifuged. All fractions were precipitated with trichloroacetic acid. The results are shown in Fig. 2. Panel *a* shows the sedimentation pattern of a *rec⁺* culture. There is a single, broad band of cells containing both [¹⁴C] leucine and [³H] thymidine radioactivity. Panel *b* shows the sedimentation pattern of a *recB⁻ recC⁻* culture. There are two bands of [¹⁴C] leucine radioactivity, corresponding to long, dividing cells and short, nondividing cells. Most of the [³H] thymidine is found in the dividing cell band. We conclude that the Rec⁻ nondividing cells are severely impaired in their ability to incorporate [³H] thymidine into TCA-insoluble material. Similar results were obtained with the *recA⁻* and with the *recA⁻ recB⁻ recC⁻* strains.

We next asked why nondividing cells are unable to synthesize DNA. Two possible reasons are either that nondividing cells do not contain DNA or that nondividing cell DNA is damaged.

To determine whether or not nondividing cells contain DNA, we uniformly labeled Rec⁺ and Rec⁻ cultures with both [¹⁴C] leucine and [³H] thymidine, treated with penicillin and centrifuged. All fractions were TCA precipitated. These results are shown in Fig. 3. In the *rec⁺* culture (panel *a*), there is a single broad band of cells containing both [¹⁴C] leucine and [³H] thymidine radio-

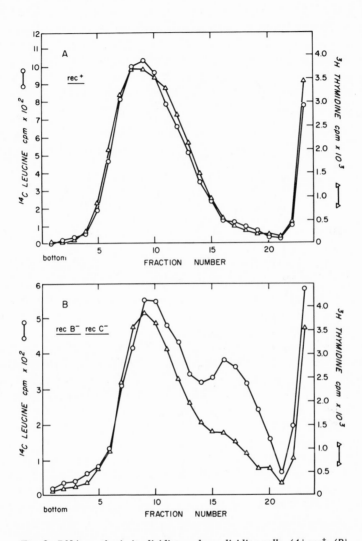

Fig. 2. DNA synthesis in dividing and nondividing cells. (A) rec⁺; (B) recB⁻ recC⁻. ○, [¹⁴C] Leucine radioactivity; △, [³H] thymidine radioactivity. Cultures were uniformly labeled with [¹⁴C] leucine, then pulse-labeled with [³H] thymidine for one-tenth of a generation, treated with penicillin, and sedimented through neutral sucrose gradients to separate dividing and nondividing cells. The gradients were collected and each fraction was analyzed for ³H and ¹⁴C radioactivity. These data are taken from Capaldo and Barbour (1975).

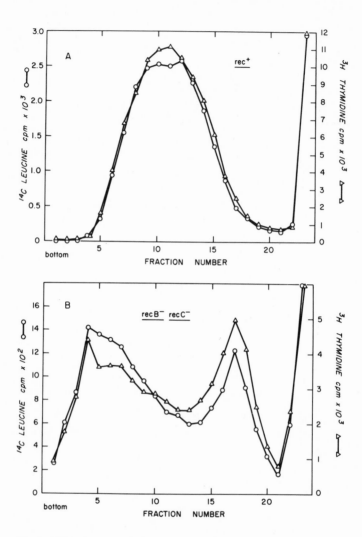

Fig. 3. DNA content of dividing and nondividing cells. (A) rec⁺; (B) recB⁻ recC⁻. ○, [¹⁴C] Leucine radioactivity; △, [³H] thymidine radio-activity. Cultures were uniformly labeled with [¹⁴C]leucine and with [³H]thymidine, treated with penicillin, and sedimented through neu-tral sucrose gradients to separate dividing and nondividing cells. The gradients were collected and each fraction was analyzed for ³H and ¹⁴C radioactivity. These data are taken from Capaldo and Barbour, (1975).

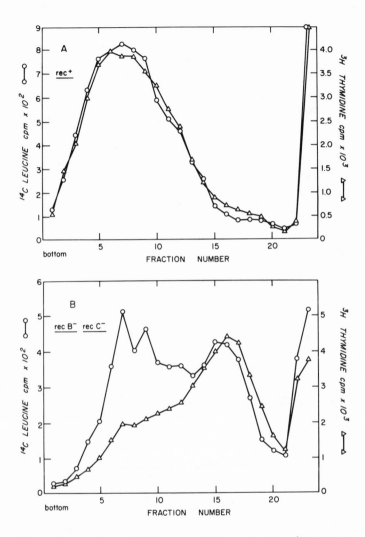

Fig. 4. Segregation of DNA into nondividing cells. (A) rec⁺; *(B) recB⁻ recC⁻.* ○, [¹⁴C] Leucine radioactivity; △, [³H] thymidine radioactivity. Cultures which had been uniformly labeled with [¹⁴C] leucine were pulse-labeled with [³H] thymidine for one generation and then grown in the presence of excess unlabeled thymidine for three generations. Each culture was treated with penicillin and sedimented through a neutral sucrose gradient to separate dividing and nondividing cells. The gradients were collected and each fraction was analyzed for ³H and ¹⁴C radioactivity. These data are taken from Capaldo and Barbour (1975).

activity. In the $recB^-recC^-$ culture (panel b) there are two bands of [^{14}C] leucine corresponding to dividing cells and to nondividing cells, and both bands contain [^{3}H] thymidine radioactivity. We conclude that the $recB^-recC^-$ nondividing cells contain DNA. Similar results were obtained with the $recA^-recB^-recC^-$ strain. However, the $recA^-$ nondividing cells appear to contain little or no DNA. Chemical measurements of the amount of DNA per cell in unfractionated cultures are consistent with these results (Capaldo and Barbour, 1975). Similar results have been reported by Hout and Schuite (1973).

Since $recB^-recC^-$ and $recA^-recB^-recC^-$ nondividing cells contain DNA but do not synthesize it, the DNA present in these cells must be synthesized in dividing cells and segregated as a function of cell division into the nondividing cells.

To test this, we uniformly labeled Rec^+ and Rec^- cultures with [^{14}C] leucine, pulsed the cultures with [^{3}H] thymidine for one generation, and chased with unlabeled thymidine for three generations. The cultures were treated with penicillin and centrifuged. All fractions were TCA precipitated. These results are shown in Fig. 4. In the rec^+ culture (panel a), there is a single broad band of cells containing both [^{14}C] leucine and [^{3}H] thymidine radioactivity. In the $recB^-$ $recC^-$ culture (panel b), there are two bands of [^{14}C] leucine radioactivity, corresponding to dividing cells and nondividing cells, and after three generations of chase, much of the [^{3}H] thymidine radioactivity has moved into the non-dividing cell band. After a six-generation chase, even more radioactivity has moved into the nondividing cell band. Identical results were obtained with the $recA^-recB^-recC^-$ strain. However, with the $recA^-$ strain, no such movement of [^{3}H] thymidine radioactivity was observed, even after a six-generation chase. This is consistent with the absence of DNA in the nondividing cells of this strain.

The products of all three rec genes are required for the repair of X-ray-induced single strand breaks in the DNA. Single-strand breaks may occur during DNA replication and perhaps also during transcription, and we and others have suggested that the inviability of Rec^- strains is a result of improper or inadequate repair of these breaks (Capaldo-Kimball and Barbour, 1971; Capaldo and Barbour, 1973; Monk et al., 1973). We have specifically looked for an accumulation of single strand breaks in dividing and nondividing cell DNA using alkaline sucrose gradients. The results are shown in Figure 5.

In panel a, $recB^-recC^-$ DNA labeled with [^{3}H] thymidine for one generation (predominantly dividing-cell DNA) is compared with ^{14}C-labeled rec^+ marker DNA run in the same gradient. The single-strand molecular weight distribution of $recB^-recC^-$ dividing-cell DNA is indistinguishable from that of rec^+ DNA. Identical results were obtained with $recA^-$ and $recA^-recB^-recC^-$ dividing-cell DNA.

The observation that DNA synthesized in dividing cells can be chased into nondividing cells provides a method of specifically labeling nondividing-cell DNA. When rec^+ cells are labeled with [^{3}H] thymidine for one generation and

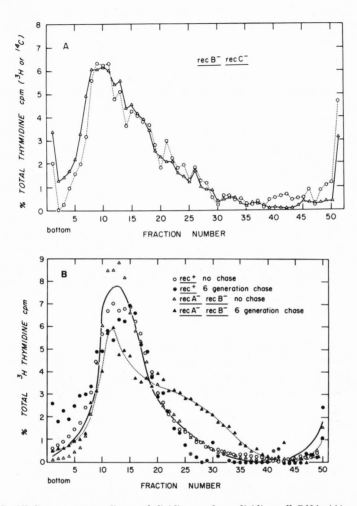

Fig. 5. Alkaline sucrose gradients of dividing- and nondividing-cell DNA. (A) recB⁻ recC⁻ dividing-cell DNA is labeled with [³H]thymidine (△) and *rec⁺* DNA, run as a marker, is labeled with [¹⁴C]thymidine (○). 100% = 1271 ¹⁴C cpm, 4508 ³H cpm. *(B)* These data are taken from four representative single-label ([³H]thymidine) gradients. For each strain, triplicate gradients containing either dividing- or nondividing-cell DNA were prepared and centrifuged at the same time using the six place SW50.1 rotor. Each experiment was done two times. In every *recA⁻ recB⁻ recC⁻*nondividing-cell gradient we observed the abnormal DNA distribution illustrated in this panel. ○,●, *rec⁺*; △,▲, *recA⁻ recB⁻ recC⁻*. Open triangles (△) represent dividing-cell DNA; filled triangles (▲) represent nondividing-cell DNA. A single curve has been drawn to fit the two *rec⁺* sets of data and the *recA⁻recB⁻recC⁻*dividing-cell DNA data. 100% = 27,993 cpm (○); 1306 cpm (●); 9686 cpm (△); 1277 cpm (▲).

chased with unlabeled thymidine for six generations there is no observable change in the single-strand molecular weight distribution of the DNA compared with DNA from unchased cells (panel *b*). However, when a *recA⁻ recB⁻recC⁻* culture is pulsed and chased in the same manner there is a significant decrease in the size of the DNA (panel *b*). We observed no increase in the amount of smaller-molecular-weight DNA when either the *recA⁻* or the *recB⁻recC⁻*cultures were chased for six generations. These results indicate that there is an accumulation of single strand breaks or gaps in the DNA of nondividing *recA⁻recB⁻recC⁻* cells.

DISCUSSION AND CONCLUSION

The products of three *rec* genes, *A, B,* and *C,* are required for normal growth and viability of *E. coli* K12 (Haefner, 1968; Hertman, 1969; Capaldo-Kimball and Barbour, 1971; Capaldo et al., 1974), and in the absence of functional DNA polymerase I they are essential for viability (Gross et al., 1971; Monk and Kinross, 1972; Smirnov et al., 1973). During normal growth of Rec⁻ mutants an event may occur in a cell which restricts the number of subsequent generations which that cell will be able to undergo (Haefner, 1968; Capaldo et al., 1974). The nature of this event is not known, nor is it known whether this event reflects a single lethal event or the accumulation of sublethal damage to a critical level. Nevertheless, Rec⁻ cultures growing exponentially contain a significant fraction of nondividing cells.

We have shown that nondividing cells of all three Rec⁻ strains are unable to synthesize normal amounts of DNA. In the *recA⁻* strain, the inability of non-dividing cells to synthesize DNA is clearly *not* the primary cause of their nonviability, since these cells do not contain DNA. However, in the *recB⁻ recC⁻* and *recA⁻recB⁻recC⁻*cultures one cannot clearly distinguish at this time whether the defect in DNA synthesis is the primary cause of the nonviability or merely a secondary effect.

A second important finding we have reported is that *recA⁻*nondividing cells lack DNA. This could result from either faulty DNA segregation or complete chromosome degradation. However, in this regard, it is significant that *recA⁻ recB⁻ recC⁻* nondividing cells contain normal amounts of DNA and further that this DNA is noticeably altered compared with viable cell DNA. That the DNA in the *recB⁻ recC⁻*nondividing cells appears normal suggests that the damage in the *recA⁻ recB⁻ recC⁻*nondividing cells is not a result of nonspecific DNA degradation caused by nonviability. Rather, we suggest that the accumulation of single-strand breaks is a direct consequence of the absence of the *recA* gene product. We propose that in a *recA* single mutant (*recB⁺ recC⁺*) these breaks are immediately attacked by exonuclease V, the product of the *recB* and *recC* genes (Tomizawa and Ogawa, 1972), and completely degraded. This would account for the absence of intermediate size pieces of DNA in our *recA⁻* alkaline sucrose gradients. This is consistent with the known mode of attack of this enzyme *in vitro* (Goldmark and Linn, 1970, 1972; MacKay and Linn, 1974). Probably the excessive DNA breakdown observed in *recA⁻* strains even under normal growth conditions (Clark et al., 1966; Howard-Flanders and Theriot, 1966; Willetts and Clark, 1969) reflects the complete degradation of chromosomes in the non-dividing cells rather than a low level of degradation in all cells of the population. The introduction of a *recB* or a *recC* mutation (resulting in a loss of exonuclease V activity) into a *recA⁻* strain blocks the excessive DNA breakdown (Willetts and Clark, 1969). We believe that as a consequence of blocking the degradation of damaged chromosomes, the initial damage, in the form of single-strand nicks or gaps, is stabilized and perhaps even accumulates. Thus we observe this DNA damage in alkaline sucrose gradients prepared from the *recA⁻ recB⁻ recC⁻* strain (Fig. 5b, filled triangles).

If breaks occur normally in the course of cell metabolism, then, because the Rec⁻ strains retain considerable viability, they must possess some mechanism for repairing these breaks. Since the products of the *recA*, *recB*, and *recC* genes on the one hand and DNA polymerase I on the other comprise the only two independent systems currently known for the repair of X-ray-induced single-strand breaks in the DNA (Town et al., 1971) it is likely that in *rec⁻* strains polymerase I provides the alternate mechanism for the repair of single strand breaks produced during cell growth. This would explain why all *polA rec* double mutants are inviable.

Thus, we postulate that the primary lethal event in all three Rec⁻ strains is the failure to adequately repair single-strand breaks or gaps in the chromosome produced during normal cell growth. These breaks may be required to allow replication (Cairns, 1963; Ogawa et al., 1968) and transcription (Pauling and Hanawalt, 1965) or may be produced as a result of discontinuous DNA replication (Sugino et al., 1972). The most striking evidence we have obtained in support of this hypothesis is the specific accumulation of DNA breaks in the dead cells of the *recA⁻ recB⁻ recC⁻* strain.

The involvement of recombination gene products in viability, studied here in *E. coli*, we believe may be pertinent to an understanding of aging processes in eukaryotic cells and organisms. Recent experiments (Epstein et al., 1973) and theories (Orgel, 1973; Yielding, 1974) suggest that processes leading to cellular aging in higher organisms may be analogous to those processes which result in the formation of lethal sectors in Rec⁻ *E. coli*. We therefore feel that study of the role of recombination in cell viability has significance as a model system for cellular aging.

REFERENCES

Barbour, S. D., Nagaishi, H., Templin, A. and Clark, A. J. (1970). *Proc. Nat. Acad. Sci. U.S.A.* **67**, 128–135.

Cairns, J. (1963). *J. Mol. Biol.* **6**, 208–213.

Capaldo, F. N. and Barbour, S. D. (1973). *J. Bacteriol.* **115**, 938–936.

Capaldo, F. N. and Barbour, S. D. (1975). *J. Mol. Biol.* **91**, 53–66.

Capaldo, F. N., Ramsey, G. and Barbour, S. D. (1974). *J. Bacteriol.* **118**, 242–249.

Capaldo-Kimball, F. and Barbour, S. D. (1971). *J. Bacteriol.* **106**, 204–212.

Clark, A. J., Chamberlin, M., Boyce, R. P. and Howard-Flanders, P. (1966). *J. Mol. Biol.* **19**, 442–454.

Epstein, J., Williams, J. R. and Little, J. B. (1973). *Proc. Nat. Acad. Sci. U.S.A.* **70**, 977–981.

Goldmark, P. J. and Linn, S. (1970). *Proc. Nat. Acad. Sci. U.S.A.* **67**, 434–441.

Goldmark, P. J. and Linn, S. (1972). *J. Biol. Chem.* **247**, 1849–1860.

Gross, J., Grunstein, J. and Witkin, E. M. (1971). *J. Mol. Biol.* **58**, 631–634.

Haefner, K. (1968). *J. Bacteriol.* **96**, 652–659.

Hertman, I. M. (1969). *Genet. Res.* **14**, 291–307.

Hout, A. and Schuite, A. (1973). *Biochim. Biophys. Acta* **308**, 366–371.

Howard-Flanders, P. and Theriot, L. (1966). *Genetics,* **53**, 1137–1150.

Lloyd, R. G. and Barbour, S. D. (1974). *Mol. Gen. Genet.* **134**, 157–171.

MacKay, V. and Linn, S. (1974). *J. Biol. Chem.* **249**, 4286–4294.

Monk, M., and Kinross, J. (1972). *J. Bacteriol.* **109**, 971–978.

Monk, M., Kinross, J. and Town, C. (1973). *J. Bacteriol.* **114**, 1014–1017.

Ogawa, T., Tomizawa, J.-I. and Fuke, M. (1968). *Proc. Nat. Acad. Sci. U.S.A.* **60**, 861–865.

Orgel, L. E. (1973). *Nature,* **243**, 441–445.

Pauling, C. and Hanawalt, P. (1965). *Proc. Nat. Acad. Sci. U.S.A.* **54**, 1728–1735.

Smirnov, G. B., Filkova, E. V., Skavronskaya, A. G., Saenko, A. S. and Sinzinis, B. I. (1973). *Mol. Gen. Genet.* **121**, 139–150.

Sugino, A., Hirose, S. and Okazaki, R. (1972). *Proc. Nat. Acad. Sci. U.S.A.* **69**, 1863–1867.

Tomizawa, J.-I. and Ogawa, H. (1972). *Nature* **239**, 14–16.

Town, C., Smith, K. C. and Kaplan, H. S. (1971). *Science* **172**, 851–854.

Willetts, N. S. and Clark, A. J. (1969). *J. Bacteriol.* **100**, 231–239.

Yielding, K. L. (1974). *Perspect. Biol. Med.* **17**, 201–208.

Author Index

(Pages 419-824 are to be found in Part B)

Subject Index